HENRY SUTTON

THE
INNOVATIVE MAN

HENRY SUTTON

THE
INNOVATIVE MAN

AUSTRALIAN
INVENTOR, SCIENTIST
& ENGINEER

BY

LORAYNE BRANCH

ISBN 978-1-925332-34-6

Cover designed by:
Benjamin Branch - Arcane Design

Image courtesy of:
Sutton Family Archives

Edited and Indexed by: M. Gregory
Tried and Trusted Indie Publishing
www.tatindiepublishing.com

Layout by:
Tried and Trusted Indie Publishing

Author Image by:
Ian Wilson Photography

For permission to reproduce parts of book or to obtain copies address the request to the author
c/o
Lorayne Branch <lorayne.branch@gmail.com>

TABLE OF CONTENTS

DEDICATION

This book is dedicated to Richard and Mary Sutton
and all the members of the Sutton family
- past, present and future.

PREFACE

"For the Love of Henry"

I am the great grand-daughter of Henry Sutton and my journey to uncover and document my great-grandfather's life and work began ten years ago. I grew up, as did all the descendants of the Sutton family, with the story about Sutton's music store, and as a little girl I was told someone in the family invented a few things, but I was never really told exactly who it was. In researching my Sutton family, I soon discovered that it was my great-grandfather Henry Sutton who was the mysterious relative I was told about. Intrigued by the few things I found online about Henry, I began to dig deeper and as I did, I started to uncover many things about my remarkable great-grandfather that had never been documented. Over time, it became apparent that what I was uncovering about Henry, his life, and inventions was an important part of Australia's history that needed to be formally recorded, and so I began collecting and collating the information I found.

My quest to learn all that I could about Henry's life has led me to all corners of the globe as very little about Henry Sutton had been recorded in the history books, journals and newspapers in Australia. As my research progressed, I soon discovered a treasure trove of information in old scientific journals in universities and libraries overseas. As I found one small clue in one journal it would lead me to another journal in another country, and so the arduous task of gathering and piecing together information was a long slow process. Early on it was clear that I had a lot to learn, not only about Henry but also to understand the technical information I was gathering, and so over the years I enlisted the help of a number of experts and historians. With the assistance of one of Henry's grandsons, Ron Sutton, I was able to travel to England and to Oxford University to research Henry and also meet up with descendants of Henry's mother Mary's family.

My task of documenting Henry's inventions was complicated. For each of his inventions I needed to do a crash course in learning and understanding the science and physics behind them, and I also needed to learn the significance of the invention and its importance and place in history. Although I learned a great deal about each invention, I was ever mindful that I was not an expert and took my leads from the various avenues and people I contacted. Albert Einstein once said "*If you can't explain it simply, you don't understand it well enough*" and this was the approach I took. So with the help of the original texts of letters and articles written by Henry and a huge amount of gained knowledge on my part, I have endeavoured to document the life and inventions of Henry Sutton in a clear and simple format so that everyone can enjoy learning about this remarkable and largely forgotten and unrecognised Australian inventor.

Throughout the book I have included many of Henry's letters, published papers, articles, and opinions along with original descriptions, words and explanations of his work written at the time. Many things Henry Sutton worked on over his lifetime were only beginning to be understood by scientists and inventors, and so clear explanations and descriptions by some people in this book may appear to be a little misleading. Writers and newspaper reporters of the time, who usually did not have a scientific background, had only limited perception of what they were actually reporting and so could only describe information as best they understood it. In fact, the various fields of science Henry Sutton worked on were sometimes a mystery even to the best men of science around the world, especially in relation to the science of electricity and wireless telegraphy.

Researching and documenting my great-grandfather's work and life has been, and continues to be an exciting and passionate journey as there are still many things to uncover and learn about his life. The encouragement and support I have received over the years has been heart-warming as I discovered along the way the many passionate people who have always loved and admired Henry. Just as events throughout Henry's life changed the course of his life, so too did researching Henry change the course of mine. Although I grew up in Melbourne, I had moved to Brisbane in 1980 but by 2013 it became necessary that I move permanently to Henry's home town in order to finish his biography, so on Christmas Eve, 2013, I moved to Ballarat in Victoria.

Henry's story is one of invention and innovation and his work is a valuable contribution to Australia's innovation history.

Lorayne Branch
Ballarat, Victoria

INTRODUCTION

Henry Sutton had achieved world acclaim for his contribution to science by the age of twenty-seven and by age thirty-four he had travelled to England to promote one of his latest inventions and to expand his scientific knowledge. His world-wide acceptance as a talented inventor and scientist had preceded his trip to England in 1890, and on his arrival he was immediately accepted into the British scientific community.

The year was 1892 when Australia's first world renowned man of science, and inventor, Henry Sutton walked into scientific history and met the legendary scientist and inventor Mr Nikola Tesla. Tesla had been invited to give his now famous lecture, 'Experiments with Alternate Currents of High Potential and High Frequency', at the London Institute of Electrical Engineers in February, and the whole of the British scientific community was excited. During the previous week, Henry had been personally introduced to Tesla through his two friends Lord Rayleigh and William Preece who felt that the two men could benefit from meeting each other, and as history would eventually show, they were right. One of the very reasons Henry had travelled to England at this time was to learn from the scientific knowledge and experience of the world's great scientists and the remarkable Mr Tesla was one of them.

On the night of Tesla's lecture it is not hard to imagine the hundreds of eager men of science, as well as the general public, vying for a seat or any standing room they could find, and right in there amongst them was Australia's young scientist and inventor. As Henry entered the lecture room, a heavy scent of tobacco emanating from the many smokers already seated must have hung over the room, the loud hum of people talking with excited expectations must have assailed his senses as he took his seat. No doubt he looked more confident than he felt, surrounded by so many notable men of science. As Henry looked around him while he waited eagerly for the lecture to begin, he must have reflected on his own work with electricity and what advances had been made with this wonder of the modern age.

His thoughts would have been broken by the sudden silence as "The Wizard', Nikola Tesla, entered the room and walked up to the lectern and began to speak. Later, this lecture would be regarded not only as a scientific awakening, but as one of those moments in history that opened new scientific directions and inevitably changed the course of many scientists' lives - none more so than that of Henry who was in awe of the master of electricity.

Henry's introduction to Tesla was a momentous event for him and, in the years that followed, the influence of their meeting and friendship had a profound effect on the course of his life as he traversed many fields of science and engineering and eventually helped give birth to the era of technology.

CHAPTER ONE

THE YOUNG INVENTOR

"The true spirit of science is to be found in the love of knowledge for its own sake."

The Duke of Argyll, 1876

A primitive tent on the goldfields of Ballarat was where Henry Sutton was born and no-one, not even his parents, could have imagined the extraordinary life he was to lead. As his parents were building a new life under the difficult and crowded conditions in the goldfields, Henry was growing up, and like his beloved country Australia, forging ahead and creating his own identity and place in the world.

Henry's parents, Richard Henry Sutton and Mary Johnson, were married in Manchester, England, on Christmas Day in 1852. Although Richard was a mechanic and Mary was the daughter of an inn keeper in Manchester, they had both come from musical backgrounds. Richard, also a native of Manchester, had spent a lot of his youth playing in orchestras in France, and Mary had learned to play the piano in England.

In 1853, Mary gave birth to their first child, Elizabeth, and it was around this time that the newspapers in England were full of articles about the Ballarat goldfields. With the English cotton industries in decline, there was a lack of available work. So, like many before and afterwards, Richard, Mary and her brother, James Johnson, were seduced by the lure of the gold rush, and sailed from England to seek their fortune in Australia. They travelled on the ship the *Invincible*, which arrived in Australia on the 26 August 1854.

Little did they know at the time that a turbulent beginning lay ahead of them when they arrived in Ballarat.

Upon arrival at Hobson's Bay in Victoria, James Johnson travelled straight to Ballarat to set up camp and search for gold. He did not know at the time that this trip would once again change the course of his life.

Richard and Mary stayed in Melbourne for two months to work and gather supplies. In November, they loaded these and their possessions onto a dray and set off along the rough road to the Ballarat goldfields. When they arrived on 21 November 1854, they promptly purchased a gold mining licence which allowed them the right to dig for gold.

Each miner at the time was allotted a piece of land 12 foot square, and so Richard and Mary, with one year-old Elizabeth, set up a tent on their allotted site on Bakery Hill in the hope of finding gold. The Sutton tent site fronted Bakery Hill Road, now known as Humffray Street, and the site backed onto what was called the Right of Way, which was an old section of St Paul's Way.

In 1855, St Paul's Church was built opposite the Sutton tent site, fronting onto the Right of Way. A decade later, due to a mine collapse on the church site, the church had to be repositioned. The church which remains today, was rebuilt in 1865 on the old Sutton tent site and is situated at 3 Humffray Street, Bakery Hill, Ballarat East.

Richard and Mary's first night in Ballarat was not a pleasant one as a great storm lashed the goldfields and blew their tent away, leaving both the family and their worldly goods exposed to the weather. This experience was a great lesson on the difficulties of living on the goldfields and would have at the time tested their resolve and ability to survive the harshness of a new and primitive country.

It was a crucial time in Australia's history when Richard, Mary and James Johnson arrived in Ballarat. The doubling of mining licence fees earlier that year had created a lot of unrest among the miners. That, and the constant harassment by government officials for miners to pay their fees, brought the angry sentiments to a head. On 29 November 1854, about 12,000 Ballarat residents and miners gathered on Bakery Hill - not far from Richard and Mary's tent. An Irishman named Peter Lalor stood strong and defiant before the crowd and asked everyone to swear allegiance to the Southern Cross flag, known today as the Eureka flag, as the symbol of the miners' rebellion. Licences were burned and Peter Lalor then led a group of miners with the flag up to the Eureka Gold Lead Mine.

Living so close to where the Eureka flag was first flown, Richard and Mary were inevitably caught up in the political and violent unrest of the rebellion and must have wondered why they had come to this wild, unruly and seemingly lawless land. Mary's brother, James, who had been in Ballarat a few months already, had become involved in the miners' plight. This led him to decide to join forces with Peter Lalor and the rebellion. On 2 December 1854, the miners banded together and built a stockade on Bakery Hill. Seeing this as a threat to the police camp, Goldfields Commissioner Robert Rede quickly sent for reinforcements. There were probably as few 120 miners within the stockade perimeter when, at 3.30 am on December 3, nearly 300 government troops attacked the stockade. James fought alongside the other miners in a bloody battle that lasted only minutes. However, this battle made a significant statement on the political landscape of Australia. What has since become known as the battle of the Eureka Stockade, stands alone as a defining point in the democracy of Australia and it changed the course of many lives.

On the night of the battle, Richard and Mary huddled together with their daughter Elizabeth in the family tent, fearing for their lives as gunshots and screams echoed across Bakery Hill. Next morning whilst still shaken and unsettled by the night's events, they were very thankful to learn that James had survived the battle.

Eventually though, the Eureka Stockade Rebellion came to be significant to Richard and Mary for a reason far removed from politics. Exactly nine months later on 4 September 1855, their first son, Henry, was born.

Life on the Ballarat goldfields was tough and demanding, so in the evenings miners would seek relief from the hard toil of the day by looking for entertainment. Using his musical instrument knowledge, Richard made a concertina to amuse and entertain his family and friends and this proved to be such a great hit among the miners that soon orders for concertinas came rolling in.

Being novices at gold mining and not very successful, Richard and Mary saw an opportunity to make some money in another way and so with Mary's urging, Richard went to Melbourne to buy musical instruments.

After loading up his wagon with an assortment of instruments, Richard took the long slow road back to Ballarat. He and Mary erected a second tent on their allotted mining site and began selling and repairing musical instruments. Their business was called R. Sutton, Musical Instrument Repairers, Bakery Hill. Richard then set about the task of making a wooden framed glass display case which was placed in the front of the tent so passing miners could see the array of musical instruments that were for sale.

The musical instruments were popular with the miners and trade soon outgrew the tent. This led Richard and Mary to move the family down the hill to the Old Plank Road, now known as Main Road Ballarat East. There Richard purchased a small wooden store for £10. The first Sutton's Music Store opened on 26 January 1856, which was Foundation Day (now called Australia Day), and by 1857, the business was called Richard Sutton, Music Seller.

Six months later, Richard and Mary's resolve was tested once again as a dreadful fire swept up the road destroying a number of shops and properties in its wake. The fire erupted at about 10.45 pm on 9 July 1857 and was soon spreading to other buildings. Richard and the other store holders would have had to wake their families and take them outside, as was the custom, to protect them in case the fire reached their stores and homes.

Such fires were a constant threat and the cause of great worry and distress in Ballarat. At such times, shopkeepers would frantically throw their wares and belongings into the street to save them, whilst many other people would help to fight the fire along with the fire brigade. Sometimes, to stop the fire from spreading, the people had to resort to desperately tearing down buildings to make a break between the stores and homes.

Like everyone else, Richard and a heavily pregnant Mary most likely threw as much of the contents of the store and home as they could out of their building just in case the fire reached them. This particular fire ended up killing two young children, clearly hitting a nerve with Richard who was so distressed that the morning after the fire, he wrote to the Editor of The Star newspaper with a suggestion on how to stop the fires spreading from building to building:

A Valuable Suggestion:

Sir,-- The disastrous fire which took place on Thursday night last, near the Plank-Road has caused me to think of a plan that would I believe, prevent many fires from being so serious in their consequences. My plan consists in breaking off the connection of the houses &c. We will suppose there are fifteen or twenty buildings all joining together, which is the case in many parts of the Main-road, and most of the buildings are built of wood: how simple it would be to put a framework of wood underneath one of these buildings, with eight small wheels cut off the trunk of a tree about 18 inches in diameter, to run on a kind of railroad, simply put on strong props well stayed to the full extent of their back ground. We will suppose there was one building so arranged in every six—don't you think it would prevent fires being so serious if there were here and there a single storey building done so? I say one storey because these would be so much easier to push back, or, with a winch fixed at the back of the premises, pull it back to the full extent of the back ground. I think breaking off the connection is the main point for consideration with respect to fires. Perhaps you will recollect the great fire at the United States School; there was a right of way between Heninway and Jones' and the Charlie Napier, and it was the open space that prevented the Charlie Napier from taking fire. Now, by pushing back an ordinary sized store, it would very likely prevent the fire from extending further. I think the cost of this framework under a moderate sized store would not exceed £35.

Sir, if you think the above remarks worthy of a space in your valuable paper, you will oblige by inserting it.

Yours respectfully, R. H. Sutton. Ballarat, 10th July 1857

The letter sheds some light on where Henry inherited some of his great ability to think outside the box. His father would later go on to patent not only a number of his own musical instrument inventions but also the machinery to make them. Although Richard's idea in the letter was in hindsight impractical, it at least posed some sort of solution to the fear of fires that constantly threatened the closely-built wooden buildings that lined the streets of Ballarat.

On the night following the fire, Mary went into labour and gave birth the next day to their third child, Alfred.

The music business grew slowly, with Richard beginning to design improvements to his concertinas and harmoniums, as well as designing and constructing the mechanical equipment to mass-produce them. The land in the section of Old Plank Road where Richard had built the first store soon came up for sale. On 8 December 1858, with the aid of the bank, Richard purchased it for around £40. The land was Lot 48, Allotment 17, block H, Main Road, which is now known as number 36A Main Road Ballarat East.

With the family growing, the wooden store was extended to incorporate extra living space, and here, another son, Walter, was born in 1859 followed by a daughter, Emilie, in 1862. The birth of their sixth child, Frederick, in 1863, meant that the family had once again out-grown living at the store. As the threat of fires was still ever-present, Richard and Mary moved their family to a miner's cottage in Scott's Parade. Richard still traded out of the store in Main Road and more than likely he also had a small stall in the front yard of the Scott's Parade cottage to catch the passing traveller's trade.

With a prospering business, Richard and Mary bought shares in the Band of Hope gold mine and by the end of 1864, Richard owned three shares while Mary had sixteen shares, each share being worth £20. Richard and Mary decided at this point that a more permanent brick store and home was needed, and so Mary cashed in her shares and they engaged James H. Spark, who was the Captain of the Ballarat Fire Station, to design their new building. Spark had designed a number of local buildings, including the Fire Station, and he, along with local builder William Cowland, contracted to build the new two-storey Sutton store.

In January 1865, the wooden music store that Richard had built was torn down, and construction of the new building quickly got underway. By late February, the roof of the new building was on and by April, the front verandah was being finished. The new brick store was completed on 6 May and opened on 14 May 1865. This new permanent building, which still remains today, marked the turning point with the Sutton family as Sutton's Music Stores would end up to be a very large part of Australia's musical history. This new store became the creative centre and platform for Henry's scientific research and inventions.

In April 1866, two of Mary's younger brothers arrived from England. Frederick and Mark Johnson arrived in Australia on the ship *The Great Britain*, much to the joy of the Sutton family. The arrival of Henry's uncles was a happy event and the family spent many hours catching up on family news and stories of England. Mary had come from a large family of nine brothers and sisters but by 1861, her mother Susannah had died followed by her father Samuel two years later. With the various industries in Manchester fallen into an even greater decline in the 1860s, Mary's younger brothers had decided to follow her and James out to Australia.

Mark Johnson quickly secured a position as a telegraph officer with the Post Office and moved to New South Wales. However, he died suddenly in April 1875 at the age of thirty four - a sad blow to Mary and the family.

Unfortunately little is known about the later life of Frederick Johnson, or his brother James after he fought at Eureka Stockade.

Family life for the Sutton family was not only full of challenges and hard work; it was also full of music and song and equally important, free thinking, learning and play. Henry was brought up to believe everyone was equal and no limitations were set on what he or his siblings wanted to do with their lives. Richard and Mary instilled in all of their children a great sense of self and encouraged them in their endeavours. All of the children were musically trained and gifted which was not surprising in light of their home environment. Family life saw an endless parade of musicians, some miners, professional entertainers and various other people from the goldfields. Impromptu concerts were held at their home and Mary and Richard's children entertained these visitors. Apart from music, all the Sutton children were very good artists, Alfred and Walter enjoyed rowing and cycling, Fred enjoyed bowls and Elizabeth and Emilie loved singing.

Richard and Mary went on to have a total of eleven children. However, Arthur Ernest Sutton who was born 1867, died as an infant, and four other children died at birth.

As Henry was growing up, his passion was science. He would spend endless hours in the library of the Ballaarat Mechanics' Institute reading everything he could and by the age of fourteen, he had read every book and journal on science in the library.

The Ballaarat Mechanics' Institute (Ballaarat being the original and still correct spelling of the institute) opened on 23 May 1859, in a temporary reading room at the Ballarat Fire Station in Barkly Street. Through a lot of effort by Peter Lalor (by then an MP), a piece of land was set aside in Sturt Street for the building of a permanent home for the Institute. Subscriptions to the Institute were payable in advance and set at £1 per annum for adults. For junior members, apprentices, and others who were under eighteen years of age the subscription was ten shillings yearly.

On 28 September 1860, in front of 10,000 Ballarat residents, the first foundation stone for the new permanent building in Sturt Street was laid.

The building, designed by Mr Charles Boykett, consisted of a reading room and library, a hall for lectures and concerts which could seat up to 1200 people, a smaller lecture room, two classrooms and a conversation room. By 1868, the space at the Institute was no longer adequate and it was decided to extend the front of the building to Sturt Street. Designs were called for and Mr J. H. Jones's design for the new building was accepted. Constructed by Mr Jonathan Coulson, the grand building which remains today in Sturt Street, was opened on 21 July 1869, by the Governor of Victoria, the Honourable Sir John Manners-Sutton (no relation).

Regardless of the constraints that living in Ballarat imposed on someone interested in science, Henry was eager to learn and the goldfields and the Institute's library became his classroom. The Institute was a beacon of light for his curious and intelligent mind, and the library helped spark his lifelong passion of scientific discoveries and inventions. He would spend many long hours there and being a studious young boy, he quickly earned the respect of many of Ballarat's citizens.

There was not one branch of science, mechanics or engineering that escaped young Henry; all subjects were explored and acted on in various degrees. Although the Institute's library was limited, he would only have to get an idea of something from one of the many books, journals or newspapers and he would work on it to try and improve an invention or find a solution to a reported problem.

Encouraged by a number of notable Ballarat citizens such as Mr James Oddie and William Bramwell Withers, Henry sought the knowledge and expertise of the many men who worked in the various industries that emerged on the goldfields. Very little escaped his curiosity and by age ten a young scientist and inventor began to emerge. Henry's father and mother nurtured his interest in science and did all they could, given their financial constraints at the time to allow Henry to follow his scientific endeavours.

Anyone who visits the Ballaarat Mechanics' Institute today, can step back in time and soak in the atmosphere of the library and peruse some of the old books and journals that young Henry read. They can sit at one of the old desks and picture a young boy of humble beginnings diligently reading and learning from the works of people like Leonardo da Vinci and Michael Faraday.

The Ballaarat Mechanics' Institute played a pivotal role in Henry's life and over many years the Institute became the backdrop to numerous events, classes, lectures and club meetings he attended over the course of his childhood and young adult life. Along with his parents, the Institute and the Ballarat School of Mines helped educate Henry's fertile and creative mind and helped to indulge his passionate interest in science, physics, mechanics, astronomy, photography and art.

Like da Vinci, one of the first things that captured Henry's attention was the flight of birds. He was not content to just watch birds flying, he started observing and making notes about how they could fly. And, unlike other young boys who sit and dream of flying, he spent much of his time studying and experimenting with the mechanics of flight.

Mary Sutton
1835 - 1894

Richard Henry Sutton
1830 - 1876

Sutton's Main Rd Music Store - opened May 1865
Alfred Sutton is on the left and Richard Henry Sutton on the right.

Ballaarat Mechanics Institute

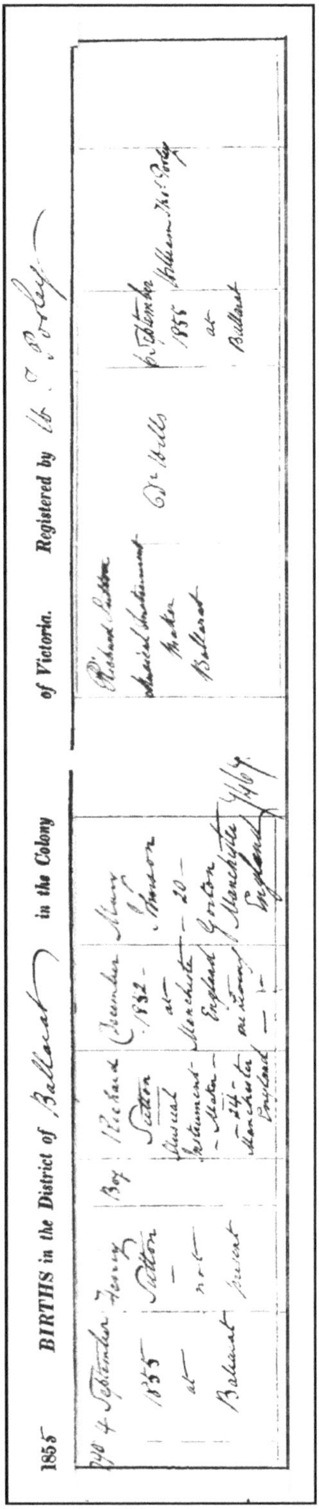
1855 BIRTHS in the District of Ballaarat in the Colony of Victoria. Registered by

CHAPTER TWO

IDEAS TAKE FLIGHT

"The subject of flight is one which ought to be approached in the most unprejudiced manner and so great and many are the advantages and benefits that would accrue to the human economy by its realisation, that it at once places itself as the highest and noblest aim for the inventive mind."

Henry Sutton 1878

At the age of ten, literally following in the footsteps of Leonardo da Vinci and others who had studied flight and flying machines, Henry spent countless hours running after birds, insects and anything that had wings and could fly, in order to observe and study their method of flight. During the course of his observations and experiments, he erected a piece of glass perpendicular to a base board and blew smoke onto the glass. As he did this, he would gently hold a bird next to the smoked glass and, as it flapped its wings, the bird would leave the pattern of motion of the wing onto the smoked glass for Henry to study.

Over the course of the next four years, Henry constructed a clockwork ornithopter that worked so perfectly and naturally in its action that a number of his boyhood friends nicknamed it '*The Goose*', albeit no doubt with a double meaning and a touch of mirth as it was very crudely constructed. An ornithopter is a flying machine which utilises flapping wings as the propelling power; it is capable of generating both thrust, the force that propels the craft forward, and lift, the force perpendicular to the direction of flight that keeps the craft airborne. The forces must be strong enough to counter the effects of drag and the weight of the craft. He constructed the ornithopter in the workshop of his father's music store, a place that would become the central hub of construction for his ideas and later inventions. The ornithopter, when wound up, would give one or two flaps and immediately dart off in a circular direction, flying in a circumference of about 12 feet. By slightly modifying the stroke of either wing, it could be made to fly right to left or left to right or, by altering the rods, it could be made to fly upward at any desired angle. It was fixed on a lever having a universal joint so it could fly in any direction at any desired angle.

Several years after Henry had finished his experiments and developed his theory on flight, he wrote to the Aeronautical Society of Great Britain and tendered his very detailed experiments and ideas. Consequently, in 1877, Henry was invited by the then Secretary of the Aeronautical Society of Great Britain, Mr F. W. Brearey, to put forward his theories and observations about flight to the Society to be formally published. He wrote and sent two papers on flight and aerial navigation, totalling some forty pages, to the Society and these were published in 1878 in the Thirteenth Annual Report of The Aeronautical Society of Great Britain. The first paper was called 'On the Flight of Birds and Aerial Navigation' and the second paper was titled, 'Second Paper on the Flight of Birds'. Brearey stated at the time that Henry had discovered the true theory of flight and as a result he was held in high regard by the Royal Aeronautical Society of Great Britain. Sadly but unsurprisingly, there was no equivalent recognition of Henry's achievements in Australia. Yet, his years of study on the subject of flight, and the publishing of his two papers, gave him the great and honourable distinction of being the first Australian to experiment with flight.

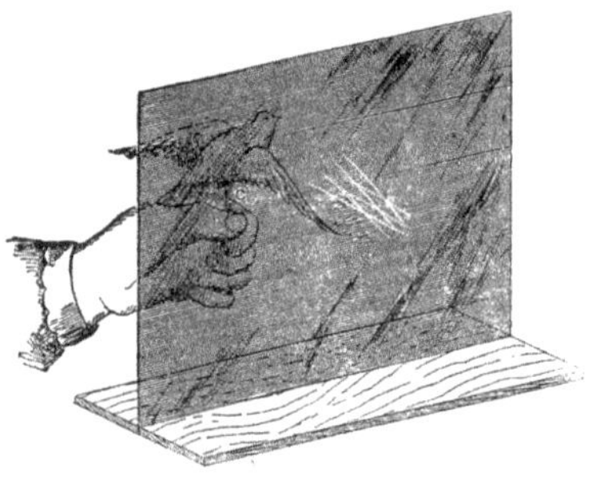

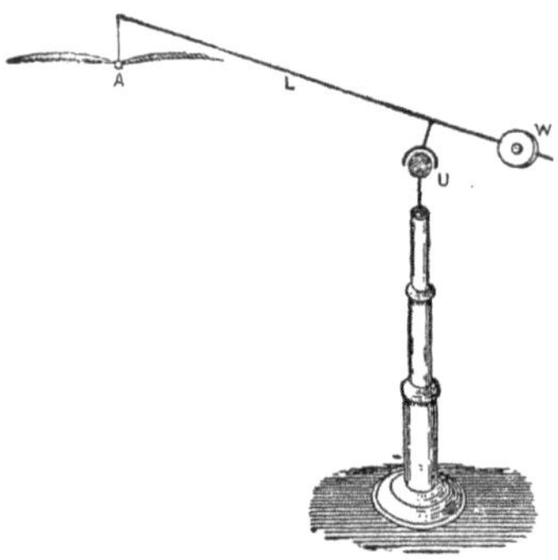

(Above) Second paper on 'The Flight of Birds' Smoked glass experiment diagram.

(Above) Ornithoper diagram on The Flight of Birds and Aerial Navigation

In 1892, some fourteen years after Henry's papers on flight were published, Octave Chanute published his now famous book *Progress in Flying Machines*. In the chapter 'Wings and Parachutes Part III', he briefly touches on Henry's observations on flight and describes Henry's ornithopter.

He also noted that Henry did not answer the important and vital question of an efficient motor. What he could not have known at the time was that Henry was only a young boy of ten when he started his theory of flight and had very limited means to develop his ideas.

Years later in 1905, Henry read an article in T*he Automotor Journal,* which reported some research on winged flight at Cambridge University by E. P. Frost, Dr F. W. H. Hutchinson and C. R. D'Esterre. Frost had been working on ornithopters since 1877. His first model had been steam-powered and quite impractical, after which he built another one with an electric motor. Now Frost had constructed one which worked with a 3hp petrol motor. Henry must have been very amused and slightly bemused by the article as most of the findings of the Cambridge trio were those he had obtained years earlier while researching his two papers. As a result, he wrote to *The Automotor Journal* stating that a clockwork ornithopter, which he had built in the 1860s in Australia, was not unlike the arrangement of Frost's ornithopter (pictured in the journal). However, Henry observed, that he and Frost had different opinions about the theory of bird flight and stated in *The Automotor Journal:*

"The prospects of flapping wing flight are at the present moment, needless to say far behind those which have been attained with aeroplanes, though of course the form of apparatus which will ultimately triumph cannot even as yet be guessed at."

Henry lived to see his childhood dream come to fruition on 17 December 1903, when Orville Wright piloted the first powered aeroplane at Kitty Hawk in North Carolina. The Wright brothers' aeroplane flew 20 feet above the ground and travelled a distance of 120 feet; the flight lasted 12 seconds. More flights were made that day and Orville's brother Wilbur Wright piloted a flight that travelled 852 feet and lasted 59 seconds.

It is clear that Henry's childhood passion, the subject of flight, was never far from his mind as he continued to read about the advances in flight over the decades and in 1909, it once again occupied his mind. The progress of flight around the world and in Australia had come a long way since his first experiments, but it is quite clear that he had kept up to date with the advances by other inventors and aviation pioneers. On 14 September 1909, *The Argus* newspaper in Melbourne published an interview with George Augustine Taylor who had, only a few months before, inaugurated the New South Wales branch of the Aerial League of Australia. This article started a series of letters to the editor of *The Argus* between Henry and Taylor in relation to a statement made by Taylor while on a visit to Melbourne:

"The rapid improvements in machines are scrapping those of practically a few months back, and to Australia a new type will not be such a difficult task, considering that the original idea of the aeroplane was invented by an Australian (Mr Hargrave a member of the New South Wales Aerial League) 15 years ago."

Needless to say, Henry, had been somewhat surprised by Taylor's statement, and he felt he had no choice but to contest this claim that the original idea of the aeroplane was invented by the Australian aviation pioneer Lawrence Hargrave.

Instead Henry wrote, *"…The original monoplane was invented by Hensen in 1843; this comprised a single plane, driven at a small angle by two propellers. In 1867 Wenham advocated the use of superimposed planes. In 1868 Stringfellow constructed a three superimposed planes' model on Wenham's plan. … Stringfellow constructed his model aeroplane 40 years ago. Wenham suggested the superimposed (box form) planes and invented the word 'aeroplane' 41 years ago. Hensen invented his monoplane aeroplane 65 years ago."*

Henry went on to say that he felt the Wright brothers had discovered principles that all others had missed and that they were, in actual fact, the true conquerors of flight. Taylor, who was a little miffed at Henry taking him to task for acknowledging Hargrave with such an auspicious credit, responded a few days later. This led both men to respond in turn and publicly debate the history of flight over the ensuing week.

[Full Text, see End Notes, Chapter Two, Nos. 1, 2 & 3.]

The editor of *The Argus* most likely feared at the time that the debate between Henry and Taylor would not end or be resolved, so on 25 September 1909, *The Argus* published a large article on 'The Conquerors of the Air'. It took up nearly half a page of the newspaper and included many drawings of the inventions by the inventors mentioned by Henry and Taylor. It also gave a short but detailed history of flight. If ending the letters and the debate was the editor's desire at the time, it certainly worked but not before two great Australian aviation pioneers had expressed their views and had educated the public on the history and advances in flight. Just over two months after the newspaper debate, on 5 December 1909, George Augustine Taylor would be the first person in Australia to achieve the flight of a heavier than air winged machine in a glider at Narrabeen, New South Wales. On the same day as Taylor's first flight, his wife Florence became the first Australian female to pilot a glider.

Prior to Taylor's achievements, Lawrence Hargrave had also studied flight and had built a box kite. He flew this successfully, for the first time, at

Stanwell Park Beach on 12 November 1894. Each of these momentous achievements would have recalled to Henry, memories from late in December 1877, when he first saw man fly in a balloon.

Professor Emanuel Jackson arrived in Ballarat from Derby, England, bringing with him what the newspapers called his "aerial contrivance". This was a hot air balloon capable of containing 20,000 cubic feet of gas and able to carry from three to four hundredweight in a basket that could accommodate up to four people. A pipe had been laid from the gas main to supply the buoyant gas to Ballarat's Eastern Oval and at 5am on New Year's Day 1878, Professor Jackson along with one very brave Ballarat citizen, a local hairdresser by the name of Drew took to the air.

Henry, along with thousands of Ballarat residents, witnessed this awe-inspiring sight as the colourful balloon rose skyward. His heart would have skipped a few beats, for even though he was a grown man, the little boy who dreamed of flight must have been filled with overwhelming excitement, a feeling he was to share with his children many years later.

On 18 March 1910, Harry Houdini came to Australia and flew his Voisin biplane at Diggers Rest near Melbourne and later that year, on 16 July 1910, Australian John Robertson Duigan flew his Pusher biplane at his property at Mia Mia, Victoria. The latter event is regarded as the first power-controlled flight of an Australian built aeroplane. However, it was not until 1911 that Henry was able to share with his children the thrill of seeing man fly.

On the 23 February, Henry woke his sons Edward and Albert very early so they could make their way to Melbourne to witness the aviator Mr J. J. Hammond flying over the city. Henry's grandson, Ron Sutton, vividly recalls his father, Albert, telling him how Henry insisting he skip school that day to go instead to the city.

It was just after 7am when the plane, shimmering in the morning light, flew over Albert Park headed for Government House. After circling there, the plane then passed over Victoria Barracks causing some soldiers to run out to join the cheering. The plane continued to fly up St Kilda Road to where Henry and his sons Edward and Albert were watching just near Princess Bridge, and then it flew towards the city over Flinders Street train station and along Collins and Bourke Streets. One can only imagine the overwhelming joy Henry would have felt in sharing this significant Australian aviation event with his sons.

Henry Sutton's contribution to manned flight had been forgotten until

1940 when the Royal Australian Air Force was sorting out their files. Published in one of the bulletins issued by the Department of Air, was a reference to a particular historic document filed in the air-force archives. The document was a copy of the Thirteenth Annual Report of the Royal Aeronautical Society of Great Britain published in 1878. News of the discovery of this, which included Henry Sutton's two papers on flight made it into the newspapers. At the time it came as a great revelation to the Australian Air Force that Australian's were experimenting with flight as far back as 1878.

However flight was not the only thing that captured Henry's attention during his early years. He was working on a number of inventions between the ages of ten and fourteen. Young Henry's idea of fun and play was to construct a torpedo. This was eight feet in length and powered by large strips of thick India rubber attached to the rod that connected to the tail or paddle end of the torpedo. In 1868, he was ready to test his torpedo and the ideal place was a large man-made water hole, named Lake Como by Ballarat residents, which bordered on the back of the music store in Main Rd. With the help of his father and brother, the torpedo was carried from the backyard of the store down to the water hole.

It is not hard to imagine a young boy's excitement as he set about preparing his invention to be tested. At the edge of the water hole, the torpedo's large rubber band was wound up and the torpedo was placed in the water. After the flat fins on the sides of the torpedo allowed it to get to the required depth, it was released. This process was repeated a number of times and the torpedo successfully travelled from between 10 to 25 yards under the water. Along with family and friends, there would have been a small contingent of Ballarat residents to cheer the successful test of the torpedo. That night Henry would have been the talk of the town, but sadly a lack of funds at the time prevented him from developing the torpedo further.

The first electric generator was invented by Michael Faraday in 1831 and it consisted of a copper disc that rotated between the poles of a magnet. It was not a dynamo, as it did not use a commutator, but it did generate a very low voltage. It was not until 1867, that Dr Werner Siemens and Charles Wheatstone independently came up with the first practical designs for a dynamo. Following on from their work in 1869, at the age of fourteen, Henry invented and constructed the world's first electric continuous current dynamo with a practical ring armature. This was two years before the Belgian scientist Zenobe T. Gramme first demonstrated his continuous current dynamo to the French Academy of Sciences in 1871. In 1869, Henry discovered that when his dynamo ran in reverse it acted like an electric motor, a fact Gramme only discovered by accident in 1873. Gramme's dynamo used the same principles as Henry's dynamo but due to Henry's age and isolation from the rest of the world, his invention and discovery was not widely reported at the time. As a result, history has credited Gramme with the invention of the first practical continuous current dynamo. In a letter published in the *The English Mechanic and World of Science* in 1879, Henry describes the difficulties associated with utilising the power of electro-magnets and then describes his continuous current dynamo/electric motor:

A New Electric Motor:

"It is generally acknowledged that the primary difficulties in the way of the successful utilisation of the immense power of electro-magnets are:-

1st Owing to the magnetic force obeying the law of inverse squares, and rendering even the most minute decrease of distance between magnetic surfaces almost in contact a terrible means of waste of power.

2nd The creation of induced currents (to the detriment of the original current) by the approach and recession of parts (i.e., electro-magnets) and make and break of current, the main difficulties being the law of inverse squares and induction from the approach and recession of movable from fixed electro-magnets. These difficulties have been held insurmountable; and yet ... I will explain an electro motor devised by myself, in which there is no approach or recession (in the strict sense) of magnets, also no make and break, but a continuous current circulating in coils of electro-magnets, also the attractive and attracted, repellent and repelled surfaces never vary their distance (yet move), but may be placed so close as to exert the greatest possible force without absolutely touching. The following drawing of side elevation will explain the principle by which these results may be secured."

"AA.-- A flat disc of soft iron, having an axis running through its centre. B.-- Bearing and support for disc AA, the disc revolving in bearing B. CC and CC'.-- Coils of insulated wire, so arranged that the disc AA is free to revolve in the coils, the disc and coils forming an electro-magnet having a core free to revolve. On a current being sent through the coils CC and CC' the disc AA becomes magnetic, having two polar surfaces N S and two neutral surfaces XX. It will now be seen that by placing the pole of any other magnet so as to act between a polar surface and one of the neutral surfaces rotation will take place, without having to break the current traversing the coils CC and CC'. DD'.-- Two electro or permanent

magnets, having their poles arranged as in drawing, the dotted portions being brought within the coils CC and CC'. The action is as follows: the S pole of disc electro-magnet is attracted by the N pole of electro or permanent magnet D', and repelled by its S pole. The N pole of disc electro-magnet is attracted by the S pole and repelled by the N pole of electro or permanent magnet D. By reversing the current in coils CC and CC', and thus changing the poles of disc, the rotation is reversed. By breaking the circuit of disc-magnet the magnets D and D' act on soft iron disc AA as a brake, and stop the rotation.

Owing to the mode by which the magnet force is utilised, the motor occupies little space laterally; therefore a number of discs can be mounted on a common axis and their combined power obtained."

[Full Text, see End Notes, Chapter Two, No. 4.]

*(*It is possible that the word "decrease" could be a typographical error and should instead read "increase".)*

The continuous current dynamo was the first machine to generate commercial quantities of power for industry and quickly led to the rapid development of the electric motor industry throughout the world. Although the continuous current dynamo was not originally designed as a motor it did become the basis for the design of all practical DC motors.

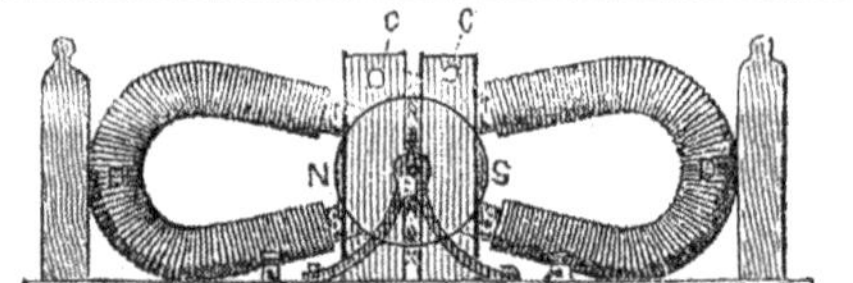

(Left) Henry Sutton's Continuous Current Dynamo diagram

Henry was a man with broad interests that included art and design as well as science. The Ballarat School of Design opened on 20 July 1870 and classes were held at the Ballaarat Mechanics' Institute. When the school first opened, there were 77 students enrolled: twelve girls and 65 boys ranging from 9-12 years of age. Henry and his brothers and sisters were some of the first students to benefit from the classes. While attending the school at five shillings a quarter, Henry not only took art classes but also trained as a draftsman - a skill that served him well throughout his life. His first teachers were Mr James Oldham and Mr Ludvig Bruun and he soon discovered his artistic talents, and art became a great love and an avenue of quiet pleasure throughout his life.

Student exhibitions were held each year and in 1872, Henry began to win prizes for his artwork - achievements that enabled him to enter his work into the Second Annual Schools of Design Exhibition in Melbourne. This was held at the Victorian State Library in late April and early May 1872. Thirty teachers and students attended, including Henry, and the government of the day gave all students cheap rail fares to enable them to attend the presentation of their prizes in Melbourne.

At this exhibition, Henry received an honourable mention for his drawing of the outline of plants. In the Eighth Class, First Division, the Judge was Mr. Chester Earles, President of the Victorian Arts Academy. Earles gave Henry an honourable mention for his drawing in the round. This drawing is of his younger sister, Emilie Sutton. This drawing is in the Sutton family archives and has been professionally preserved.

Henry's sister Emilie. Drawing by Henry Sutton, circa 1872.

Another student who competed alongside Henry in the 1872 exhibition was Frederick McCubbin who later went on to become one of Australia's greatest artists. Coincidently, years later, Henry's niece, the Australian artist Hilda Rix Nicholas, was to train under the guidance of Frederick McCubbin who had become well known in the Heidelberg School, one of the more important periods in Australian art.

In 1873, Henry started attending the Ballarat West School of Design where his teacher was once again Mr Ludvig Bruun. At the Ballarat Student Exhibition which was held on 18 June 1873 at the Ballaarat Mechanics' Institute, Henry won First Prize for Shaded Ornamental Drawing from the Flat. The prize for this artwork was four volumes of Cassell's Technical Educator which, given Henry's interests, was a perfect prize. He also won the Best Drawing award for Shaded Figure Drawing from the Round, one year's free scholarship to the Ballarat School of Design, and First Prize for Shaded Figure Drawing from the Flat. A month later, Henry won Second Prize in the Second Class Ornament Division at the Third Annual School of Design Exhibition in Melbourne, held at the Victorian State Library. He was able to accept the prize in person as the government gave all students who entered free travel to attend the event.

Henry's great love of art and education was highlighted when an exhibition was held in Ballarat in September the same year. Complaints were made against the decision of the judges, appointed by the Agricultural and Pastoral Society, regarding the validity of a prize won by a Mr Adamson who was associated with the School of Design. Feeling outraged at the content of the complaints, Henry felt compelled to express his views on the matter in a passionate letter to the Editor of the *Ballarat Courier*. The following section of transcript from his letter highlights his passion for art and education. It is also clear from the letter why in the years to come Henry would become such an exceptional teacher:

"Does it sound not reasonable that because a person has qualified himself for a certain position he is debarred from taking a higher, No Sir, it is against all common sense. There is too much of this now-a-days. It is at this point that our Victorian technological institutions fail us. They educate up to a certain point, and there stop. There have been many men lost from this cause to the world, who would have left their mark on it. They would have reached a certain position, and from the want of proper encouragement have retired. Now, Sir, are we to allow this in our schools of design, most decidedly not! If we wish to inspire an artistic taste in our population we must encourage artistic advancement."

Such a strong point of view and profound words from a young man showed great strength of character and conviction beyond his years.

In 1874, Henry once again attended the Ballarat City/ Ballarat West School of Design at the Mechanics' Institute. His teachers were Mr James Oldham, Mr Turnbull, Mr Percy Oakden and Mr Charles Douglas Figgis. The latter was a very well-known architect who designed many of Ballarat's historic buildings. He was the teacher most instrumental in honing Henry's excellent drafting skills. In June 1874, Henry won a Bronze medal for general excellence in all subjects, a Special Certificate for drawing from nature and designs, and a First Prize for figure drawing from the flat: Human figure drawing in crayon. A month later, at the Fourth annual schools of design exhibition at the Industrial Museum in Melbourne, Henry won the Extra Prize for figure drawing, senior. This drawing is still in the Sutton family archives. The drawing is in a very fragile condition and still has the Extra Prize notification from the exhibition attached to it. Henry also received a special mention for his entry in drawing from nature in the round, sixth class senior. Examinations were held for the students while they were there and Henry won Second Prize for colour drawing and also for free hand drawing.

On the 11 November 1874, at the Ballarat West School of Design Exhibition, the committee of the Mechanics Institute awarded Henry a Silver Medal for efficiency in drawing from nature as well as design. Designed by Mr Ludvig Bruun and made by Mr Thomas Gerrard, the medal was displayed in Gerrard's shop window in Sturt Street, Ballarat, prior to it being presented to Henry at the Mechanics' Institute. From 13 to 23 December 1875, the Exhibition of the Schools of Design was held at the Public Library in Melbourne at which Henry received the Second Prize for Coloured Designs and a Second Prize for Colour in Examinations at the exhibition.

During 1875, Henry once again attended the Ballarat West School of Design and exhibited his work and won a number of awards. At the Annual Exhibition held at the Ballaarat Mechanics Institute on the 29th of October, Henry won a Special Certificate and Scholarship for twelve months for his original drawing, "From Australian wild plants, leaves; and flowers; a lady's lace handkerchief; collars." . The judges considered Henry's artwork fully carried out the object for which he schools of design were started.

Later that year at the Victorian Schools of Design Exhibition in Melbourne, Henry came 3rd in the Coloured Design section and also sat for Examinations in Colour and came 2nd in that.

Henry's 1875 intricate drawings of a lady's lace handkerchief and other ladies lace dress wear designs were taken from the textile fabric pattern 'Australian Wild Plants, Leaves and Flowers'. The drawings in pencil which have survived the years, are part of the Sutton family archive and consist of two lady's lace cravat designs, a lady's lace dress front handkerchief design and a lady's lace collar design. On the 2 June 1876, Henry submitted his drawings to Mr Batten of the Ballaarat Mechanics Institute, to be exhibited in the Ballarat Mechanics Institute Exhibition of Fine Arts which was due to open on the 15 June 1876. In the catalogue for the exhibition Henry's drawing is listed under Drawings Exhibited by Pupils of the Mechanics Institute and Ballarat East Schools of Design with item number 227. Although his drawings were mentioned in the papers, it is not known if he was awarded any prize for them. In all, over the years, Henry won over thirty prizes in art and design, including the prestigious Silver Medal in 1874.

There are a number of family stories based around Henry's drawings and paintings. One style of drawing in particular fascinates all the family members who report having seen them. Henry created a number of beautiful drawings using thousands and thousands of minute dots; a style of art that would have taken a great deal of concentration and patience to create. He continued throughout his life to enjoy art and painting and many of his paintings hung on the walls of the family home and the family music stores. In the music store in Elizabeth Street, Melbourne, the customers could enjoy a display of Henry's paintings. They took up the whole of the left wall on the second floor. These paintings remained there until the store closed in the 1980s and one former Suttons Music Store employee said they were taken to the tip at that time. Sadly only a few of Henry's drawings from the 1870s remain with the family.

Music was also a big part of family life in the Sutton household. All of the children were taught to play the piano and violin, along with many other musical instruments. Elizabeth and Emilie were not only very good musicians, they were also exceptional singers and took part in many Ballarat Harmonic Society soirees. Elizabeth had a beautiful mezzo soprano voice which enthralled audiences at many concerts in Ballarat and around Victoria. Alfred, having a rich bass voice, joined the Ballarat Liedertafel Choir which later sang with a young lady called Nellie Mitchell, who was at the time known as Mrs Nellie Armstrong, and is remembered today as Dame Nellie Melba. Frederick was also musically gifted and became President of the Ballarat Lyric Orchestra and President of the South Street Society in Ballarat. Sutton's music stores became patrons of the South Street Society and each year the Sutton Shield and the Sutton Cup became highly sought-after prizes by bands and contestants all over Australia.

As for Henry, he not only played the piano and the violin, but was a composer as well. One particular lovely piece of music he composed was a waltz known as "Sutton Waltz". It quickly became a well-known musical piece around Ballarat. However, no copies of musical compositions by Henry have yet been found either in Ballarat or in the family archive.

Music was always a big part of Henry's life and the family tradition of gathering around the piano and singing continued through the generations. Henry's son Edward's "Sutton Piano" still remains in the family, proudly sitting in one of Henry's great grandchildren's homes waiting for the next generation of the Sutton family to learn how to play.

Henry grew up a quiet but restless child who was fascinated by everything, but especially by science. Richard and Mary found it hard to keep up with his strong desire to learn about the world and science. In his very early years, Henry was schooled by his mother but from about ages ten to twelve Henry attended state school. Unfortunately, due to poor records of that time, it is not known which school in Ballarat Henry attended. In 1869, a school teacher named Mr John Finlay opened up a private school on a seventy-four acre property just outside of Ballarat which was called Gracefield College.

Gracefield was advertised as an 'Agricultural Training, Grammar and Commercial School' and for Henry and his brother Alfred, it was the perfect place to continue their education. Mary, who was often pregnant and busy with the younger Sutton children as well as helping with the music store, was no doubt pleased, or maybe even relieved, to send her two boys off to boarding school. Keeping young and active young boys occupied and out of trouble on Ballarat's goldfields would have been hard and so continuing their education at Gracefield was a great solution all round.

Henry and Alfred were some of the first students of this new private college which had both borders and day students. Mr Finlay was a creative and somewhat eccentric teacher and the college combined scholastic tuition with horticultural pursuits which Henry and the other students enjoyed greatly. The Sutton brothers, like the majority of students, boarded at the college during the week. Each Monday morning they would be picked up from outside the North Grant Hotel on the corner of Bridge and Peel Streets, Ballarat East, and from there they would be taken by two wheeled horse and buggy cabs to Bradbury's, on the road to Melbourne at Brown Hill. From there, they would walk about a mile and a half to the school. On Saturday mornings they would meet the cabs again at Bradbury's at Brown Hill and spend the weekend with their family. The school day would start with morning prayers and classes consisted of various practical lessons and handicrafts, gardening and farming.

Henry and Alfred had a great time at Gracefield. After the afternoon classes, Mr Finlay would entertain the boys with lots of stories or with his fiddle and the boys would do square dances. Mrs Finlay was a great cook and the boys were fed very well.

In the students' free time, the boys were left to explore and play and as the school was set in acres of nature's unspoiled splendour, the school grounds were a great source of adventure and fun. Henry and Alfred were already great friends with many of the students, and they revelled in the freedom from family life and the pressures of the hustle and bustle of the fast growing city of Ballarat. Fun for the boys at Gracefield College was running around the paddocks and forest and playing games such as robbers and police, follow the leader, hare and hounds and various sports. Alfred Sutton attended Gracefield College from 1869-1872 and Henry Sutton from 1869 -1871. Henry then went on to become a student of the Ballarat School of Mines.

The Ballarat School of Mines was opened by the school's first president Sir Redmond Barry, on the 26 October 1870, in a refurbished court house at the top of Lydiard Street South, Ballarat. By January 1871, a prospectus of classes on offer at the School was available for all prospective students and classes began on 23 January 1871. Mr John Phillips was elected Master of the school and Mr Joseph Flude who was a pupil of the eminent scientist Michael Faraday, offered his services to teach the students inorganic chemistry and metallurgy. Flude also helped establish the school's laboratory which opened on 30 May 1871 at a cost of £300. The metallurgical laboratory was later opened early in 1872. At first, most of the students were adult men who were employed at the various mines in the district, but the number of younger students steadily increased over time. The School of Mines was originally set up to educate and train men in the specialised fields of the mining industry but later expanded courses in other areas such as Electricity and Telegraphy.

Henry first enrolled as a student at the Ballarat School of Mines in January 1872 and attended Mr Flude's chemistry classes on Tuesdays and Thursdays from 7pm to 9pm. At the time of his first enrolment, the school consisted of two buildings: the old courthouse which was dedicated to lecture rooms and contained a museum, and a second building that housed the metallurgy and assaying rooms. The school's lecture hall was about 50ft by 25ft with two classrooms, offices and store rooms as well as a room at the back of the hall that housed a library and museum. The second building contained the chemical laboratory, and a metallurgical laboratory which was considered at the time to be the most complete laboratory in the colony. There was also a weighing room, workshop and fuel shed. In the mining and land survey school section, some more work needed to be done as it did not have a full set of the instruments needed; the mining and practical engineering school and the mineralogical department had very few minerals for students to study. From these humble beginnings resources were slowly obtained and the school eventually grew as funding became available.

In the school's first year, there had been a total of forty students enrolled. By 1872, the total number of students enrolled was sixty-six and in 1873, there were fifty nine students. In 1873, classes began on 13 January and the subjects on offer to students were mathematics, natural philosophy, mining and land surveying, mechanical engineering, chemistry, metallurgy and assaying; each class was two hours in duration.

The opening of the Ballarat School of Mines beckoned in a new era in Henry's life. At last he had a place where he could, in practical terms, expand his knowledge and he enthusiastically embraced the school as a wonderful arena for education and scientific learning. In 1872, Henry was enrolled in the Chemistry and Assaying class and in 1873 he enrolled in the Chemistry class again. From the very few records available, what is known is that Henry had an "O" next to his enrolment listing in the school register. It is thought that this might have meant "Open" and that he could attend any class he wanted to in the school year. Given the variety of classes available this would have kept a budding young scientist and inventor totally engaged in his studies.

On 28 May 1873, His Excellency the Governor of Victoria, Sir George Bowen, and Lady Bowen visited the Ballarat School of Mines. Ten to fifteen thousand people greeted them at the Ballarat train station and later that afternoon the Governor took a tour of the school. Undoubtedly, Henry was in class at the school that day and the Governor's visit would have been a highlight in his young life. A few months later, he met the Governor again when Sir George Bowen presented him with prizes for his art in Melbourne on the 12 July 1873.

Between working daily at the family's music store and on his own personal scientific research and inventions, Henry attended various classes at the Ballarat School of Mines all through the 1870s and was still enrolled as a student in the chemistry class there in 1881. At a meeting on 14 September 1881, at the school, the Mayor of Ballarat, Cr John Noble Wilson honoured Henry Sutton's achievements as a student of the school and bestowed upon him a three year scholarship. After the mayor finished his speech, Mr James Oddie presented Henry with the scholarship and the crowd erupted in great applause. After the presentation, Henry gave an hour long lecture explaining the nature of electricity and the mode of its operation. His lecture was well delivered and met with great applause from the audience.

Henry was now 21 years old and the scholarship was seen as great encouragement for him to continue his study and work. The scholarship gave him the freedom to use all the facilities of the school in his scientific endeavours and when working on his many inventions. This in turn bought acclaim to Henry and the school.

During the next few years Henry was busy experimenting and working on a number of inventions, all of which had great benefit - not only to the school but to the world as well. He invented and demonstrated and also gave lectures on many of his inventions while he was a student at the school. In 1878, his telephones; in 1881, his mercurial vacuum pump; in 1882, a new invention for volatilising all metals; his new motor, and his secondary battery just to name a few. What Henry learned during his time as a student at the school was invaluable. The knowledge gained served him well as his scientific research continued, and his body of knowledge was reflected in many of his letters and papers years later.

(Below) Ballarat School of Mines circa 1870
Images courtesy of Federation University Historical Collection

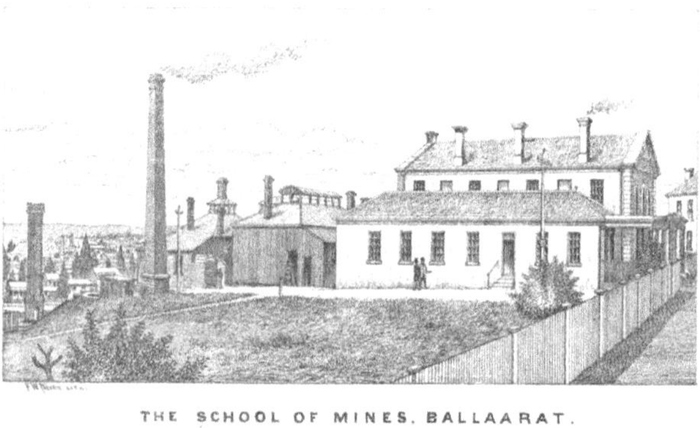

THE SCHOOL OF MINES, BALLAARAT.

CHAPTER THREE

HENRY AND THE TELEPHONE

"The inventor... looks upon the world and is not contented with things as they are. He wants to improve whatever he sees, he wants to benefit the world; he is haunted by an idea"

Alexander Graham Bell.

The invention of the telephone in 1876 came at a crucial time in Henry's life. During the previous year his father Richard had been unwell for quite a while. Rumours that he was dead had spread around Ballarat and people started coming into the music store to give their condolences to the family. Each time Mary explained that Richard, in spite of his recent ill health, was very much alive. Finally, frustrated with the continuing rumour and the procession of people into the store, Mary sent Henry down to the office of the Ballarat Courier newspaper. He was to tell them in no uncertain terms, that although Richard had been unwell he was not dead! So, on the 1 July 1875, the *Ballarat Courier* published the following article in their newspaper. "*The report which was freely circulated in Ballarat yesterday, that Mr R. H. Sutton of the Main road, was dead is, we are requested to state, untrue, as that gentleman is much better in health, though he has lately been very unwell.*"

However, just over a year later Richard Sutton became gravely ill. While most of the family were in church, they received word that he had suddenly taken a turn for the worse. They rushed home and were just in time to witness his death. Richard Henry Sutton died on 24 September 1876 with all of his family by his side. His death certificate states that he died of dropsy but it was his heart that suddenly gave in. His funeral was attended by many of Ballarat's residents and the service officiated by Reverend Joseph Albiston at the Barkly Street Methodist Church. Richard was laid to rest in the Old Ballarat Cemetery on the 26 September 1876. His death was a great blow for Mary and the Sutton family; only months before they had celebrated Elizabeth's marriage to Henry Finch Rix, and now Mary and the Sutton children had to somehow carry on Richard's legacy regardless of their enormous grief. There was little time for any of the family to grieve they had a business to run and important decisions to make.

Although Henry worked at the family's music store each day, his scientific endeavours occasionally took up much of his time. This meant that the huge task of running of the music store largely fell on Alfred Sutton's shoulders. This was a tall ask for a young man of nineteen but Alfred admirably and successfully assisted Mary with the running of the rapidly expanding family business. Decisions needed to be made quickly regarding the future of the business and they decided to open a larger store. Sturt Street by this time was fast becoming the new shopping and business district of Ballarat and it made sense to open the new store there in order to take advantage of the flourishing trade. To this end, Henry organised the rental of Mr Finlayson's premises at 31 Sturt St, Ballarat. The new store opened on 18 November 1876, under the name R. H. Sutton and Co. while the music store on Main Road remained as the family home and workshop for the business.

The announcement of Alexander Graham Bell's new invention, the telephone, was first bought to the public's attention on 25 June 1876 at the Centennial Exhibition held in Philadelphia, USA. The invention of a device that enabled people to talk to one another over long distances immediately caught Henry's attention and so, excited about the great possibilities this invention might bring, he set straight to work experimenting with the new device.

Within six months of Bell's first public demonstration of the telephone, Henry had designed and constructed at least twenty different telephones, sixteen of which were subsequently patented by others overseas. Henry first set up telephone lines between the three floors of the Main Street store and by mid-1877, he had set up a system which ran between the music store at 36 Main Road, Ballarat East, and the new store at 31 Sturt Street, Ballarat. This being the first telephone system in Australia.

Not long after he began to investigate the telephone, Henry wrote to Alexander Graham Bell. This started a lifelong personal and professional association between the two inventors. It is more than likely that it was Henry's association with Dr. Bell and other scientists around the world that led to sixteen of his telephones being patented by others. Henry freely shared his knowledge and designs in the hope that they would benefit the advancement of the telephone, clearly understanding its possibilities as a means of communication in the future. In the following article, Henry gave some insight into one of his creative and unique methods for testing the resonance in telephones, which was based on his great knowledge of music and musical instruments.

Testing for Resonance:

"I have no doubt it will interest a number of your readers to hear of a little fact I discovered during some telephonic experiments. I was endeavouring to improve the acoustic properties of the telephone by the adaption of a soundboard, as in musical instrument construction, and was testing various boards with the object of choosing the most suitable in this manner, shown in section.

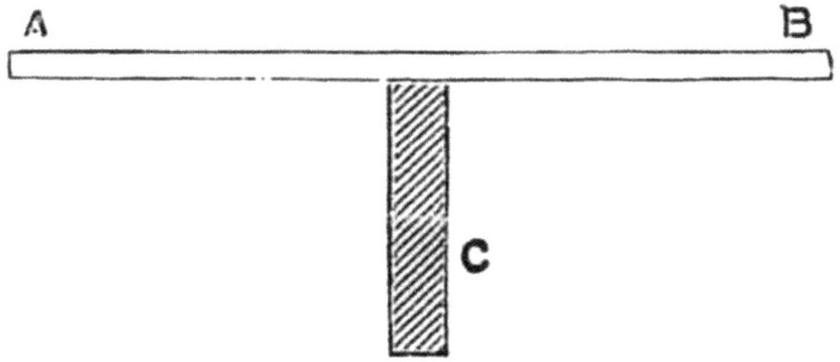

A B represents a section of soundboard, C represents a rod of wood (beech) fixed to centre of board; the board is thus held as in Chaldni's experiments with vibrating plates. Now, by taking the rod, C, between the thumb and fingers, and simply applying a tuning fork to end of rod, the capability (if any) of the board for resonance is immediately made appreciable to the ear. I tested many boards in this manner, and was surprised at the variation between apparently similar boards as reinforcers of sound; in fact, by perfect imitation of form, thickness, &c., and matching the grain, I could never get two alike as reinforcers, but by testing in the manner explained boards having the most dissimilar form, appearance and grain would produce similar results; in fact, I was using a most perfect method of choosing soundboard resonators by means of resonance itself...

Now, it occurred to me during these experiments that the old Cremona violin-makers utilised some such principle (i.e., testing for resonance) in the selection of the wood for their incomparable instruments, and not by any peculiar method of matching grain, form, &c., nor yet by using the most seasoned woods. Modern times have shown that most perfect use of these latter methods do not produce violins anything like equalling the efforts of the Cremona makers of last century. The reason is sufficiently obvious--- the capability of any board as a resonator depends on the elasticity of the fibre, which may vary considerably in boards of apparently similar grain and season.

Acting on this clue I selected out of a number of pieces of wood in the manner explained those having the best resonating power. From these I constructed a sound-box, and certainly I was not quite prepared for the result. This box, although roughly constructed out of common pieces of wood lying about, is as a resonator far superior to any of the many violins I have at my command (some 30); when a vibrating fork is placed on it the volume is extraordinary. Not only this, but it has a delightfully pure tone, and is exceedingly rich in the more valuable harmonics. I have not yet tried its effect on vibrating strings, but I opine it must be similar.

You can see that it is quite possible by this means to select every individual portion of a violin, before putting together, by resonance, which evidently must be the most perfect principle. It is nothing more nor less than selecting for resonance by resonance."

(H Sutton, English Mechanic and World of Science, No 711, Nov 8, 1878, p222.)

Henry often had people over to the family home when conducting tests on his telephones, one of which was placed in the basement while another was placed on the third storey in the family living room. On one such night, his family and friends were entertained by his rendition of some very well-known songs such as 'Rule Britannia', 'Little Brown Jug' and 'Paddle Your Own Canoe'. The songs were sung into the telephone and were heard clearly at the other end of the line and from there, a chorus was sung in response. The voices of the singers were clear and recognisable, as were the sounds of animal noises, cooeeing and laughter. On other occasions, Henry played the piano and one time he played the waltz he had composed. It was easily recognised by all as 'The Sutton Waltz'.

Late in 1877 and early 1878, newspapers began to print many reports of experiments carried out with telephones constructed by a number of different people in Victoria and Australia. By January 1878, Henry was making modifications to his telephones almost daily; particularly after a public experiment involving other telephone researchers. At the time, he discovered some of his telephones were less powerful than others. Within a week, Henry had made a number of enhancements to his devices and in so doing made world history. However, due to isolation from the world, his telephone improvements bought him no fame. One of the reports published at the time described the unique design of Henry's new double compound telephone. It was the first telephone in the world with what is known today as a telephone handset. The first public test of the design was on 13 February 1878, at the Main Road store and from the 15 February, reports of Henry's new telephone began to be published in newspapers around Australia, such as this account in *The Argus*:

"Some interesting experiments in telephony were made a few evenings ago at Mr H. Sutton's music warehouse, in Sturt Street, in the presence of Mr Oddie, Mr Bechervaise, the post-master, Mr Blandford of the telegraph office, and several other gentlemen. The instruments experimented with were two of Professor Bell's portable telephones (made by Mr Sutton), from which splendid results were obtained, and a new form of telephone, described by its inventor, Mr H. Sutton, as a compound telephone, and the general opinion of those who witnessed the experiments is, that in spite of the adverse circumstances under which they took place, this telephone is a great improvement on Professor Bell's. In all the experiments a resistance equal to 30 miles of ordinary telegraph wire was used in the circuit. It appears while experimenting seven months ago with the horse-shoe form of permanent magnet, the poles of which were surrounded by convolutions of fine wire and various diaphragms, Mr Sutton found that when subjecting a vibrating diaphragm to the action of both poles of the magnet at the same time, instead of the entire magnetic force being utilised in involving electricity, that part of the force was neutralized in the diaphragms, which seemed to act in the same way as the keeper of a permanent magnet. He found by presenting only one pole of the same magnet to the diaphragm that a far stronger current was evolved, and knowing the action is reciprocated between each pole of the same magnet, the idea was suggested to place a coil of wire on both ends of a bar magnet. The compound form of telephone is the outcome of these and other experiments. This instrument is different from Bell's portable telephone, in that it consists of a curved magnet, surrounded by coils of wire at both ends. Each coil has a separate diaphragm, the curve of the magnet being so arranged that one diaphragm comes opposite the mouth of the operator, whilst the other diaphragm reaches the ear. The following good results are obtainable from this arrangement:-

1st That he can both speak and listen at the same moment.

2nd That it involves mere intensity of sound.

3rd the timbre of sounds transmitted are reproduced more perfectly.

4th When used as a transmitting instrument, the results obtained from all other telephones of the ordinary form in the same circuit are considerably improved.

5th The instrument is mounted on an adjustable stand, which leaves both hands of the operator free.

In Professor Bell's arrangement, two telephones are necessary, one held to the ear, another to the mouth, both hands being thus engaged. Mr Sutton's telephone being compound, opens up a new field for experimental research, and we understand that he is still engaged in experiments which may still further improve the application of telephonic science."

James Oddie
Photograph courtesy of Federation University Historical Collection

On Saturday 16 February 1878, Mr James Oddie, who was a great mentor and advocate of Henry's scientific talents, organised a dinner and telephonic séance which was held at Craig's Hotel, Ballarat. To highlight Henry's work with telephones, Mr Oddie invited twenty two gentlemen involved and interested in telephony to attend the dinner. The guest list included Mr William Bechervaise from the Telegraph Office in Ballarat and Mr Peter R. Challen from the Melbourne Telegraph Office. At this séance, Mr Challen was to learn firsthand of the success of Henry's telephones. He commented later, that up until then he'd thought that he had been the first in the Southern Hemisphere to experiment widely in the construction and use of the telephone. He gallantly complimented Henry on his successes and hoped that they might be mutually helpful to each other, for which Henry expressed his gratitude.

The evening was spent discussing advances in the telephone and conducting various telephone experiments. For these, wires had been run into the hotel from the main telegraph line outside and had been placed around the dining hall and along the passages to two other rooms in the hotel. For part of the time, the group divided up and gathered around telephones attached to the wires. Through coil distances ranging from 98 to 200 miles, conversations were held and songs sung. Mr P. Cazaly and Mr J. Robson added vocal and instrumental music to the charm of the evening.

Mr Bechervaise drew versions of private telephone boxes at telegraph stations, where one day soon people could talk to someone a thousand miles away. It should be noted that when the first official telephone service began in Ballarat in February 1883, Mr James Oddie became its first subscriber and received the telephone number "*1*".

The next morning, Henry and a number of gentlemen gathered at the Ballarat Telegraph Office and took part in some further telephonic experiments which were arranged by Mr Bechervaise and Mr Challen. Four telephones, one of which was Henry's, were placed in a circuit and communication was held with the Melbourne Telegraph Office. A quartet sang 'Home Sweet Home', which the Melbourne man said he heard very distinctly. As the experiment progressed, someone else entered the Melbourne office and the operators prevailed upon them to sing 'Sunny Days' after which 'The Brother's Lullaby' was clearly heard on the Ballarat telephones. In Ballarat, a music box was also placed on the circuit and this was heard in Melbourne, and the proceedings closed with 'God Save the Queen' sung in Ballarat and then repeated in Melbourne.

News of Henry's new telephone spread quickly and on 20 February 1878, several representatives of the Press were invited to witness a demonstration of Henry's new double compound telephone at his home in Main Road. All present expressed their surprise at how well the telephone worked and one journalist present reported: "I*t was found that the double compound telephone would transmit spoken words from a distance of about 8ft, though not always very distinctly, singing however could be heard up to 14ft away. The new instrument promises to develop into something of great utility and it is credible to Ballarat that some improvements have already been made here.*" Henry went on to exhibit two of his double compound telephones at The Juvenile Industrial Exhibition which was being held in Ballarat. The Governor of Victoria, Sir George Bowen attended the exhibition.

Local citizens of Ballarat were so intrigued by the constant reports on the new invention that one employee of the London Chartered Bank on the corner of Sturt and Lydiard Streets (now the site of the Commonwealth Bank) connected a line of string from the second floor of the Bank across the street to the second floor window of the Union Bank on the opposite side of Lydiard St. Then, employees of both banks tried to talk to each other using tin cans attached at ends of the line. This must have proved quite an amusing sight early in the morning in the centre of the city of Ballarat and inevitably the event would have created an amusing topic of conversation around town. It is not known who the employees were or if their experiment was successful, but certainly it would have amused Henry at the time. This attempt by the bank staff to talk, mimicking a telephone, may well have been one of the first times in Australia that people used tin cans on the ends of a string to talk to each other. The fun of this experiment went on to become a part of everyday play for children growing up in Australia and is still a source of great amusement. Such was the local interest in the new invention at the time that preliminary steps were taken to form a Telephonic Club in Ballarat.

On 27 August 1878, a number of gentlemen from the Telegraph Office attended another series of telephonic and microphonic experiments that were held between Ballarat and Geelong. The experiments conducted by Henry and Mr W. J. Thomas of Geelong were designed to hear what results could be obtained through the use of a microphone made by Thomas. Although the men's telephones worked well, the microphone did not. When it was in the circuit, the sounds were less distinct but the strains of a concertina playing in Geelong were still reproduced in Ballarat.

Henry put his telephones to good use while he was still a student at the Ballarat School of Mines. He set up a telephone system there in late 1877, making it the first educational institution in Australia and perhaps the world to have a telephone. The system went from the Registrar's office to the School's laboratory to test other people's telephone inventions. Later in 1884, when Henry was a lecturer at the School, he set up a telephonic circuit for use by the students which was laid out between the engine-house and the workshops.

In April 1877, the Ballarat Fire Station decided it would install a telegraph line from their Station on the corner of Barkly Street and East Street in Ballarat East to the Ballarat City Fire Station in Sturt Street. At the time, the telegraph line ran along the train line from Melbourne to the Ballarat train station, then from there up Lydiard Street on the left hand side and then down the left-hand side of Sturt and Bridge Streets and over to Main Road Ballarat East. The telegraph line then ran along the left side of Main Road to the old wooden Ballarat Post Office, which was at the time between Humffray and Barkly Streets. In order to connect the two fire stations, it was necessary to install a number of new telegraph posts from the corner of

Lydiard and Sturt Streets up to the Ballarat City Fire Station. This of course took time, as the new telegraph posts had to be installed for a total of five blocks up to the corner of Sturt and Raglan Streets.

During this time, Henry worked on several types of telephones. When the telegraph poles were finished being installed Mr Bechervaise and Mr Blandford from the Ballarat Post and Telegraph Office connected the telegraph/telephone line along the route so that the two stations could be connected. Exactly one year after the decision to install a telegraph line, the line was connected to the tower of the Ballarat Fire Station.

The line from the station ran directly from the tower, across East Street and across the land opposite and connected straight to the Ballarat Post Office buildings telegraph pole on Main Rd. From there, the line connected to the established telegraph line which ran to the corner of Sturt and Lydiard Streets, and then it was connected to the new line of telegraph poles up Sturt Street to Raglan Street and was connected to the tower of the Ballarat City Fire Station

Henry generously donated two of his double compound telephones, one for each of the fire stations. These were attached to a wall high on the tower at both places. At the Ballarat Fire Station, the mounting board from the installation in 1878 is still there with the faint outline of the shape of Henry's telephone.

Ballarat City Fire Station
Picture courtesy of Federation
University Historical Collection

Ballarat Fire Station
Photograph courtesy of the
Ballarat Fire Brigade Archives

On 7 April 1878, Henry connected the telephone system between the Ballarat City Fire Station and the Ballarat Fire Station making them the first fire stations in Australia to have a telephone system. Henry, Captain Williams of the Ballarat City Fire Station and Captain George Morris of the Ballarat Fire Station and several other officers and gentlemen were present when the first telephone conversation took place. The telegraph line and Henry's telephones were an invaluable gift, not only to the two fire stations but also to the residents of Ballarat and it was not long before they proved to be a very useful life-saving addition to the two stations.

Henry's telephones were in use at the fire stations until September 1881, when the stations were connected to the Ballarat exchange and the stations installed Edison Bell telephones. Henry continued to design new telephones, but he did not patent any of them. However, a number of years later, on 26 October 1886, he lodged the following patent for a method of telephonic transmission through gas or water pipes:

PATENT: "Improvement in Electric Circuits for Telephonic Purposes."
"Designed to enable the usual battery attached to the transmitter to be dispensed with, and in place thereof the necessary electric current is derived from a circuit composed of a gas main or service pipe and a water main or service pipe, such circuit being connected with a telephonic transmitter known as a battery transmitter, or the circuit may be formed of either gas or water main when connected with an earth plate capable of conducting electricity."

To date, none of Henry's telephones have been discovered but it would certainly of great historical interest to find one. Henry's specialty and expertise was his telephone receivers and the knowledge and experience that he gained through this work proved invaluable for his future inventions and discoveries.

In 1998, the Australian poet Timoshenko John Aslanides wrote a book of poems called *AnniVersaries* in which he wrote a poem for each day of the year. One of these, a poem called 'Inventions', paid homage to Henry and his telephones and inventions.

CHAPTER FOUR

A SERIES OF BRIGHT INVENTIONS

"We live in an age of wonders, and there is no telling where the electricity of the future may come from."

Henry Sutton 1879

The quest for a safe efficient electric light globe dates as far back as 1810 when Sir Humphry Davy invented the arc lamp. Many obstacles were overcome along the way including the necessity to find an effective filament and a way to produce a perfect vacuum, as well as the need for the light globe to be safe and economical. The race was on all over the world, but it would take a number of years before solutions started to be found.

Henry was not in the race per se he was merely trying to perfect something of great importance, knowing that the successful invention of a safe and effective light globe would benefit everyone. It was during the course of his research that Henry began collaborating with Alexander Graham Bell regarding the properties of carbon.

Once he established that carbon manila fibre would provide the most effective filament, he turned his attention to perfecting a method of creating a perfect vacuum. After constructing several forms of mercurial vacuum pumps, Henry hit upon a design that created the closest condition to a perfect vacuum inside a glass bulb that was possible at the time, and so the Sutton incandescent light globe was successfully completed in January 1880.

From as early as 1878, reports appeared in newspapers and scientific journals that Thomas Edison in the United States and Joseph Swan in Britain were both working on the light globe. However, Henry was working alone and was unaware how close they both were to achieving that goal.

On the 4 November, 1879, Thomas Alva Edison lodged his patent for the Edison electric light globe. The first public announcement and public demonstration of his incandescent light bulb occurred on 31 December, 1879, at Menlo Park in the United States.

News travelled fast around the world. Henry may have successfully completed his light globe before Edison but he had not had the opportunity to publicly demonstrate it to anyone. On finding out about Edison's demonstration, Henry quickly contacted Mr R.L.J. Ellery who was the Government Astronomer and the President of the Royal Society of Victoria. He then travelled to Melbourne, and demonstrated his light globe to Ellery on 16 January 1880, just sixteen days after Edison's announcement and public demonstration.

Although Henry received no credit for his achievement, the significance of his success in producing the light globe should not be lost. This quiet young man worked alone, isolated from the rest of the world and with very little funds, unlike Edison who had enormous funds at his disposal and a significant number of assistants to help him. Edison and his assistants had conducted over two thousand experiments just to find an effective filament; needless to say, Henry had no such luxury but had still completed an effective incandescent light globe at the same time.

In light of the announcement by Edison and subsequently by Joseph Swan, Henry's achievement was barely noticed which must have been disappointing for him. The only consolation would have been that he did not do all that hard work for material gain. Henry Sutton merely wanted everyone to be able to benefit from what he had invented. A chance soon came to do that close to his home.

By May 1882, the City of Ballarat was considering options to change the city's Sugg gas-powered lamps to electricity because the Sugg system was failing to solve the problem of adequately lighting the streets of the city. Henry Sutton gave this problem much thought and painstakingly calculated how much it would cost, first by gas and then by a number of different systems of electricity. His research, recorded in detail, took into account such things as candle-power, gas used per hour and horse power generated by a dynamo.

Henry confirmed Councillor Claxton's statement that a twelve horse-power gas engine would light up Sturt St. It would power enough electric lamps on a circuit, to have a lamp at every cross street from Grenville to Lyons Streets, and only need a relatively thin conductor. The consumption of the gas by the engine would be about 250 cubic feet per hour, the same quantity that should be burned in the combined Grenville and Armstrong Street Sugg Systems.

Of the various systems of electric lighting then available, Henry considered that although a number of them would be suitable, but there were disadvantages with others.

The Siemens system was too expensive for Ballarat and the Australian Electric Company system needed improving as their candle-power was below standard. Although the Gramme and Siemens systems were suitable, Henry felt at the time that the Weston system which was already in use in Australia, was far superior and more suited for Ballarat's needs. He calculated that the cost of the dynamo, five lamps and the required electrical accessories to complete the circuit would be about £350, and there was also a ten light machine if required.

Henry considered that Sturt Street was especially suitable to the electric light and that if the City Council were to adopt it, they would also be able to light the City Hall internally by aid of incandescent lamps, the electricity being supplied from the street circuit. Unfortunately, the Council did not act on Henry's information and advice and continued to dally with the problem of the Sugg lamps and electricity. But it did not stop Henry from pursuing the advancement of electric lighting in Ballarat, and later that year a business opportunity presented itself and he quickly jumped on board.

The Australian Electric Company, which was based in Melbourne, was now venturing beyond Melbourne and was beginning to expand into the country cities and especially into the mines of Victoria. The company had invited the public to buy shares in order to provide the means to expand the business. A few well know Ballarat residents took up the offer and the company setup a local committee to oversee the installation of electric lighting in Ballarat for the company. The committee which was formed in early November 1882 consisted of the Mayor of Ballarat, Mr Frederick Moses Claxton; Mr James Oddie, Council Member; Thomas Houlden Thompson, Mr William H Barnard and under the urging of many Ballarat residents and Mr Oddie, Henry Sutton was appointed to the committee as his expertise was highly respected.

The Australian Electric Company had secured the exclusive patent rights to the use of Joseph Swan's incandescent light globes for the lighting of halls, theatres, shops, businesses and in mines. Mr John Ditchburn Junior was the

Ballarat agent. It was great satisfaction that only days later, the company had secured the contract to install electric lighting in the Band of Hope and Albion gold mine.

In an effort to demonstrate the advantages of electric lighting to the people and businesses of Ballarat, the company and the committee set up electric lighting in Jago's Boot Palace and Mr Pojoy's, Unicorn Hotel in Sturt Street. The installation was powered by one of the company's dynamo-machines which was driven by an eight horse power portable engine, supplied by A.G. Corbett and Co. The installation of electric lighting was aided immensely by Henry Sutton and on 25 November 1882, the two premises were successfully and spectacularly illuminated for all of Ballarat to see. To say the demonstration was a roaring success would be an understatement.

A few days after the success of the electrical lighting display, the Australian Electric Company's electrical engineer, Mr G.W. Selby, gave a lecture on electric lighting at the Ballarat Royal Exchange Room. During the lecture Selby took the opportunity to praise Henry for his expert assistance with the display and then went on to congratulate him on the success of his recent battery invention. He stated it was the best of its kind that he had ever seen. The battery which will be covered later in this book, had by that time, bought Henry world acclaim and Selby's statement was greeted with great applause from the audience.

In late December 1882, the Australian Electric Company was busy installing a powerful dynamo-machine at the Band of Hope and Albion's number seven mine. The dynamo was to be driven by a 12 inch cylinder steam-engine and would be powerful enough to furnish 120 lights around the mine. Initially there would be 41 lights above ground to light up the mining site and about 46 in other parts of the mine. An arc light was to be suspended from the top of the poppet-head at the number seven shaft.

As part of the local committee, Henry must have been very satisfied to finally bring electric light to the mine that both his mother and father had bought shares in all those years ago. Richard and Mary's shares in the Band of Hope and Albion mine helped pay for the building of the family business over the years and it would have brought great joy to Henry to be involved with the historic installation of electric lighting there.

On 1 February 1883 at 8pm, the lights were switched on and thousands turned out to witness the event. The arc light at the top of the poppet-head was so bright that a newspaper could be read 300 yards away.

At the conclusion of the lighting display at the mine, the company representative expressed the hope that soon there would be electric lighting in the Ballarat Hospital and the Benevolent Asylum and encouraged the members of the electric light installation committee to take the matter up. A few days later, Barnard, who was on the committee, announced that he was now the local agent for the company and was now able to supply the necessary equipment to install electric lighting in Ballarat.

Just as it is now in Ballarat, residents had various ideas of how to fund electric lighting in the city. A year had passed since Henry proposed the best way to light the streets and now that a company had set up business in Ballarat their hopes were high. At the council meeting on 7 May 1883, the subject was once again put to the council. At the meeting Cr. Smith asked in a facetious manor, "*How about the electric light?*" Cr. Thompson was quick to reply and stated back, "*Give me £1,000, and I will quickly show you the electric light.*" It was clear to everyone over time that unless funds were made available, Ballarat streets would not get electric lighting.

The years rolled by and when Henry left for England in 1890, the city of Ballarat still had no electric street lights. Three years later, when Henry returned home, he must have lamented that situation was unchanged. One could only imagine the long frustration that he and the residents of Ballarat endured as the wheels of politics turned slowly, holding progress back as the city was trying hard to move forward.

Finally in January 1894, the Ballarat Electric Supply Company was formed by Oddie and a number of leading Ballarat gentlemen. On 26 January 1895, the streets of Ballarat were finally illuminated by electricity, nearly thirteen years after Henry's proposal. Today there are approximately 10,000 street lights in the City of Ballarat, but the memory of the Sugg lamps still remains as the replicas of those beautiful lamps still adorn the heart of the city.

Life could not have been all work for Henry; during this time, a lovely young lady by the name of Elizabeth Ellen Wyatt caught his eye.

Both the Sutton and Wyatt families were members of the Wesleyan Church in Lydiard Street Ballarat. Henry, having been a member of the Lydiard Street Mutual Improvement Society since 1872, attended many social gatherings that were based around the church and its various associations. Elizabeth was a friend of Henry's youngest sister Emilie. Both of them sang in the Lydiard Street Wesleyan Choir and both gave solo performances at various concerts in and around Ballarat.

Henry and Elizabeth's affection for each other grew over time and eventually Henry fell in love with the lovely and talented dark haired beauty and she accepted his proposal. On the 6th September 1881, two days after his twenty sixth birthday, Henry Sutton and Elizabeth Wyatt were married by the Wesleyan church minister, Mr John Cope. Henry's best friend Thomas Grundy was a witness and many of Ballarat's finest attended the ceremony.

A year later, on the 13 September 1882, Henry's brother Alfred married Miss Eliza Ann Treloar. Richard and Mary's children were growing up and one by one they were leaving the family home. Walter, who was a school teacher, went on to marry Mary Elizabeth Ruddock in 1887, and young Frederick married Elizabeth Jago in 1888.

Henry and his new wife set up home at 74 Doveton Street, Ballarat and Elizabeth soon became pregnant. However, the joy of the impending birth gave way to sadness when on the 24 March 1882, Elizabeth's brother William Wyatt suddenly died of epilepsy, a condition from which he had suffered for several years. It was not an easy time for the Sutton or Wyatt families, but sorrow became joy again when on the 25 June 1882, Henry and Elizabeth's first son Arthur Ernest was born.

After his marriage to Elizabeth, Henry put aside the invention of the light globe and focused on one of the other inventions he had designed, the vacuum pump. At a meeting at the Royal Society of Victoria on 13 December 1881, the President, Mr R. L. J. Ellery, announced that a short paper had been received from Mr H. Sutton of Ballarat, describing a new vacuum apparatus for the rapid production of high vacuum suitable for electric lighting by the incandescent system.

Paper Read: Description of Vacuum Apparatus with Diagram.

"This apparatus is the outcome of several forms of vacuum apparatus I have devised for the rapid production of high vacuum suitable for electric lighting by the incandescent system, or investigations such as Mr Crookes, has recently given to science.

A mercury bottle (m) when held up, allows mercury to rise in the tube (A), filling reservoir (B), and rising in branch pipe (C) isolates lamp globe (H) from reservoir (B). The mercury passes by outlet pipe (D) filling vessel (G) till it overflows by pipe (E) into vessel (F); (G) is a glass tube of large diameter, having a glass stopper in each end, both stoppers being covered by a mercury jacket; the lower half of this tube is filled with broken pumice, soaked in sulphuric acid to absorb any moisture, the upper half being filled with pumice, gilded with gold leaf to absorb traces of mercury vapour. On lowering mercury bottle, the mercury descends in reservoir (B) and pipe (C), leaving a Torcillian vacuum in (B), the mercury in (D) being supported by atmospheric pressure. On lowering (m) sufficiently, communication is established between reservoir (B) and vessel to be exhausted (H) by means of branch pipe (C) and dissecting apparatus (G), the air in (H) then expanding over into (B); the mercury bottle (m) is then raised again, and the mercury rising in (B) drives out the expanded air by outlet pipe (D). This pipe is bent back on itself at the lower end; this is to prevent the ingress of any air that might collect on end of pipe in vessel (I), alternate raising and lowering the mercury bottle (m) producing a vacuum."

[Full Text, see End Notes, Chapter Four, No. 1.]

A few months later, on the 17th March 1882, the E*nglish Mechanic* and *World of Science Journal* published a letter from Henry in which he describes one of his vacuum pumps and some experiments he made. He also referred to the innate problems associated with the Sprengel vacuum pump and the problems that he solved in relation to it.

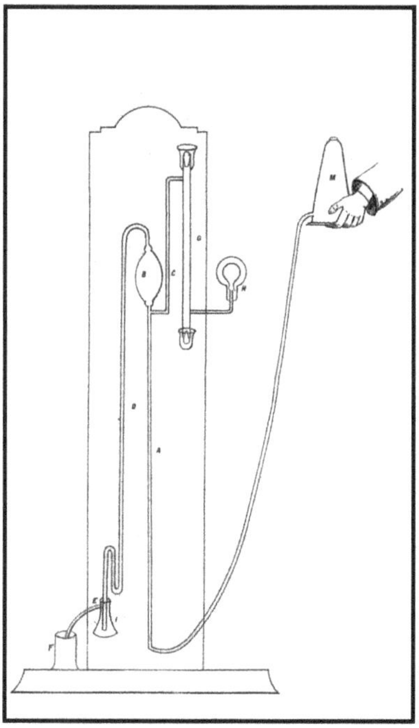

Vacuum Pump diagram
by Henry Sutton 1881

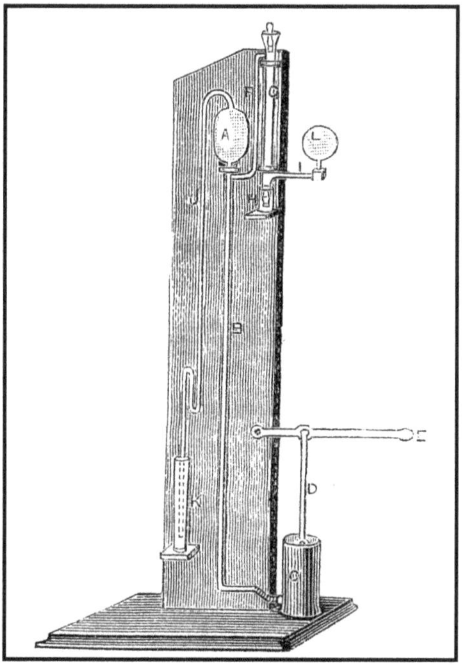

Sutton's Apparatus for the Rapid
Production of High Vacua. 1882

The letter stated:

"Sutton's apparatus for the rapid production of high vacua."

"I have just received my English Mechanic for Nov. 11, 1881, in which, under the heading "Obtaining and Measuring High Vacua," there is a description of Prof. Rood's improved Sprengel pump: its limit is stated to be arrived at "when the leakage of the apparatus is equal to the amount of air that can be removed in a given time."

I send you a description and a sketch of a vacuum pump invented by myself some time ago, and which, you will see by its construction, has a higher limit still, as the amount of air that can be removed in a given time exceeds the leakage many hundred-fold; in short, the air can be withdrawn more rapidly than it can leak in. Its construction is simple; there are no stop-cocks, nor valves; it is not liable to derangement, and will do in a few minutes that which takes hours on the Sprengel principle."

[Full Text, see End Notes, Chapter Four, No. 2.]

Henry made a number of different vacuum pumps and in 1881, he had donated one that worked by water jet to the Ballarat School of Mines and this was put into daily use by the students. He never patented his vacuum pumps, but once again chose to let the world make use of them for the benefit of all.

His vacuum pump proved to be an invaluable tool in the medical field, in hospitals in Britain and to industries around the world. Joseph Swan and Thomas Edison, both of whom had also invented light globes, quickly took notice of Henry's new vacuum pump and it was subsequently utilised by both men in the production of Swan and Edison incandescent globes in England.

The great merit of Henry's vacuum pump was that it was inexpensive and certain in operation and created so perfect a vacuum that even long after his death in 1912, his vacuum pump was still being used in London hospitals and in various industries.

In 1880, the Government Astronomer and President of the Royal Society of Victoria, Mr R. L. J. Ellery, wrote to Henry to ask him if he could improve the design of the batteries used by the Victorian Torpedo Corps. Ellery, who was also the Captain of the corps at that time, was no longer satisfied with the Leclancé batteries the corps were using to fire the torpedoes and mines in Port Phillip and felt a new style of battery would serve the corps better. The Leclancé battery had been patented in 1866 by the French scientist Georges Leclancé.

Henry naturally took on the challenge, but before he could complete his new design, a tragic event occurred in Port Philip Bay.

Shortly after 5pm on the 7 March 1881, a huge explosion occurred as the corps ship, the *Cerberus*, was laying out mines just off Queenscliff; five men were tragically killed. The news of the explosion and the subsequent inquiry would have weighed heavily on Henry's mind as he went about inventing his new battery for Ellery and the corps. The inquiry into the explosion was held and many questions were asked, regarding the electric current and batteries used, in the attempt to find out what had gone wrong. Eventually the inquiry concluded that the premature explosion of the mine was caused by an electrical current but could not establish just how it occurred.

In August 1881, Henry demonstrated his new battery to Ellery who immediately knew it was a great step forward in the evolution of batteries. They immediately made plans to protect the invention, even though Henry insisted on giving his invention free to the world as he believed that the fruits of science belonged to, and should benefit, everyone.

It was decided that Henry would write two papers disclosing the details of his new battery. One was sent to the Royal Society in London and the other was put under lock and key at the Melbourne Observatory, to be read immediately after the Royal Society in London received Henry's original copy.

It did not take long for hints and speculation of Henry's new invention to circulate and on the 9th December 1881, one paper was read by Ellery in his role as the President of the Royal Society of Victoria. He later stated that, "*Henry's discovery will revolutionise the present mode of using electricity as a motor.*" The original paper was read a few days later on the 15 December, by the President of the Royal Society in London, Mr William H. Spottiswoode. Henry's paper and subsequent supplementary notes to both societies went on to describe his experiments and his new form of secondary battery.

Henry's paper presented to the Royal Society of Victoria:

"The Great utility of some thoroughly practical method of conserving electric force has caused a great deal of attention to be applied to the subject. No system of electric supply can be considered perfect until some means is used to store the force generated that it may be drawn off equally and regularly, and this whether the generator be on or off.

If we take as an example of electric supply the present systems of electric lighting, it is at once seen, should an accident or stoppage take place in the machinery generating the current, the whole of the apparatus, such as lamps or motive machines, would be influenced. Should there be a reservoir of electricity between the generator and the apparatus for utilising the force, this inconvenience could not occur, as the reservoir would then supply the necessary force."

[Full Texts, see End Notes, Chapter Four, Nos. 3 & 4.]

An easy and simple description of Henry's new battery appeared in the *London Graphic* newspaper on 4 February 1882.

"(A) Is a flat copper case with a lid (B) of paraffined wood, and a slip (C) of the same waterproof material fitted into its bottom. In each of these pieces of wood is a central groove for the reception of an amalgamated plate of lead (E), i.e., a plate of lead coated with mercury to prevent local chemical action. Through the top (B) projects a glass tube, which serves both for the inlet of the battery solution and for ascertaining the condition of that solution, as will presently be explained.

The cell is filled with sulphate of copper (common bluestone) dissolved in water, with the addition of sulphuric acid. When this cell is subjected to a primary current, the decomposed oxygen from the solution forms a coating of peroxide on the lead plate, while the hydrogen replaces the copper salt, and metallic copper is deposited on the containing cell. As this action goes on, the solution loses its blue colour, to regain it when the operation is reversed, and when the cell is giving out the secondary current. We now understand the use of the little glass tube, through which the colour of the fluid can be seen. This colour test shows unerringly when the operation of charging is complete. Mr. Sutton gives his invention to the world untrammelled by any patent rights and its obvious importance will speedily cause it to be submitted to rigorous tests. To give an idea to its power, the inventor states that a cell, four inches deep by four inches internal diameter, gave a current which for two hours red-heated an inch of wire (No. 28 gauge)"

A personal telegram dated 4 January 1882, was sent to Henry from Mr Spottiswoode the President of the Royal Society London, which read:

"Dear Sir, I have the greatest pleasure in receiving the addition to your former paper on 'Storage Batteries,' and have put it forward for reading at our next meeting. Your former communication created much interest among our fellows. I hope soon to send you printed copies of it.

With best wishes for success in the continuation of your researches.

I remain, &c., W. Spottiswoode."

News immediately spread about Henry's new battery and reports were

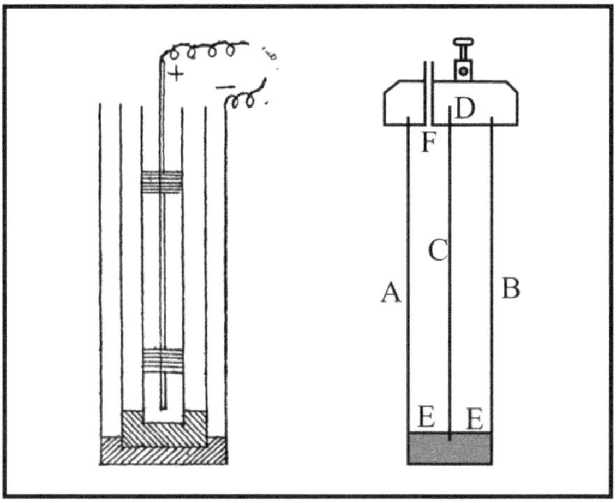

Battery Cell diagrams by Henry Sutton 1882

published in newspapers and journals in Australia and all over the world. He received world acclaim for his new invention, so much so that Thomas Edison was to state at the time that it was the best battery in the world. A copy of a news article on Henry's battery was kept by Edison and is in the Edison Papers online. Henry gave his first lecture on his new battery on the 14 December 1881 at the fourth meeting of The Chemical and Electrical Industry Society which was held at the Ballarat School of Mines. The President, Mr James Oddie, Henry's long time mentor and advocate, opened the meeting and after Henry had talked on the history of batteries and the power of electricity, he demonstrated his new battery and conducted other experiments to the delight of the audience. The evening was concluded by a statement that Mr Sutton had produced a battery that was "second to none".

On the night, Henry donated a six cell battery to the Ballarat School of Mines.

On 8 February 1882, The Chemical and Electrical Industry Society had met again in Ballarat and Mr Oddie informed the members that at a recent meeting of the Royal Society London, Mr Sutton's paper on his storage battery was bought prominently forward and that accounts of it had appeared in several scientific journals. News of Henry's success was met with repeated applause. However, two days later it was reported that the Royal Society of Victoria had seemed to take only minor notice of Henry's achievement when all over the world such publications as *Nature, English Mechanic, World of Science and Iron,* just to name a few, were praising Henry's new battery, and many journals and newspapers noted the fact that Henry had not patented his battery, instead he gave it free to science and experimentalists wherever science is cultivated.

News of the Royal Society of Victoria's seeming indifference to Henry's battery sparked outrage in Ballarat and drove one citizen to write to the Editor of the *Ballarat Courier* expressing his personal outrage. After chastising the Society, the outraged citizen went on to state the following:

"The fact that Mr Sutton is a modest unassuming man… may mitigate against his becoming famous in the ranks of scientists, and may have influenced his judges against him… The fact that Ballarat has the honour of being the seat of the discovery may have raised the jealousy of the Melbourne savants,… With a citizen like Mr Sutton among us, we have more to be proud of than many other cities in the world; and I think it only due to such men, when their lights are shaded by the petty jealousies or ignorance of others, that their fellow citizens should honour them before the world… I think therefore the people of Ballarat should show their recognition of the value of Mr Sutton's discoveries and the honour of having him as a fellow citizen."

Although a naturally humble man, Henry quickly responded by writing a letter to the Editor of the *Ballarat Courier* to quell the growing outrage of the people of Ballarat. He explained that the Royal Society of London received his paper before the Royal Society of Victoria and the Victorian Society had not yet had a chance to appraise it. The announcement of his battery had to be timed with both societies in order to make the wide-reaching announcement of his giving the battery freely to the world.

The citizen who expressed his outrage was right in one respect. Little fuss was made in Australia regarding Henry's battery, while the rest of the world did take notice. However, his battery did cause the world to focus on the city of Ballarat in a new way. Now, it was not only known for the discovery of gold, but also as the new centre of scientific study and achievement.

On the 4 April 1882, Henry demonstrated his secondary battery to a large audience at the Ballarat School of Mines. It was a proud moment for him as his fame was bringing great recognition to the school and to Ballarat.

In a very short time, many scientists around the world began experimenting with the new battery, including the notable scientist, Joseph Swan. Just over a month after Henry's paper was published, Swan took the opportunity to give a lecture on the results he had obtained with the Sutton cell. The tests enabled him to realise all the advantages of the cell and its economy and durability. Swan immediately utilised Henry's battery in his lighting installations in mines around England and Wales. The stored electrical energy and the lamp were all in one portable unit, and the lamp was so well guarded as to render it a very perfect safety lamp. In a brilliantly lit room, Henry's battery did not display to its fullest advantage. However, in a mine, the light was far superior to what miners were accustomed to. For lighting in railway carriages, Swan thought the most feasible arrangement would be to have a small dynamo and a set of store cells for each carriage. He went on to say that if ever railway carriages or tramcars are to be driven by electricity, the electric storage of energy must play an important part in bringing this about.

Henry's new battery was embraced around the world. It could not only store electricity, but was much smaller, lighter and more economical. However, it was only after a lot of pressure from the scientific community that the humble Henry was persuaded to enter his battery in the 1882 Crystal Palace Electrical Exhibition in London. There it was considered by many scientists to be far superior to the batteries produced by Gaston Planté or Camille Faure, who also exhibited their batteries at the exhibition.

Historically it was believed that Henry had not lodged a patent for his battery, but in actual fact a patent was lodged in England on 21 July 1882 by the well-known English patent attorney, John Henry Johnson. The records have it listed under Johnson's name, on Henry's behalf. The late lodgement of the patent in England would appear to be due to either pressure from the scientific community or the exhibition of his battery at the Crystal Palace. As usual, Henry had no intention of commercially benefitting from his invention and regardless of the patent continued to allow the world to benefit from its uses.

[Sutton Battery Patent: Full Text, see Patents, Chapter Four, No.1]

On the strength of the worldwide success of Henry's new battery, societies around the world wished to attract into their membership the new man of science from Ballarat. On 6 July 1882, Henry, now a noted scientist in his own right, was nominated to become a member the Royal Society of Victoria, by the society's president. The following week, on 13 July 1882, at the monthly meeting of the society, he was formally elected as a Member.

Henry could only be classed as an Associate Member of the society as he had no formal qualifications. However, his acceptance into the membership of the Royal Society of Victoria was the first formal acknowledgement in Australia of his innovative talents and work, and this achievement would have pleased the citizens of Ballarat.

The following year, on 24 May 1883, Henry was elected as an Associate Member of the Society of Telegraph Engineers and Electricians in England which was the leading electrical society of the world. This would have been a source of celebration in the Sutton household as it was a great honour to be ranked among the most eminent scientists in the world. Henry then went on to become a member of the Society of Chemical and Electrical Industry.

In 1884, following his acceptance of membership of the British society, and due to his tenure as lecturer of Applied Electricity and Magnetism at the Ballarat School of Mines, Mr Louis Adolphe Cochery, the Minister of the Post and Telegraph Office in France, invited Henry to become a member of the Societé Internationale des Electriciens. Cochery had recently formed the societé and the invitation was a prestigious honour. The societé comprised many notable men of science both in France and around the world and had been formed to coincide with the Electrical Exhibition in Philadelphia which was due to begin on 2 September that year. Even though the invitation was a great honour, Henry chose not become a member of the French Societé.

Between 1883 and 1884, Electrical Societies from America, France, Belgium, Russia and Britain all sent offers of membership to Henry. Of these, he chose to only become a member of the British Society, which is known today as the IET - the Institution of Engineering and Technology.

Around the time of Henry's battery invention, gas shares around the world were losing value as many investors were promoting the promising future of electric lighting. Electricity was being seen as the power source of the future, with the potential to be utilised in many areas. Investment was needed to develop the equipment and batteries to provide the power. Henry's battery could store electricity and was economical and small and could be used on trains, in the mines and in various industries too numerous to mention.

The success of Henry's battery reached New York. He had two years to lodge a patent for it in the United States and many companies and investors there wanted to ensure that if he did, that they would get in on the ground floor and gain monetary advantage from it.

In 1882, ten powerful business men - industrialists and financiers - ruled the world of finance in the United States. These men were known at the time as "The Kings of Wall Street".

Early in the year, Henry received a letter from an American consortium stating that, not with-standing Henry's paper to the Royal Society in

London and the several electric storage systems he had since devised and offered free to the world, they would be very happy to find the money to join him in business. The consortium's letter stated that if he preferred, they would buy his patent outright from him. Amongst the company's principles were Mr Cyrus Field, and Mr Jay Gould (two of the Kings of Wall Street).

To some people this would have been a dream come true. Henry could have stated any price for his patent, or made any kind of business deal and been well rewarded financially. Of all his inventions to date, this was the one that could have made him a lot of money.

Henry Sutton, however, was a man of his word and he wrote back to the Kings of Wall Street with the very firm answer that he was not going to go back on his word to the Royal Society and the world and that his battery would remain free to be used by everyone. A decision that Henry would have grappled with and at times regretted as the years went on, but his act of scientific philanthropy was, at the time, revered all over the world.

Just after Henry invented his battery, the question arose of sourcing electrical power for the new venture of Melbourne's tramway system. In August 1882, a series of letters appeared in the Melbourne *Argus* newspaper debating the question of the best form of electrical power for trams. After a few days of reading the lively correspondence, Henry decided to weigh into the debate between Mr F.C. Rowan of the Australasian Tramway and Engineering Agency and a reader of the newspaper about a number of different forms of batteries and the form in which trams may be made to work using electricity:

"Sir:- I have read with interest the letters of 'Alleinhandel' and Mr Rowan, and feel it necessary to point out several errors in 'Alleinhandel's' letter of 23rd inst.

In reference to the Sellon-Volcmar battery, 60lb. is the weight given per horse-power per hour. This is incorrect, and represents the weight of metal composing the elements; to this should be added the weight of case and solution, fully 20 lb. more; total 80lb. per horse-power per hour. Nothing has been publically heard of this battery since the Crystal Palace Exhibition, so that up to the present we cannot accept it as superior to either Faure's or my own, both of which are at present unsuitable for tramway work.

'Alleinhandel' states that the nature of the metallic parts used by the Sellon-Volcmar Co. is a secret. This may be so to some, but is well known to a reliable authority in Melbourne as being identical with a metallic paste

discovered and used for the purpose by myself prior to the Sellon-Volcmar patents. I can assure 'Alleindhandel' that were the Sellon-Volcmar Co.'s secret exploded the value of the battery would be considerably lowered.

Two facts I give. 1St. This metallic paste costs about 2s. a pound to make, 2nd Minimum used for the same purpose by Faure costs about 5d. a pound so that Faure's can be made at least one fourth the price of Sellon-Volcmar.

The great objection in the latter is the necessity for replacing the position plates, estimated every 15 months, so that any tramway or other company, by spending £2,000 on them, would have to further spend £1,000 every 15 months to keep them in order.

I thoroughly agree with Mr Rowan that up to the present time no system has been published which enables electricity to be used economically, or with any advantage as a motive power for trams. The thing seemingly will never be done by any system in which the electric current reaches the tramcar through cables, overhead wires or rails, the difficulty in the peculiar insulation required being apparently insurmountable.

The authority Mr Rowan quotes, stating that the early possibility of using electric energy for the purpose will be in the discovery of portable storage batteries, seems to me perfectly right in his conclusions; and there is no doubt from the importance of the subject, and the number of investigators now attacking the problem, that we may soon hope to see the matter solved, and if things do not very much alter, solved in Victoria."

Henry Sutton

At this time, as apparent by this letter, no viable system for running trams on electricity had been found. However, whilst Henry's letter did predict that trams would run on electricity, he was wrong in thinking that power would never be delivered to trams by overhead wires. Melbourne did eventually get electric trams that began running in 1889, but owing to disputes, that particular venture failed and it would not be until 5 May 1906, that electric trams would became a permanent fixture on the Melbourne landscape.

As well as considering the delivery of electrical power above the ground, Henry also gave some thought to its delivery underground. On 12 December 1882, at the New Australasian Mine in Creswick, Victoria, there was a disastrous mining accident in which 22 men died. A year later it still weighed heavily on everyone's minds. Henry was no exception and he wrote on the value and possibilities of electricity in mines:

The Value of Electricity to the Miner

"First and most important, we have light. Electricity is capable of producing a light which cannot vitiate the atmosphere - a light produced by a quadruple interchange of energy. We have the heat of the furnace converted into mechanical force by the steam engine; then this force converted into electricity by the dynamo-electric machine; then the electric current converted into heat by the resistance to its passage in the carbon of the lamp, this heat then rendered the carbon so hot as to emit light. The miner's candle supports itself by withdrawing oxygen from the air at the miner's expense. With the electric light, the miner can work in pure air. Electricity can be applied to indicate the presence of fire, and thus becomes a fire alarm; it can also be made to indicate the presence of water, and then becomes a flood alarm. Had an automatic alarm been in use at the New Australasian Mine at the time of the disastrous accident which cost the lives of twenty two men, not a single life would have been lost. For the men in the most remote parts of the mine would have been instantaneously warned as to the velocity of electricity through copper wire is one hundred and eleven thousand seven hundred and eighty miles per second. The presence of foul air can be detected and the warning sent. The whole of these danger detection and warning signals are capable of being automatically controlled by electric force: it is therefore worthy of the best attention of our Legislature. As a method for the transference of energy electricity is fraught with the greatest importance to the miner. In our leading mines a great deal of work is done through the aid of winzes, no inconsiderable amount of time and labour being spent in hauling by common windlass placed at the mouth of a winze. Now this may be economically obviated by an electrometer doing the work of winding. A motor of this description would weigh something like fifty pounds per horse-power, the electric current being supplied from a generator on the surface, and the force transmitted down the shaft through conductors. Of the energy spent in driving the generator, fifty per cent may be realised in the motor."

Henry's electrical fire and flood alarms were simple inventions and easy to install, but it would take many years before the legal requirement for effective safety alarms in mines to become a matter of serious attention both for mining companies and for government legislature in Australia.

Although fire alarms in general were slow to be implemented in Australia at the time of the mining disaster, they were however being introduced into mines, fire stations and hospitals in cities around the world.

After discussions with Henry in December 1883, James Oddie asked him if he could invent a fire alarm system for Ballarat's Benevolent Asylum. Henry soon came up with a new system, which many experts of that time considered to be a superior fire alarm.

The most current version of fire alarm on the market in the early 1880s was the Bright system which had been exhibited at the Paris Electrical Exhibition in 1881, and had already been installed by the London Fire Brigade. The Bright alarm had a "closed circuit" system and the battery had to be kept working day and night. It had a history of sounding false alarms and this was a serious problem which plagued the London Fire Brigade.

Henry's fire alarm was an "open circuit" system and the battery power was only used when the alarm was in use, and in this system it was impossible for the alarm to send a false signal.

In order for Henry's fire alarm system to be installed at the Benevolent Asylum, many obstacles needed to be overcome such as cost, the use of the telegraph line and making a connection to the Ballarat City fire station in Sturt Street. At that time, the telegraph poles only reached as far as the Ballarat City fire station on the corner of Sturt and Raglan Streets and the Benevolent Asylum was a number of city blocks away in Ascot Street. Unfortunately, the cost and other considerations involved in setting up Henry's fire alarm system meant it was never installed.

It would not be until June 1886, that the Ballarat Fire Brigade finally decided to install Ballarat's first fire alarm system. By this time, the fire station's telephone system had been connected to Ballarat's telephone exchange and some residents could call the fire station directly. However, the growth of many large trees obstructed the view from the station's towers and the many residents who could not telephone were still at risk. By the 7 January 1886, information and plans on the Bright alarm system had been sent by the Melbourne supplier, Mr J Wilson, and the brigade recommended adopting the system for Ballarat.

The system of alarms was purchased for £150 but not even the supplier could shed any light on how to put them together and install them.

By January 1887, many members of the Ballarat's fire brigade had expressed their disapproval of the Bright fire alarm system and at the monthly meeting on the 7 January, Captain Trotman, announced that the alarms were a "white elephant". There were many other criticisms made by other members of the brigade who had read the reports, from England and Melbourne, of all the problems with the Bright alarm system including the frequently triggered false alarms. Over the ensuing weeks these views sparked much public debate about the system's cost and usefulness.

That Henry had had designed a fire alarm system years before, which was considered by experts to be a better system, was publicly bought to the attention of residents via the local newspapers.

In an attempt to quell the adverse opinions of the Bright system, and put the equipment to use, the Brigade had little choice but to seek Henry's advice on how to install it.

Henry respectfully inspected the Bright system for the brigade and, regardless of the fact that they were not going to use his system, agreed to put the purchased alarm system together. Before the installation could begin, a number of factors needed to be resolved, including permissions from the Postmaster General to use the telegraph lines.

The installation began and it took the combined talents of Henry, Mr Bechervaise from the Ballarat Telegraph Office and two of his workers, Mr Crisp and Mr Burbridge, a year to complete the process. It would not be until April 1888, that the fire alarm system was trialled with alarm boxes placed in Humffray Street, near the North Grant Hotel, and in Bridge Street.

By November 1888, the Bright fire alarm system was up and running albeit with a few mechanical issues and some vandalism by some Ballarat youths and the foreseeable problems of false alarms.

On the 4 March 1889, the Ballarat Fire Brigade sent a letter of thanks expressing the brigade's gratitude to Henry and the other men for their valuable services in the connection with the erection of Ballarat's fire alarm system.

1883 Swan Edison Light Globe
(Above)
(Image courtesy of Queensland University Physics Museum)

Vacuum Pump believed to be made by Henry Sutton circa. 1882 (Right)

The Kings of Wall Street
Image courtesy of The Miriam and Ira D. Wallach Division of Art, Prints and Photographs: Print Collection, The New York Public Library. (1882).

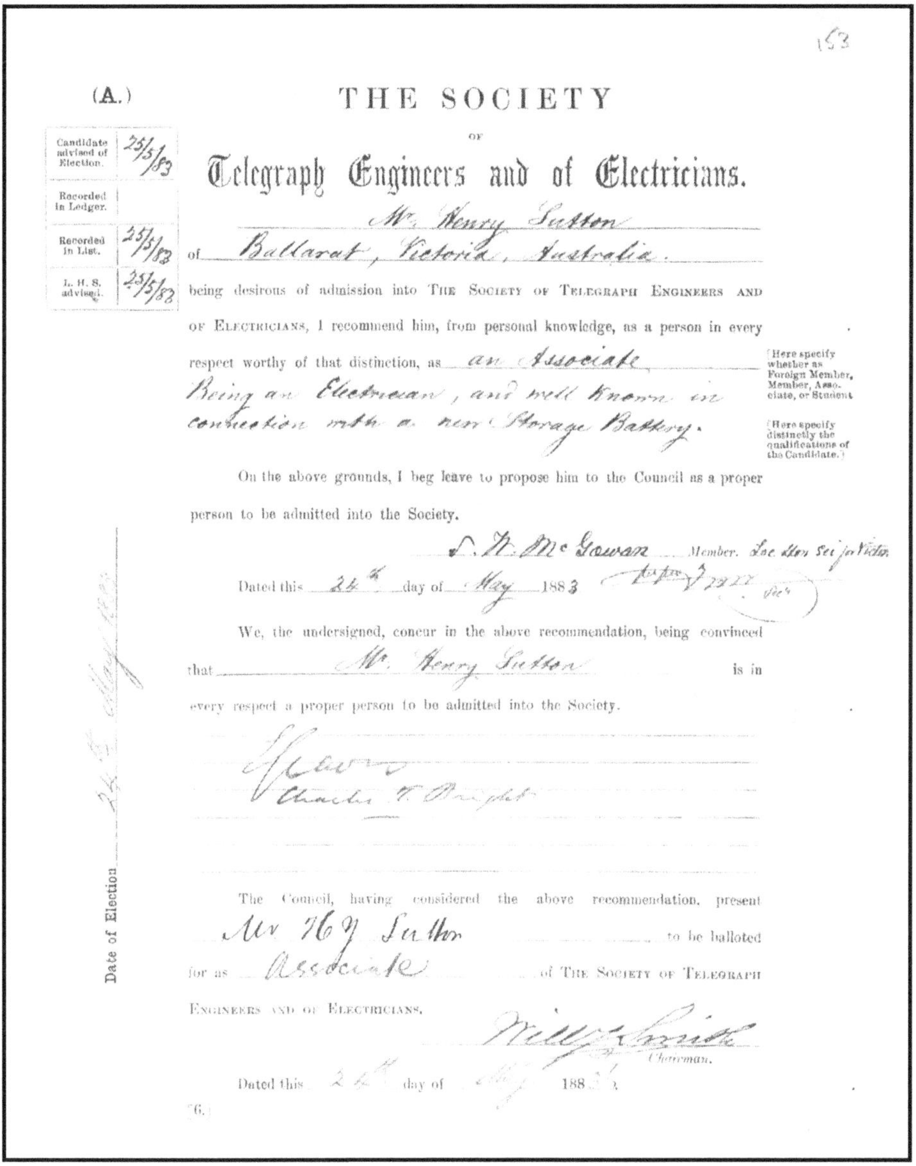

153

(A.)

THE SOCIETY

OF

Telegraph Engineers and of Electricians.

Candidate advised of Election.	25/5/83
Recorded in Ledger.	
Recorded in List.	25/5/83
L. H. S. advised.	25/5/83

Mr. Henry Sutton

of Ballarat, Victoria, Australia.

being desirous of admission into THE SOCIETY OF TELEGRAPH ENGINEERS AND OF ELECTRICIANS, I recommend him, from personal knowledge, as a person in every respect worthy of that distinction, as an Associate (Here specify whether as Foreign Member, Member, Associate, or Student)

Being an Electrician, and well known in connection with a new Storage Battery. (Here specify distinctly the qualifications of the Candidate.)

On the above grounds, I beg leave to propose him to the Council as a proper person to be admitted into the Society.

S. W. McGowan Member. Loc. Hon Sec for Victo.

Dated this 24th day of May 1883

We, the undersigned, concur in the above recommendation, being convinced that Mr. Henry Sutton is in every respect a proper person to be admitted into the Society.

Charles T. Bright

Date of Election 24th May 1883

The Council, having considered the above recommendation, present Mr H. Sutton to be balloted for as Associate of THE SOCIETY OF TELEGRAPH ENGINEERS AND OF ELECTRICIANS.

Willy Smith
Chairman.

Dated this 24th day of May 1883

(6.)

Copy of Henry Sutton's Admissions Record of Membership
to
The Society of Telegraph Engineers and Electricians
London, England.
1883
(Image courtesy of the Institute of Engineering and Technology, Savoy Place, London.)

CHAPTER FIVE

HENRY'S MICROSCOPE AND OTHER INTERESTS

"Somewhere, something incredible is waiting to be known."

Carl Sagan

It is not known exactly when Henry started his studies with a microscope but it is clear by the content of one of his letters to the *English Mechanic and Mirror of Science Journal* that by 1870 he had been studying with one for a number of years. In his letter published in the journal, Henry gives advice on the best methods of mounting microscopic objects, referring to a method that was used in Canada at the time.

On 12 October 1883, it was proposed at a meeting at the Ballarat School of Mines to establish, in connection with the Field Club and Science Society, a class in microscopy to be presided over by Mr W. H. Wooster, who was a member of the Microscopical Society of Victoria. Because Henry was actively engaged in the subject, he was eager to share his knowledge with Wooster as he continued his studies on the subject.

Henry had, by this time, turned his attention from the study of plants and insects to the study of germs and diseases and had acquired a microscope that could magnify objects 2500 times. The microscope enabled him to take photographs of objects which were invisible to the naked eye and was an invaluable asset to his study of germs. He was well aware of the fragility of the human body and the threat germs were to human existence and knew of the hard work of men of science to try and alleviate the suffering of diseases. People such as Louis Pasteur and Robert Koch had bought the fight of germs to the forefront of science and although some progress had been made with some diseases, the race was still on to find causes and cures for others.

Interest in the study of germs became personal for Henry when, in 1883, his sister Emilie was diagnosed with tuberculosis. This tragic news brought home how, when it came to germs, no one was immune. At the time of Emilie's diagnosis, Henry's writing about this subject is etched to its core with sadness as he laments on the beauty and the fragility of the human condition:

"Perhaps, beyond everything, the welfare of our bodies should be our first care; for with unhealthy frames what are we fitted for? Absolutely nothing: our whole energy is built on the state of our system. We are truly, as the ancient sage writes, 'fearfully and wonderfully made.' Look at our muscular system, the ropes and pulleys of our osseous being. How well and how fitly designed is everything! And then that intricate and beautiful nervous system, comprised of motor and sensory nerves, the former transmitting the will to the various organs, the later conveying the sense of touch whether it be pain or pleasure; also that most marvellous optical instrument, the eye, with its lenses, refracting humors and beautiful retina taking up and analysing the multitudinous vibrations of light.

Then our ear, with its tympanum malleolus, and apparently clumsy but exceedingly sensitive system for the transmission of sound waves. The most complicated piece of mechanism ever constructed by man is nothing to the human body, and yet how easily the work of man falls into disrepair! How much care, therefore, should we not take of ourselves! Disease, that fell enemy of ours, always overtakes us, and I suppose ever will; for we can never hope for the elixir of life and universal youth."

Henry Sutton (Alias Delta): Scientific Notes, The Microscope and Disease: "The Touchstone" Ballarat, 28 December, 1883, p. 7.

Sadly, on 9 August 1884, Emilie Sutton died of tuberculosis at only 22 years of age. Her funeral service was held at the Wesleyan Church in Lydiard Street on the 11 August, and was presided over by the Rev. Thomas Williams. The church organist Mr C Eyres played the *Dead March* out of respect to Emilie and in the evening the choir, of which she had been a member, sang Pope's well known ode "Vital Spark". The funeral procession to the cemetery was attended by her family, her dear friends and by many leading citizens of Ballarat. Emilie Sutton was buried with her father, Richard, in the Old Ballarat Cemetery. Today, over half a million new cases of tuberculosis are recorded each year around the world and can be successfully treated, but any hopes for eradication of this disease seem as far away now as it did then.

Photograph of Emilie Sutton
Taken for her singing debut, 1880.

After Emilie's death, Henry focused more heavily on his studies into germs and at the time of the outbreak of Asiatic cholera in Spain, early in 1885, Henry obtained a slide of the *"Comma Bacillus"*. The cholera germ had only been identified two years before in Germany by Professor Robert Koch of the Berlin University. In Australia there was an outbreak of this disease on the ship *Dorunda* carrying over 470 people from England to Queensland. When the ship docked in Townsville, Queensland on 10 December, 1885, it was confirmed that cholera had broken out on board the ship. Under orders from the Colonial Secretary, the *Dorunda* was ordered to proceed directly to Peel Island in Morton Bay, where it anchored offshore on 14 December. Newspapers around Australia carried alarming news of the outbreak and deaths on board.

The cholera outbreak and the threat to Australia clearly spurred Henry into action for on the night of 18 December, he set about the task of trying to photograph the cholera germ from the slide he had received. At the time, this was not an easy task and it would take great skill and patience. Only after many hours of trial and error did he succeed in taking a number of photographs. Clearly understanding their importance, he wrote to the Editor of *The Argus* newspaper in Melbourne including with his letter copies of his photographs of the germ. *The Argus* published these on 28 December 1885 in order to announce the significance of the achievement to the medical fraternity. In the letter Henry also described the difficulties in photographing the germ and offered his assistance:

"The process of photographing very minute bodies of this class with the microscope is one of great difficulty. The little germs, the images of which you will find in the enclosed prints, are magnified 1,000 times linear. You may readily recognise them from the impurities of the slide by their comma shape. They are a little indistinct, caused by the germs being visible only when stained, and the staining not being properly differentiated from the medium in which the germ is mounted… When I tell you the mere focusing of the microscope occupied about two hours, you can understand the difficulty in photographing this class of object. Should the method adopted interest any of our medical readers, I will be happy to supply any information in my power."

This was most likely the first time the cholera germ had been photographed in Australia and it would have been an invaluable service to the Australian medical fraternity and to the study of the germ which plagued the world.

In February, 1889, Henry took a photograph of the tongue of a house-fly, the photograph had been taken under his microscope and had been magnified 2,500 linear. That photograph, which he put on display in the family's music store in Sturt Street, was a source of great interest to the public.

During the time Henry was in England, he attended a number of lectures related to microscopy. On one particular occasion at the London Camera Club on 7 April 1892, he attended a lecture on the subject of "Microscopy and photo-microscopy" given by one of the world's experts, Mr T. Charters White, who had just published a book on the subject. Question time after the lecture was very lively and many questions were asked by Henry and the audience.

Henry continued to study microscopy all through his life, including the subject of photo-micrography for which many advances in photographing germs and microscopic objects had been made.

Henry had quite an eclectic interest in the various fields of science and he always felt the need to educate people in a variety of scientific and engineering fields. This is demonstrated by the number of published articles written for the general public in 1883 and 1884. On the subject of criminal science, he wrote about the processes of forgery and photo manipulation. In relation to steam locomotives, he explained the latest inventions in steam engines, and on the subject of mineral extraction he discusses some of his own experiments based around Andrew Crosse's past experiments he found in a book of his life titled, Life of Crosse. Henry's expanse of knowledge knew no boundaries and it is clear he enjoyed passing on the fruits of his studies.

[For full texts of subjects, see End Notes, Chapter Five, Nos. 1, 2 & 3.]

One particular event late in 1883 occupied a few weeks of Henry's time. On 20 May that year, the volcano on the Indonesian island of Krakatoa suddenly erupted. The series of eruptions that followed on 26 August was nothing short of catastrophic and had a profound effect around the world. The last eruption opened up fissures in the walls of the volcano that allowed the ocean to pour into the magma chamber, creating an explosion of super-heated steam that destroyed most of the island. The explosion was heard in Australia 3500 km away and was the loudest-ever sound in recorded history.

The eruptions sent huge ash clouds up into the earth's atmosphere and these travelled around the world, even as far away as the United Kingdom. In the months that followed there was a lot of speculation worldwide about why the sunsets were so vividly colourful and strange.

While it is now understand that it was sunlight reflecting from the ash particles that created the peculiar sunsets, at the time the spectacle baffled many of the world's scientists and astronomers and gave rise to many theories about the causes. Even Australia's government astronomers such as Mr Ellery struggled to explain the phenomenon to the inquiring public, and by December 1883 all sorts of theories were being bandied about.

Every explanation offered was the subject of a great deal of interest and debate and each was likely to be challenged. In response to a reader's letter in *The Ballarat Courier*, Henry put forward information which was published in *The Touchstone* on 7 December 1883, under the title, "The Peculiar Sunset".

'I noticed a letter in the Ballarat Courier of Wednesday, in which the writer endeavours to explain the cause of our highly-coloured sunsets. I would not call your attention to the matter, only that the knowledge of some of our most able men is incidentally denied. I refer to the writer's novel announcement to the effect

that there are no skilled spectroscopists out here. Now, this is an error; for both Mr Ellery, the Government astronomer for Victoria, and Mr Russell, filling the same office in New South Wales, are highly skilled in the use of the spectroscope.

We have in our very midst a gentleman who has been under the tuition of the greatest living authorities on spectroscopy, namely, Professor Smith, of the School of Mines, who, I believe, has studied under Roscoe and Bunsen, than whom two higher exponents of both the art and science of spectroscopy it is impossible to find. It may interest the writer of the letter in "The Courier" to know that actual observation with the spectroscope led Mr Ellery to conclude the presence of watery vapour.

To the theory that we attached the remnants of Biela's comet, it is only necessary to say, the amount of matter representing the remains of Biela's comet would not refract the sun's light sufficiently. According to Proctor, millions of cubic miles of the comet's tail may be compressed into an ordinary sized hat. Starlight is not appreciably dimmed after passing through three or four millions of miles of tail.

The peculiar sunsets are seen over some sixty millions of miles of the earth's surface, so that the matter of Biela's comet, as last seen, would be too tenuous to refract the sun's light to the extent observable over this immense area'.

[For full text, see End Notes, Chapter Five, No. 4],

As The Touchstone's science writer Henry also went on to write some further explanations of the phenomenon, but these did not satisfy the original letter writer, a Mr Alexander from Allendale, Victoria.

Alexander thought differently to Henry's possible explanations and re-stated in his response letter, that the peculiar sunsets were due to the residue of Biela's Comet. Henry felt otherwise, and was drawn into a detailed debate in the newspaper with Alexander.

As the two correspondents responded to each other's letters, it was clear that Alexander was not going to relent and the subject was not going to be resolved. However, what he failed to realise was that, in his letters, Henry was putting forward not just his own opinions but those of various notable scientists and astronomers that he felt were being ignored.

It was Alexander's last letter and his personal attack on Henry's intellect that was to finally end the debate. Alexander was soon to learn that very few people win a debate with Henry, who responded in a sarcastic but polite manner.

"By your last issue I see Mr E. J. Alexander has satisfied himself that I have an

easily satisfied mind. Now, there are certain minds which, owing to a combination of dullness and stubbornness, are never satisfied. I feel truly grateful to Mr Alexander for so kindly placing me out of this class."

This letter was somewhat humiliating for poor Alexander and at the end of it, Henry stated that he must now withdraw from the discussion. It took a lot to rankle his normal steady and polite resolve but clearly, Alexander had over stepped the mark by questioning his position and information on the subject.

[Full Texts, see End Notes, Chapter Five, Nos. 4, 5, 6 & 7.]

Even the very learned Henry Sutton was found to be wrong when he stated that he doubted that the peculiar sunsets were due to volcanic ash, however he had allowed for the possibility that it was. The debate with Alexander showed that no matter what went on around the world, human beings always want to learn and understand their environment.

Eventually it became clear to astronomers and scientists and the public that the sunsets were caused by the ash cloud that encircled the earth after the eruption of Krakatoa. However before then, the newspapers and journals all over the world had debated the reasons for the phenomenon just like they did in Ballarat. For a period of about three years, the peculiar sunsets were captured by artists around the world such as Turner, and today their paintings are a lasting reminder of the effects mother-nature can inflict on the world.

CHAPTER SIX

PASSING ON KNOWLEDGE

"I never teach my pupils, I only attempt to provide the conditions in which they can learn"

Albert Einstein.

Thomas Bury, a well-known Ballarat journalist who wrote a column in *The Ballarat Courier* under the pseudonym 'Tom Touchstone', was a great admirer of Henry Sutton. He put the suggestion to the people of Ballarat that every effort should be taken to induce their local, eminent scientist and electrician, to form an electricity class at the Ballarat School of Mines.

Mr James Oddie agreed to the suggestion and took up the challenge. After many meetings over a number of months, he finally persuaded Henry to accept the offer to lecture at the school. By April 1883, steps were underway to fit out a laboratory for the new class in Applied Electricity and Magnetism.

Some of the equipment needed for the class consisted of, a resistance coil, lathes, tools, two dynamos and measuring instruments, all of which were ordered from Melbourne. A twelve horse powered Clerk gas engine had also been ordered for the electrical laboratory and classroom. This engine, which was to come from England, would not arrive in time for the start of the classes but when it did arrive, the electric lighting for the class room and laboratory was set up by Henry.

Just prior to Henry's first class, the Government Astronomer and President of the Royal Society of Victoria, Mr Ellery, wrote to the School of Mines stating he was very pleased that Henry would be an attached officer of the school.

On Monday 21 May, the first class in Australia of Applied Electricity and Magnetism commenced with ten students. Henry's classroom was in a separate small building at the rear of the laboratories and it measured 33 feet x 16 feet. The classes were of one hour duration and the cost of attending was £2/2s which covered the six month term. Students of Henry's first class comprised of a member of the City Council, a well-known solicitor, a mining investor, a mine manager and others.

The text book adopted by Henry was titled *Elementary Lessons in Electricity and Magnetism* by Silvanus Phillips Thompson which had been first published in 1881. Henry began the class with a twenty minute lecture on 'Statical Electricity' after which the students commenced their studies by constructing their own electroscopes. Most scientific instruments were very expensive at the time. They could only be obtained from England or Europe and took many months to arrive in Australia.

From the beginning, it was Henry's plan that the students should learn to make their own equipment. This was a new approach to teaching, but it proved a very good foundation for the students to learn and understand their trade and the instruments and equipment they would use. As the term's classes continued, the students constructed their own dynamos and telephones.

Early in 1884, Henry set up a telephonic circuit between the engine-house and workshops at the School of Mines in order for his students to test their telephones. Over the next few years Henry's practice of teaching the students to construct the tools of their trade began to expand and from these humble beginnings, the first commercial production of scientific instruments began in Australia.

With Henry's classes now up and running, Mr W. D. McKee chairman of the Head Teachers of the Ballarat District arranged a meeting with Henry, the Vice President of the Ballarat School of Mines, James Oddie and the members of the Boards of Advice and the head teachers of the school district. At the meeting, which was held on the 1 June 1883, McKee read from a letter written by Mr Andrew Berry Registrar of The Ballarat School of Mines in which he had notified the district school councils that the School of Mines had engaged the services of Mr Henry Sutton as a lecturer in applied electricity and magnetism. Oddie informed the attendees that arrangements had been made with the School and Mr Sutton for a free one hour lesson to be held each Friday afternoon for senior state scholars in the district. Henry then went on to explain that the new state school classes would consist of twenty-six students and that the class would be a six month introductory course of instruction. At the end of the term, those sufficiently advanced could be drafted into the evening classes of the more advanced students. Henry then informed the guests regarding the benefits and advantages of the class and for the students. It was resolved that the lectures will be at 4pm on Fridays and that the students chosen should have their parent's permission and be willing to commit to the six month course.

The number of students expected to enrol quickly expanded; one teacher was eager to let the whole of his sixth class attend which amounted to between thirty and forty children, but the School of Mines could not permit such a large class as the space available was insufficient for all of the prospective students. On the 27 July 1883, fifty-five students, representing eleven schools, attended Henry's introductory class. Due to the large attendance, a request was made to the head teachers of each school to select a limited number of the more advanced pupils for admission to the class. This request brought down the number of eligible students to that required and the classes proved to be a great success.

By September there was so much interest in Henry's class that it was decided that the scientific notes of the class would be regularly published in the local papers. The first report by the class was, to no-one's surprise, on the subject and uses of electric lighting:

First Students' Report Published in "The Ballarat Courier", 22 September, 1883:

"One of the uses of Electricity"

"New uses and applications are daily being found for the electric light. A very novel one, and one that must be of great use on some occasions to the medical profession was brought up at the meeting of the Academy des Sciences recently. It was called a medical light apparatus. It is intended for being used for illuminating natural cavities, or a deeply seated operating field, and being very light is applied to the forehead of the surgeon who can thus direct his light at will and instantaneously, by slightly turning his head. The apparatus consists of a small Swan incandescent lamp contained in the hollow of a metallic cylinder between a reflector and a small convex lens. It furnishes a very intense light, the field of which can be altered by a slight displacement of the lens. The source of the electricity is a small bichromate of potash battery."

Annual reports on the class continued to be submitted by Henry to the School of Mines and to British scientific journals for a number of years.

The first stage examination in Electricity and Magnetism and the first half-yearly examinations in Electricity for the State School Classes were conducted by Mr Ellery on the 7 December. Twenty-two students presented for the examination and seven students successfully passed. Two of these were, David Avery and William Henry Sharwood, (who in 1927 became a Commonwealth Crown Solicitor). David Avery excelled under Henry's tuition and went on to become an excellent student of The Ballarat School of Mines. At the student's reunion on 9 February 1884 at the School of Mines, Henry presented a gold medal in the form of a Maltese cross to Master David Avery who occupied first position in literary and practical work combined. David also received a one year scholarship to the School of Mines.

In July 1884, a new building had been erected at the Ballarat School of Mines and Henry's classes also made use of this larger structure. The building had a ground floor and an upper floor, was lit by day by large windows and three large window lanterns (raised skylight windows) in the roof. Two thirds of the ground floor was an addition to the chemical laboratory and the other third of the floor was set up for the electricity and magnetism class.

The twelve horsepower, Clerk gas engine made by Sterne and Co. of Manchester had arrived and the whole school had electric lighting which Henry had helped install.

Henry's classroom now consisted of a number of new devices and was the only place in Australia at the time to have a Thermopile (an instrument for measuring small amounts of heat). Other new equipment included various types of galvanometers, a small motor about half a horsepower which was driven by one of Henry's batteries and a Weston dynamo. A visitor to the class witnessed one of the students in the process of making his own Tangent Galvanometer and another student putting the finishing touches to the wiring of an eight commutator, Siemens Alteneck Armature for a shunt dynamo.

By mid-1885, although Henry had done all that he could to attract students to his class at the School of Mines, the number of students dropped. The school decided to combine the adult class with the state school class and Henry was happy to do this and forgo the extra fees as it would save the school about £30 a year.

Just before the start of the combined classes, an odd thing happened and Henry, confused by the incident, wrote a letter of complaint to the School of Mines Council. His letter was in regards to the attendance of students to one of his classes.

To have a great influx of school boys from Mr Lowther's Errard Street Common School was amazing enough, but when only two boys showed up for the next class, Henry had no choice but to complain as the low attendance had a marked effect on the success of the class.

The school made inquiries about what had occurred and it turned out that Cr. Martell had recently addressed the boys at the Common School on the advantages to be gained from attending the School of Mines.

The interest from the state school boys was so pronounced that many of them banded together and left their school to attend Henry's class. At the following session when only two boys showed up, it was because all the boys that left their school on the previous occasion had been reprimanded by Lowther for not asking permission to leave the school grounds. The classes resumed as normal after the boys were disciplined, but it did show that the young boys were very interested in what Henry could teach them.

That Henry was an exceptional teacher was soon realised by Mr Ellery and Professor, Henry Martyn Andrew, who was a member of the Royal Society of Victoria and for a brief time a lecturer at Melbourne University. Both men considered Henry to be one of the best lecturers the Ballarat School of Mines had employed. His contribution to the school and its reputation was obvious to everyone, including Mr Oddie who had been instrumental in securing Henry's services. The high regard and respect Mr Oddie had for Henry would in the coming years play a pivotal role in changing the course of Henry's life.

Outside of the classroom, Henry would also give lectures at the School of Mines on the electrical transmission of power and on his various inventions. The lectures would attract both men and women and various organisations such as the Ballarat Ironworkers' Association who would pay to hear his lectures. On one occasion in August 1885, Henry gave a lecture to the Ironworker's Association, but it was his experiment at the end of the lecture which amazed the audience. He demonstrated a new invention by Sir Oliver Lodge called an 'Electrostatic Precipitator' which is a filtration device that removes fine particles from smoke. Henry wowed the audience by condensing smoke into a confined space, making it virtually disappear before the audience's eyes.

Henry gave lectures on subjects as diverse as telephony, the electrical transmission of power, electric lighting and batteries, just to name a few. He also gave lectures on his many inventions of which many were donated to the school.

In 1886, The Ballarat School of Mines became affiliated with Melbourne University. While this sparked a new era in the history of the school, it unfortunately highlighted the academic attitudes and prejudices of the day. Regardless of Henry's worldwide reputation as a true man of science and his notable success as a lecturer at the school, he did not have a formal degree. This fact was not relevant to the scientific world or academics overseas, but in Australia it was clearly an issue and had been for quite a while. Only a few men of Australian science such as the President of the Royal Society of Victoria, Mr Ellery, fellow society member Professor Andrew and men who worked in the field of electrical science such as Mr. Challen, were champions of Henry's genius. The most established men of science and academia in Australia ignored Henry and their attitude towards him was highlighted by the fact that during Henry's tenure as a lecturer at the School of Mines, the University of Melbourne refused to acknowledge or promote that Henry was a lecturer there. This brings into focus the fears of the Ballarat citizen who once criticized the Melbourne establishment for ignoring Henry's scientific success. You could argue that there may have been a certain amount of jealousy and embarrassment at play; clearly the university frowned upon what they perceived and regarded as uneducated and unqualified men taking such a high profile and public prominence at the institution. The attitudes of the day transferred into the record and history books, so much so that if it were not for the great Ballarat historian William Bramwell Withers, Henry's early achievements would never have been officially recorded in Australia.

There is a certain irony, and one might say a certain smug satisfaction in the fact that in 2012, Australian science journalist Robyn Williams gave a speech at the graduation ceremony of Melbourne University students. On the closing year of 2012, which was the centenary of Henry's death, Williams educated the graduating students on some of Henry's scientific achievements. He also quoted Professor Mark Dodgson, who is the Professor of Technology and Innovation at the University of Queensland Business School, who stated that Henry Sutton is *"One of the greatest inventors in history."* One can only hope that these words reverberated down the hallowed corridors of Melbourne University and shook the ghosts of the university's scientific and academic men of the past.

The attitudes of others did not bother Henry in the slightest: his only focus was on science and this only served to showcase what he felt as a young man when he wrote about education in his letter to the editor of the *Ballarat Courier* in September 1873. Henry's style of teaching, although unorthodox was simply to educate his students and to give them a sound basis on which to expand their knowledge and so ensure they became proficient in their chosen fields and hopefully go on to greater things. For Henry, his students were the promise and future of Australia, science being a very honourable and noble path for any budding student or man to tread.

Henry's teaching style was creative and interesting; full of the drama and excitement the new field of electrical science brought to the world. Almost daily new inventions related to the now flourishing electrical industry were created and Henry was eager for his students to learn about their quickly evolving future occupations. To gain the experience to fulfil the professional requirements of their chosen field, the students needed the most up to date equipment. This posed a problem for the finances of the school and Henry was soon held to account by the schools committee for the extravagance of his classes. For Henry it was not a question of money, it was about the future and the new frontiers his students would bring to the world of Electrical Science.

During his years as a lecturer at the Ballarat School of Mines, Henry worked on a number of improvements to inventions, many of which were used by his students and the School of Mines. Previous to his employment there, Henry had already donated two of his telephones, one of his batteries and a vacuum pump. Another item donated was a new and improved method of cheaply and effectively sterilising bottles or containers in a rapid and practical manner. There is also an ammeter, now part of the University of Ballarat's historical collection which is believed to be Henry's, and also in the collection is one of his vacuum pumps. Other items donated by Henry to the School of Mines were the 13th Annual Report of the Aeronautical Society, which included copies of his two papers on flight, copies of his two papers on his electrical storage battery from the Royal Society, and a copy of *Volume 1, Pluttner on the Blowpipe* which was needed to help educate the students.

During his time as lecturer, Henry continued working at the music store and on various inventions. In 1886, Elizabeth once again became pregnant and on 16 March 1887, their second child Reginald was born. Tragically, though, little Reggie suffered from the inability to absorb proteins and calories. He died of malnutrition on 20 May 1887, at the age of only two months. One year later on 25 March 1888, Elizabeth gave birth to another son, Percival, and the sad loss of Reginald somehow eased. Soon all was going well for Henry and Elizabeth again, and Arthur was growing up fast. Arthur was now six years old and Percival was a bonnie baby and a delight to his parents, but on the 10 December 1888, young Percival contracted meningitis. For twenty days little Percy struggled for life, but sadly on the 30 December he lost the battle and died. Life had dealt a cruel blow in the Sutton family, a blow from which Elizabeth found it hard to recover, and in the years to come that heavy toll would become more evident.

In 1884 a visitor to the school stated, *"As the Ballarat School of Mines provides the means to properly develop whatever of genius there is about a youth, I see no reason why a Copernicus, a Descartes, a Faraday or an Edison should not be raised up as a memorial of its educational power"*, little did they know the Institution already had one.

Henry Sutton lectured at the Ballarat School of Mines from May 1883 to December 1889. He had spent a total of seventeen years at the school, first as a student then as a lecturer of Applied Electricity and Magnetism. Henry's class was the first of its type in Australia and in many ways was the catalyst that forged the way for the new and various fields of science and technology that followed. His legacy at the Ballarat School of Mines has remained and his reputation as a great inventor, scientist, engineer and teacher has carried forward into the folklore of not only the school's but also Australia's history.

By 1889 it was clear to Henry that it was time to expand his own knowledge and the time had come to venture out of Ballarat and make his own mark on the world. The mid 1880s was a busy time for Henry; he was tirelessly working on a number of other inventions including a half-tone photographic process, colour photography, a facsimile or fax machine as it is known today and telephotography or vision at a distance, known today as television. We'll return to each of these inventions later in the book, as each has significance and a relationship to other inventions or events through Henry's life and will be best understood in those contexts.

From a young age, Henry had many interests and hobbies and while he was a student and lecturer at the Ballarat School of Mines he was a founding member or member of a number of Clubs and Societies in Ballarat. The Ballarat Field Club and Science Society was formed in June 1882 with James Oddie as the first President and Henry as a founding member. Within months the club had about eighty members because of the variety of subjects, in many fields of interest that were offered. The club was founded for the purpose of investigating and discussing the natural sciences and at first confined its attention chiefly to geological and mineralogical subjects. On the 15 August 1882, the club elected as members a number of notable people: Ferdinand Von Mueller, Robert L.J. Ellery, Mr Foord, James Cosmo Newbery, Professor Frederick McCoy, Professor Henry Martyn Andrew, and William Charles Kernot. Lectures were delivered, papers read and discussed and periodical excursions were made to elucidate special physiographical points or for the collection of specimens.

In the first year of the formation of the club/society, Henry's classes in electricity and magnetism began at the Ballarat School of Mines. At the club's first Annual Meeting on 26 June 1883, Mr Oddie expressed his hope that Henry's new classes would attract fresh blood and fresh subjects for study and discussion. During the first twelve months of the club's existence, there had been twenty six meetings, eighteen of which were held at the school and eight in the field. The topics of the meetings included geology, zoology, botany, astronomy, chemistry, microscopical observations and electricity and all members were encouraged to contribute.

At the Annual Meeting of the Field and Science Club in 1883, discussion was held on the subject of procuring, as soon as possible, a high class telescope for astronomical observations. This would be for the use of members and for the serious study of astronomy. Up until this time there were only three telescopes in Ballarat. One had been recently purchased by *"The Ballarat Boys' Astronomical Telescope Company"* for use by young students in the new astronomy classes at the Ballarat School of Mines. A second telescope was owned by Mr John Wall, while Henry owned a reflector telescope. All were excellent instruments but not adequate enough for the long term study of astronomy.

The subject of the telescope had first been discussed at a club meeting the previous March when James Oddie became aware that a powerful telescope was for sale in Melbourne. The opportunity of using an effective telescope would excite and interest many people and an organised club of observers would, in time, make observations from Ballarat that were of importance to astronomers everywhere. During this time, a favourable offer for the purchase of the telescope had been made to Oddie from Government Astronomer Ellery.

On 11 May 1886, the Ballarat Observatory was officially opened on Mount Pleasant - just outside the city of Ballarat. A 10 foot long, 12½ inch Newtonian reflector telescope was unveiled by Mr James Oddie and Captain Henry Evans Baker, the maker of the telescope.

At the time of the opening, the building was known as the Oddie Observatory, as Oddie had personally funded the construction at a cost of £4000. The new observatory operated largely as a teaching tool for the students at the Ballarat School of Mines, but was also for use by the local Astronomical Society and for the amusement of Ballarat citizens.

After the opening, and in order for the observatory to have the best equipment, Mr Oddie employed Captain Baker to construct a great telescope. On the 2 May 1888, the observatory had another milestone opening when new buildings were completed and the observatory director, Captain Henry Baker, unveiled his newly built, 16 foot long, 26 inch diameter Newtonian reflector telescope which has since been named 'The Baker'.

Mr Oddie was in England at the time of this later occasion and so could not attend, but Mr John Wall announced that Mr Oddie had sent word that he was gifting the observatory to the city of Ballarat. About 90 people attended this second opening including Henry and his wife Elizabeth. To mark the occasion and to officially record the day, Henry set up his camera to photograph the day's events. He gathered all the attendees at the front of the observatory and took a group photograph. This still exists as an historical photograph of the day. Henry processed his photographs and posted a copy of the group shot to Mr Oddie in England and another to the Government Astronomer, Ellery, in Melbourne. Ellery's copy of the photograph is now in Victoria's Melbourne Museum Collection.

Upon Mr Oddie's return to Ballarat, he made arrangements to install electric lighting at the observatory. The electricity was to be generated by a dynamo driven by a small four-horse powered steam engine and stored in a 28 cell storage battery. This gave enough power to light the observatory for three weeks. The fine display of electric lighting revealed on 28 September 1888 was installed by Baker. Henry, who had previously inspected the installation, complimented him and when the observatory was opened up to the public and the evening's festivities began, he went on to entertain the guests with a brief but instructive talk on the theory of electricity.

Henry Sutton went on to give many lectures and demonstrations on a variety of subjects to the Ballarat Field and Science Club as well as to the Astronomical Society where he was a member from 1882-1889. In 1883, Pon's Comet was to be a celestial visitor and its arrival was highly anticipated by many astronomers including Henry who wrote in the Ballarat newspaper The Touchstone:

"So we are shortly to have another celestial visitor, in the shape of Pon's comet, which, according to the Government Astronomer, will be visible here in a few weeks.

Well, I am sure if gazing be any criterion of welcome, our old friend will receive a hearty reception, provided only that a little more consideration be given to those of us who would really like to turn astronomers for the occasion, without becoming martyrs to science through rising from a warm bed at half-past three in the morning. It is earnestly to be wished that Pon's Comet will be a little more accommodating in this respect than last year's wonderful apparition. Comets are strange bodies, in that they usually have a tail. They are not unique in this respect, as there are celestial bodies of another order having this appendage. It is a peculiar and very significant fact that astronomers are unable to account for the tail. There is no lack of theories, but there is not one truly satisfactory; evidently, a tale requires unfolding in this direction. Pon's Comet, the reappearance of which was first observed by Mr Brooks, by aid of the telescope, when at an enormous distance, is one of a group, all having calculated periods of about seventy years.

It was last observed in 1812, and is expected to be visible to the naked eye by the end of December. It does not seem likely to become a brilliant object; but still we don't know."

The 1880s were the boom years for Sutton's music store. The business had, over the years, forged a very good reputation state wide and with commerce booming in Victoria due to fortunes made on the goldfields, it was time to expand.

In order to capitalise on the large Melbourne market, the four Sutton brothers decided they needed to open a Melbourne store. After a lot of discussion within the family, Mary Sutton agreed to loan her four sons £250 each to finance the venture and so the firm of *Sutton Brothers* was formed. The existing Ballarat music store would continue to be run by Mary Sutton, and the new company would trade as an independent business. This separation of companies would protect *R. H. Sutton and Co.* if the business venture failed.

It was decided that Alfred Sutton would move to Melbourne with his family and set up the new store. Finding a suitable location to trade from did not take long and the new store opened on the 3 February 1887, at 87 Elizabeth Street which was on the corner of Little Collins Street.

Under Alfred's management the Melbourne store prospered and he soon needed more help. As a result, Walter Sutton gave up his teaching career at Golden Point School in Ballarat to help grow the business in Melbourne. Henry and Frederick continued to assist Mary with the business in Ballarat, which by this time had expanded to include agencies all over the state of Victoria. It is hard to imagine how Henry found the time to help with the music store but he contributed to the day to day running, although he probably did more on some days than others.

Government Astronomer and President of the Royal Society of Victoria, Robert Lewis John Ellery.
Image courtesy of State Library Victoria

Ballarat journalist - Thomas Bury, aka 'Tom Touchstone'.
Image courtesy of the Ballarat Historical Society

Electricity classroom at the School of Mines.
Image courtesy of Federation University Historical Collection.

The 1888 official opening of the Baker Telescope
Ballarat Observatory
Image courtesy of Museums Victoria.

CHAPTER SEVEN

SUTTON'S PHOTOGRAPHIC PROCESS 1881-1899

"Photographs open doors into the past, but they also allow a look into the future"

Sally Mann

Henry's love of photography started when he was quite young and by the 1880s it had become one of his favourite hobbies and a subject to which he turned his inventive talents. Not satisfied with what was available to him, he designed and constructed much of his own photographic equipment including cameras, photometers and lenses. He had constructed his own tripod with a mechanism that would allow him to rotate his camera in any direction. Over the course of time, he became one of the founding members, and President, of the first photographic club in Ballarat and went on to invent a new half-tone photographic process that he patented worldwide. This invention and patent altered the course of Henry's life immensely and eventually led him to England to start up a new business.

On the 2 June 1886, a meeting was held by a number of photography enthusiasts in Ballarat, with the aim of forming Ballarat's first photographic club. The meeting was held at the Ballarat School of Mines and Henry Sutton was called to chair the meeting. During the course of the evening it was decided to name the new club the Ballarat Amateur Photographic Association. The objective of the association being, to promote the study of the science and art of photography. Both ladies and gentlemen could apply to be members as long as long as they did not make a living out of photography. It was then decided to formally request permission to hold the association's monthly meetings at the Ballarat School of Mines. Membership was subject to a joining fee of ten shillings and a subsequent annual subscription of the same amount. Meetings were to be held on the fourth Monday of each month. At the meeting, the following officers were elected: President, Professor Alfred Mica Smith; Vice President, Mr Henry Sutton; Secretary, Mr W.E. Burbridge and Treasurer, Mr H. Crisp, who later developed the Crisp photographic process.

At the next meeting on the 23 July 1886, a letter was read from the Council of the School of Mines granting the photographic association permission to hold their meetings at the school. The club was given the use of the school library as well as permission for a room at the school to be converted into a dark room. During the meeting a committee was appointed to arrange the fitting out of the dark room and members exhibited their cameras and other photographic apparatus. Members Figgis, Fitzpatrick and Donaldson displayed a number of negatives and prints, Mr Oldham exhibited from his private collection a number of reproductions by various mechanical processes and Henry Sutton presented his newly invented photometer, an appliance for measuring the power of chemical rays in light. News of this new photometer was published in *The Argus* newspaper in Melbourne on 13 July that year and was reported to be of great value to photographers, particularly as it enabled them to arrive at a correct estimate of the exposure their plates would require in any light.

A year later, on 29 June 1887, Professor Alfred Mica Smith presented the first annual report of the Ballarat Amateur Photographic Association. During this meeting, Professor Smith made the first public announcement of Henry's new photographic process, which he stated, would be more than of mere commercial importance to the world. Henry had been working on a number of new photographic processes since 1881. One of these was colour photography and one was a new type of half-tone photographic process. The idea for the latter had come about years earlier when he felt that a new, faster and cheaper way of processing and printing photographs for newspapers could be found. Henry, obviously not shy of a challenge, set about working on that very issue.

On 20 October 1887, Henry Sutton submitted his first patent application for his new half tone process in Victoria. Over the next few months he lodged a total of fifteen patent applications, five in Australia and ten in various countries including Britain and the United States. Initial accounts of his new process were published in newspapers and journals both in Australia and overseas and they solicited a lot of interest from people in the fields of photography and publishing. By November, Henry had entered a number of his half tone photographs at an exhibition in the Geelong Jubilee Juvenile and Industrial Exhibition. Some of these were entered in the Fine Arts section. Henry's photographs of scenery were greatly admired, one titled *A Bit of Nature* was a Victorian bush scene and was described as '*A gum tree has thrown its thin branches aslant a streamlet, amid thick growths of grass and bushes.*' The second photograph *The Fair Poachers* was described as '*A young girl fishing on the banks of Lake Wendouree, in the neighbourhood of overhanging trees and plants, whose forms are partly reflected in the glassy water.*'

Patent:

"An improved process of converting a photographic image on a gelatine surface into a relief or intaglio printing surface."

"Be it known that I, Henry Sutton, a subject of the Queen of Great Britain, residing at Ballarat, in the British Colony of Victoria, music-seller, have invented an Improved Process of Converting a Photograph Image on a Gelatine Surface into a Relief or Intaglio Printing Surface; and I do hereby declare that the following is a full, clear, and exact description of the invention, which will enable others skilled in the art to which it appertains to make and use the same."
[Full Text, see Patents, Chapter Seven, No. 1.]

When particulars of Henry's new half tone process were published, the news spread quickly. Further articles appeared in newspapers all around Australia and New Zealand. Buoyed by the interest, Henry gave a demonstration in Ballarat to a number of scientific men and representatives from local and Melbourne newspapers on 30 May 1888. The demonstration was held in his laboratory at his residence at number 74 Doveton Street, Ballarat and the people present were Professor Mica Smith, Mr Burbridge, Mr Bechervaise of the Ballarat telegraph office, Mr Challen from the Melbourne telegraph office, Mr James Curtis, Mr F. W. Niven, Mr James Oldham and representatives from the Ballarat newspapers and the Melbourne *Argus*. Up until this point, only a couple of people had been privy to Henry's new process.

The Government Astronomer and President of The Royal Society of Victoria, Mr R. L. J. Ellery, who had been aware of Henry's work, sent his apologies as he could not make it to the demonstration. Professor Smith had also been privy to Henry's work and was especially delighted with it. Mr

Niven a well-known local printer and publisher had given Henry some help in earlier experiments. He was pleased that Henry had taken the research up, as he had the scientific knowledge to do so.

It had become Henry's practice to keep someone informed of his research and inventions, as a safeguard in case something happened to him. When Oldham stated, 'That art and science were becoming more linked together more so than ever before,' he referred to the splendid pictures in the art journals obtained by the etching process. He went on to say that Henry's discovery was very important in a commercial sense. It would be of great advantage to have the newspapers illustrated with the events of the previous day.

Ellery, who had prior to the public demonstration been privy to a personal demonstration of Henry's process, had previously stated:

"I believe the process will be found of the greatest value for ordinary and cheap illustration for which woodcuts are now largely used. The chief advantages are, in my opinion, the short space of time required between obtaining a photograph and the issue of productions from the printing press, the accuracy of the results and the exceeding cheapness of the reproductions. It only requires, I believe, to be put to the practical test to prove itself a most valuable (commercially) process."

Many photographs were printed that evening using the new half-tone process, one of which was of the American actress, Mary Anderson, who was currently performing in Australia.

ELECTRO - PHOTOTYPY.

Letterpress Metal Block, untouched by acid or tool. Produced by Henry Sutton, as described at Camera Club Conference. From a Photograph by Mendelsohn, Victoria.

The photograph, which was originally taken by the Melbourne photographer Mendelsohn, was then published in a supplement in T*he Ballarat Courier* on 30 May 1888. This was the first half-tone photograph published in a Ballarat newspaper and one of the first few published in Australian newspapers in 1888 but unfortunately the supplement it was published in no longer exists.

The photographic reproduction that Henry made of that photograph, still has the distinction of being the first published using a half-tone process wholly invented and produced by an Australian. The same photograph that was published in the Ballarat Courier supplement was used a few years later in the advertising of the Sutton-type process in England.

The article below, published in the Melbourne Argus on 31 May 1888, was one of the first to appear that related to Henry's process:

"*Mr Henry Sutton, patentee of the Sutton type instantaneous photo-engraving process gave an illustration of his patent at his laboratory, Ballarat, today in the presence of a number of gentlemen interested in engraving and printing. In Sutton type the actual engraving of the picture taken in the camera is purely automatic. It is affected by the action of light, and no mechanical or artistic skill is required. In the process a photograph of a view is taken in almost the ordinary way. The negative on being removed from the camera is developed in the usual way, and then the only original operation of the Sutton type comes into play. The negative is placed upon a metal tray and heated, and as the gelatine film melts, it appears as an engraving. Mr Sutton, in his experiments, discovered that when a photographic image, bearing a pattern, is melted, the power of capillary action is brought to work, and this power he found in the negative to be proportionate to the action of light. The effect of this melting and the operation of capillary power is that each little line and dot automatically graduates itself in proportion to the action of the light, and becomes in gradation of size an exact counterpart of the half-tones of the original. The electrotyping from the photoengraving is done in the usual way. Sutton type forms the link between the cameo and the electrotype. The process is marvellously rapid, exceedingly cheap and very accurate. A negative, as soon as properly developed, can be converted into an engraving ready for the electrotyper in from three to seven minutes. The cost, as compared with present processes, is trifling, and its accuracy is undeniable. Some negatives were engraved during Mr Sutton's demonstration today, and he exhibited a number of engravings produced by this process, some of them from instantaneous photographs.*"

In 1889, the Ballarat Fine Art Gallery gave Henry permission to photograph a number of oil paintings, one of which was the painting *Ajax and Cassandra* by Solomon J. Solomon that the gallery had recently purchased and put on display. The photograph of this powerful and inspiring painting was developed and printed using his new half-tone process and was one of many photographs he exhibited while he was in England. Just as the painting *Ajax and Cassandra* had inspired Henry, this same painting inspired a young painter, Norman Lindsay, around the same time.

Thomas Bury, who wrote under the alias 'Tom Touchstone' for *The Ballarat Courier*, was a great admirer of Henry's work and wrote many articles about it. One that he wrote about, the new photographic process, caused quite a stir with Ballarat resident, Mr Charles Gorsuch, a survivor of the American Civil War who had immigrated to Australia and settled in Ballarat in about 1865. Gorsuch wrote to the newspaper chastising Tom for gushing over Henry's new invention which he was of the opinion was nothing new. This small letter started a barrage of published letters in the Ballarat Courier titled 'Gorsuch v. Sutton' which lasted for two weeks. The letters between Touchstone and Gorsuch went from debating the merits of Henry's new process to a rather lively but friendly and amusing personal tit for tat between the two correspondents. These letters created a lot of interest and debate in the wider community about Henry's process and greatly amused not only Henry but all of the residents of Ballarat.

Before long, information on Henry's new faster and cheaper process began being published in journals all over the world which gave rise to many additional inquiries from overseas.

On the 1 June 1888, Henry received, by mail from America, an offer from a company to purchase the United States patent rights to his photographic process. It is unclear what eventuated regarding this particular offer though.

Over the next year, Henry gave demonstrations to Australian publishers and newspapers. As overseas interest increased and photographers around the world were experimenting with the new process, Henry was encouraged to go forward and exploit the true value of his new process and that was to bring it into commercial use worldwide.

By September 1889, Henry's new process had been the subject of endless talk in the printing, publishing and photographic world and so the time was ripe to take commercial advantage of the interest. More accounts and praise were published in the Australian newspapers and his new process which had been called 'Suttontype Patent Automatic Engraving' was now being advertised in the Melbourne Punch newspaper. Henry set up a small commercial production facility on one of the upper floors in Sutton's music store at 131 Elizabeth Street, Melbourne, and a number of advertisements included published photographs produced by the new process. Many people felt that Henry was destined to make a fortune from this invention. Henry also knew it was time to venture out into the world and to gain more scientific knowledge and used this opportunity to do so. It was his hope to meet and learn from his scientific peers personally, one of his long held dreams.

Encouragement to take his invention to the commercial market overseas came from many sources but it was with the aid and business help of James Oddie that Henry took the brave step that would alter the course of his life and take him to seek his fortune in the world. In May 1887, Oddie, who was well apprised on Henry's process, was on a trip to England where he visited the London Royal Academy exhibition. While there he purchased an official illustrated catalogue and decided to call upon the publishers to show them examples of Henry's photographs printed by his new process. The publishers freely admitted that the photographs were much better than they had previously produced by a similar process. By 1889, Oddie had taken another trip to Europe and England and while there he talked to many contacts about Henry's new process, one of whom was the owner of the London *Graphic* newspaper. The people in the *Graphic* office had promised that they would assist Oddie as they had known him for many years. After this meeting, Oddie reported that it was probable that the Sutton-type process would be used for daily engraving in the London *Graphic*.

Henry had originally intended to travel to London in 1887 but due to work and family commitments and the deaths of two of his infant children he was unable to go. However, by the end of 1889, Henry was finally free to leave to try and make a success of his photographic process.

In January 1890, just prior to Henry leaving for England, a meeting of citizens was convened at City Hall by the Mayors of the city and town of Ballarat to devise a means for showing their appreciation of the services of Henry both as a citizen and scientist. There was a good attendance and Mayor William Little occupied the chair, a committee was formed and Mr John Whitburn Rail was appointed secretary.

At 8pm on 3 February 1890, a farewell dinner was held at the Ballarat Town Hall to pay homage to Henry Sutton. At the event, Henry was seated to the right of Mayor Little and to his left was Colonel Smith MLA. The vice-chairs were occupied by Mr John Robson and Mr William Bechervaise.

Over the evening many compliments were given to Henry and his scientific genius. The Hon. David Ham MLC spoke first to honour Henry and then told of how he now had undying fame. Colonel Smith said, '*Ballarat has been exceedingly fortunate on being the birthplace of one of the greatest scientists in the world.*'

Bechervaise stated that he would have gone a thousand miles to honour Henry and was sure that he would make his mark in the world. Professor Mica Smith said, 'as a citizen, a scientist and a man, they had in Mr Sutton a grand example of the youth of Australia.' Mr J W Kirton MLA remarked that Henry's parents deserved credit for his success. Where most young men were involved with sport, Henry's talents led him in a different direction which was encouraged and supported by all that knew him.

Robson announced that he had received a telegram from James Oddie who was in England. The telegram stated that he had placed the sum of £500 towards defraying Henry's expenses for his overseas trip. Mr Oddie had on a number of occasions offered Henry large sums of money to further his scientific career but Henry had always declined his offers and, true to form, he also declined this offer with thanks, declaring that he liked Mr Oddie too much to take his money.

Ballarat was proud of Henry Sutton and his extraordinary progress in electrical and scientific research and his marvellous inventive facility. In light of this, the organising committee, the Mayor of Ballarat City, William Little and the Mayor of Ballarat Town, Edward Murphy, had decided to present an illuminated address to Henry. This was the highest compliment the citizens of Ballarat could pay to one of their own. After all the speeches, J. W. Rail read the illuminated address to the audience, and then formally presented it to Henry. For the rest of the evening it was put on display for all to see. The evening finished with a thunderous applause from the many guests. Henry hardly knew how to thank them as he had not realised how much the people of Ballarat admired him. The illuminated address presented to Henry was, over a century later, donated to the Ballarat Historical Society by the Sargent family.

The text of the illuminated address is reproduced on the next page.

Mr Henry Sutton

Dear Sir,

On learning that you were about to leave this city for several years, (perhaps forever) a number of citizens, representing all sections of the community were desirous that you should not be allowed to depart, in your usual unobtrusive manner, without their having the opportunity of letting you feel that though it was not your habit to mix with the hurly burly of city life, yet you have won the lasting esteem of all who knew you intimately in this your birthplace.

While you so assiduously confined yourself to your long series of scientific experiments in your laboratory, and so patiently burned the midnight oil, they have not been totally unobservant of the results of that labour, which has won for you European honours and fame.

These citizens are naturally, and perhaps somewhat selfishly, proud of the scientific triumphs which a Ballarat boy_ born and bred and educated in our midst_ has achieved; and they feel assured that, if your life be spared, you are certain to win still greater triumphs in Europe and America, thus reflecting more glory on your birthplace and it's institutions which enabled you to cultivate your natural genius.

While regretting that that you are leaving their midst, they feel somewhat assured that it is for the best, because in Europe & America (in closer touch with the leading scientists of the world) you will have a much better chance of cultivating your intellectual faculties, and of developing that inventive genius which has already been so richly productive.

Though you are going to seek "fresh woods and pastures new" yet your Ballarat friends will often hear of you, for, from the nature of your services in the scientific world yours will be a name that men will not let die.

On behalf of your many admirers, we heartily wish you, Mr Sutton and your boy a pleasant voyage to the mother land and increased health and happiness and presenting you with this valedictory address,

We, by their desire, subscribe ourselves.

William Little Mayor Ballarat City.
Edward Murphy Mayor Ballarat Town.
John. W. Rail Hon. Secretary.

Ballarat 3rd February 1890.

To

Mr Henry Sutton

Dear Sir,

On learning that you were about to leave this City for several years, perhaps for ever a number of citizens, representing all sections of the community were desirous that you should not be allowed to depart, in your usual unobtrusive manner, without their having an opportunity of letting you feel that though it was not your habit to mix with the hurly burly of city life, yet you have won the lasting esteem of all who knew you intimately in this your birthplace.

While you so assiduously confined yourself to your long series of scientific experiments in your laboratory, and so patiently burned the midnight oil, they have not been totally unobservant of the results of that labor, which has won for you European honors and fame.

These citizens are naturally, and, perhaps somewhat selfishly, proud of the scientific triumphs which a Ballarat boy - born, bred and educated in our midst - has achieved; and they feel assured that, if your life be spared, you are certain to win still greater triumphs in Europe and America, thus reflecting more glory on your birthplace, and its institutions which enabled you to cultivate your natural genius.

While regretting that you are leaving their midst, they feel somewhat assured that it is for the best, because in Europe & America (in closer touch with the leading scientists of the world) you will have a much better chance of cultivating your intellectual faculties, and of developing that inventive genius which has already been so richly productive.

Though you are going to seek "fresh woods and pastures new" yet your Ballarat friends will often hear of you, for, from the nature of your services in the scientific world yours will be a name that men will not let die.

On behalf of your many admirers, we heartily wish you, Mr Sutton and your boy a pleasant voyage to the mother land and increased health and happiness and presenting you with this valedictory address,

We, by their desire, subscribe ourselves.

William Little Mayor Ballarat City

Edward Murphy Mayor Ballarat Town

Hon. Sec.

Ballarat 3rd Feby 1890.

The illuminated address presented to Henry Sutton in 1890.
Photograph courtesy of the Sutton Family Archives,
Illuminated Address was donated to the
Ballarat Historical Society by the Sargent family.

Inspired by Henry and his unqualified genius, local Ballarat poet, Mona Marie, wrote a farewell poem in his honour. This poem was published in the *Ballarat Evening Post* on 3 February 1890 and was later published in her book of poems titled, *Ballarat Chimes.*

Ballarat's Farewell
To Her Child of Genius, Henry Sutton, Scientist
By Mona Marie

As pass from us the summer hours so thou art passing now,
With the dawning ray of deathless fame upon thy classic brow-
With the hope exultant in thy breast that other climes, as well
As this, thy native clime, will own thy genius. "Fare thee well."

But, oh, we trust, like summer hours, thou'lt midst us come again;
For love of country springs not forth from thy fond heart in vain.
And this, thy native city, shall with pride it's bosom swell,
When, crowned with fame, she welcomes thee- her Sutton. "Fare thee well."

Excelsior! Young eagle rise on wings of genius soar.
Command the elements, and realms of science vast explore;
And so thy name, like music heard in a pearl-lined ocean shell,
Shall from obscurity be drawn to heights of fame, "Farewell!"

When naught but ocean billows heave around thy vessel's lee,
When o'er thee shines the spangled skies, beneath the starlit sea,
Let mem'ry wing it's backward flight to Austral's shores, where dwell
Staunch friends, who wish thee now success, and bid thee now "Farewell!"

This poem was the first of a number of poems inspired by Henry's achievements. The great Australian poet, Les Murray, wrote a poem called *The Tube* about Henry's invention of his television.

After months of planning and preparing, Henry Sutton and his ever faithful and patient wife, Elizabeth, and young son, Arthur, set off for Melbourne's Williamstown Pier in Hobson's Bay. They went aboard the aptly-named ship Ballarat which set sail for England on 7 February 1890, carrying with it all the hopes and dreams of Australia's young inventor.

CHAPTER EIGHT

A YOUNG INVENTOR SPREADS HIS WINGS

"Success is to be measured not so much by the position that one has reached in life as by the obstacles which he has overcome while trying to succeed"

Booker T Washington

To understand the significance of Henry's invention and his trip to England, a brief history on not only photograph publishing and its processes but also the history of newspaper publishing both in Australia and overseas will help put it into perspective.

The 1890s were the time of great change for publishing. The number of illustrations in newspapers, books and magazines had increased due to the adoption of half-tone photographic processes and mechanised reproduction. Photomechanical processes consisted of line processes which could only produce linear originals and half-tone processes which captured the tones of paintings and photographs. These half-tone processes became the catalyst for the birth of the new business of photo-journalism which changed the face of newspapers all around the world. Prior to this, most publications used hand-carved block work for their images or retouched photographs. Since the early 1880s, experimenters such as Meisenbach from Germany, Frederick Ives from the US, Polish born Leon Warnerke (real name Wladyslaw Malachowski) and Joseph Swan in England had all produced half-tone processes and by 1888, newspapers, magazines and books were publishing half-tone photographs alongside their wood engravings. The process of producing these half-tone images for publication was slow and expensive, which limited their use. This made the market ripe for someone to find a way to produce the images both quickly and cheaply. Henry's process was much faster to produce and the cost was a quarter of the price of other processes that were on the market, which made it a viable competitor.

Although utilised to some degree in Australia, there were a number of reasons why Henry chose to heavily commercialise his half-tone photographic process overseas. A major reason being that Australia did not have the population to make producing and publishing half-tone photographs in newspapers and journals commercially viable. Consequently, until Henry left for England in 1890, half-tone photographs were sparingly used throughout the various publications in Australia. The financial outlay for publishers to purchase the equipment required was a huge deterrent to producing and printing the photographs on a large scale. Slowly, with the advent of more viable and economic photo-mechanical processes like Henry's becoming available in Australia, the tide began to turn. Instead of waiting a week for wood block engravings to be produced, it eventually became sound business sense to utilise half-tone photographs that could be produced on a daily basis. It would not be until the turn of the century that half-tone photographs started to become more main-stream in Australian publications.

Henry Sutton left the shores of his beloved country, Australia, not knowing if or when he might return. He left, not only because he intended to exhibit his photographic process at the Columbian Exposition in Chicago, scheduled to be held in 1893, but he also intended to travel the world to meet and learn from the world's greatest scientists to expand his scientific knowledge. His first aim was to travel to England to see how he could bring his photographic process to the commercial market, after which he intended to meet and learn from the best scientists in England and then travel to Europe and America to do the same.

Henry's quest was ambitious, but for a true man of science it was a necessary journey to take to fulfil his inexhaustible need for knowledge.

After arriving in England, Henry, Elizabeth and their young son Arthur found accommodation in Clapham just near Clapham Commons and wrote back to his family to tell them of his safe arrival. A short time later, Henry and his family moved into a lovely three story terrace house at 116 Powerscroft Road, Clapton, London.

Henry's reputation as a gifted man of science preceded his visit to England and there he was greeted and accepted as an equal among the giants of the scientific world. This acceptance opened up many doors for him which eventually led him down many avenues of scientific interest.

Not long after arriving in England he began the long, arduous task of promoting and demonstrating his new process and by May 1890, articles had begun appearing in journals praising it. The editor of the photographic journal *Photographic Answers* stated:

"By Mr Sutton's process, the production of curved blocks to suit rotary presses is quite as easy as of flat ones; therefore with this method there is at least a reasonable chance of getting decent half-tone blocks in papers of large circulation, without the grid-iron effect being so painfully obvious."

Mr W. T. Wilkinson, one of the world's leading experts, wrote about Henry's process in various British and American journals of photography:

"The ordinary method of producing printing blocks in half-tone can be produced with certainty in a few hours. These blocks for quick printing are wonderfully deep and the process ingenious and simple."

Henry arrived in London just as the great Baring financial crisis began in Britain. His attempts to get his process on the commercial market were hindered as interest in the process as well as finance proved difficult to secure.

For investors caught up in the financial crisis, money was scarce. This meant that there was little incentive for them to consider funding new businesses or inventions. Regardless, Henry kept plugging away, such was the belief that his photographic process could be a commercial success for investors. However, it soon became clear that he needed to do more to bring awareness of the financial benefits of his process to the publishing world. He had to find a new marketing direction to bring his process into commercial production.

In January 1891, Henry became a member of the London Camera Club, which at this time was a social, scientific and artistic centre for amateur photographers and other interested people. He was introduced to the Camera Social Club, which was a very prestigious institution and was home to some very notable photographic scientists. Among them were the President (later Sir) Captain William de W. Abney, Lord Rayleigh, Major J. Fortune Nott, Charles Vernon Boys and Leon Warnerke. During Henry's time at the London Camera Club, he not only attended, but gave, many lectures. One that he presented at the club on 22 October 1891, was on reversal in photography. It was met with great interest and he described in vivid detail some experiments that he undertook while photographing a thunderstorm in Ballarat a few years before.

Reversal:

"In Australia we have some of the finest electric storms, and there I have seen many flashes, the impression of which, on the eye, was of greater duration than would be due to persistence of vision.

On one occasion I was fortunate in observing a brilliant thread, suspended between two clouds, for several seconds…

One evening I was observing a display of electrical discharges in a large cloud situated on the northern horizon; these discharges were accompanied by sympathetic discharges in a cloud on the southern horizon. The northern horizon cloud was approaching, and I concluded that, should the clouds be at different altitudes, a difference of air currents might enable the one to overtake the other. A more rapid movement of the northern cloud was soon noticed; a camera fixed up and carefully focused on the southern cloud, by aid of the local discharges. All being in readiness and the shutter of slide drawn, but cap still on the lens, I waited events. When the northern cloud reached the zenith, it began to show considerable symptoms of electric excitement at the edges; knowing this to be my opportunity, the lens was uncapped, and in about half a minute a pair of brilliant flashes darted from the zenith to the southern cloud. I estimated the duration at over one second.

The plate was immediately developed, but not a sign of the flashes. I expected to find a pair of curly black streaks, right across the plate; instead, there appeared a very fair cloud negative, and the black outlines of some buildings low down in the field. These buildings enabled me to verify the position of the camera, which being attended to, proved the field must have included the flashes."

[For expanded version of Text, see End Notes, Chapter Eight, No. 1.]

Many of the lectures and papers on photography that Henry presented were published in British and American photographic journals, and reading them gives an interesting insight into his vast knowledge of photography.

From a young age, Henry had designed and built much of his photographic equipment, including the lenses. His knowledge of the processes was often the subject of lectures at the Ballarat Photographic Club meetings. In one such lecture that Henry gave on photographic optics in 1886, he highlighted the depth of his knowledge as he described various lenses that were available, including those of his own design.

While in England, Henry endeavoured to try and commercially market his lenses. This timed beautifully with the opening of the testing facilities at Kew Observatory in London in 1891. After many a trial and error with prospective manufacturers in England, Henry became frustrated and disappointed because of their technical and mechanical inability to produce his type of lens. In the end, because the testing at Kew Observatory was so stringent, he was unable to gain the necessary certificate of examination that he needed to commercially produce his lenses. A member of the Edinburgh Photographic Society, Mr John Traill Taylor, often displayed at camera club meetings, photographs taken by Henry using his specially designed lenses.

It was not long before Henry was added to the list of presenters at the camera club's annual conference. He presented his paper and lecture on the subject of "Electro-Phototypy, A New Method of Photo-Mechanical Printing" at the camera club's Fifth Annual Conference which was held on 7 April 1891, at the Society of Arts. Electro-Phototypy/Photypy was the official scientific name given to Henry's new process by the English authorities. Although he was not a prominent listing on the official program for the conference, his paper was the talk of the evening. It was met with great applause and became the subject of great interest and conversation for quite a while. As reports on the conference began to circulate, photographic journals all over the world wrote again about Henry's success, re-invigorating interest in his process.

Henry made many contacts during his time in England as he had become very well known for his secondary battery and other inventions. Due to his fame and social connections, Henry and Elizabeth were personally invited by the Mayor of London, the Right Honourable Sir Joseph Savory, to attend a Royal event at Guildhall. On the 10 July 1891, Henry had the honour of being a guest at the reception of His Imperial Majesty the German Emperor, Kaiser Wilhelm. In attendance at Guildhall were numerous members of Britain's Royal family, the Duke of Cambridge with Princess Mary, the Prince and Princess of Wales, Lord and Lady Salisbury, the Duke and Duchess of Edinburgh and numerous other titled gentry.

Henry was in familiar company as the notable engineer and politician George Swinburne had also been invited to the event.

In the midst of the Royal event, Henry wondered how on earth a young man, born in Ballarat, could have risen to such heights as to end up in such illustrious company. The next day he excitedly wrote back to his family in Ballarat, painting a very colourful word picture of the wonderful event he had been so privileged to attend. He described in great detail what it was like mingling amongst the many royals and the Emperor and Empress.

Regardless of the financial crisis, by the end of 1891 Henry had successfully promoted his photographic process and attracted investors who could bring it into commercial production. Within one week of issue, all the shares on offer were taken up. To have gained financial backing at this time was a great testament to Henry's determination, perseverance and hard work. The speed at which the shares were sold highlights the confidence of the business community in the new process. At the time of the Barings crisis, Henry stated that his photographic process was the only invention in England to gain financial backing from investors.

On the morning of 16 November 1891, at the office of Henry's London accountants, Wallington and Company, a number of influential gentlemen in the printing and publishing trade were invited to witness a demonstration, given by Henry, of his photographic process. He began by taking a photograph of one of the invited guests which would later be published in a journal. Then, he invited Mr Dellagana, electrotypers William Richardson and Alfred Isger, Carl Hentschel, George Meisenbach, Leon Warnerke and John Traill Taylor into the dark-room to see the photograph being developed, a process that occupied about fifteen minutes. When the men emerged from the dark-room, Henry put the plate into an ordinary dish on top of a small stove and heated it for about three and a half minutes, while his audience eagerly stood on their seats and leant over one another's shoulders so that they would not miss anything of the secret of turning this simple piece of glass into a mould for the electrotyper. When this process was completed, the plate was handed around the room for all to see and then it was given to the editor of the English journal *The Printing World* who immediately placed it in a column of one of the page forms and sent it to press to be published in the next issue of the journal.

After the demonstration, Henry was subjected to a great deal of questioning from some of the gentlemen who attended, including Mr F. Plummer from *Waterlow and Sons* and a representative from *The Autotype Company.* He answered all their questions to their satisfaction and in doing so the men could not find any fault with his process. Henry told them that his process could execute a block in four hours at a uniform charge of two pence per square inch, yielding a profit of forty per cent, and they could not argue with that.

The question as to whether the process could be adapted for rotary printing was actually answered by Mr William B. Marr, who was attending the demonstration on behalf of the publishing firm of Unwin Bros., one of the first companies to use the process commercially. Having received a number of photographic plates from Henry to publish, Mr Marr stated in *The Printing World,* 1891, that, *"The plates could be as easily printed from on a rotary machine as a page letterpress."*

"The inventor, Mr Sutton, deserved great credit for his simple and ingenious method of making a process block," declared the editor of *The Printing World,* summing up the general feeling of the meeting.

The gentlemen were all informed that the capital for the syndicate was already subscribed and that very shortly they would hear of the new British company working the patents. News of Henry's very successful public demonstration and formation of the syndicate in London on 16 November, reached Australia and by January 1892, reports were being published in newspapers all over Australia.

Sutton's Process Syndicate Limited became a registered company on 18 November 1891; the capital raised was £15,000 and shares were listed at a cost of £1 each. Henry became the majority shareholder agreeing to 10,000 shares in the company in return for the selling of his patents. Such was his belief that the venture would be a great success, he accepted the promise of ongoing royalties instead of a monetary payment. Under the agreement, Henry was to keep current all of the fifteen patents that had been lodged around the world related to his process. Henry was the head of the company, the company's accountant was Mr William Peet, Mr Charles Wallington was the Managing Director and the company solicitor was Mr Gerrish Foster. All company business was run from the Wallington and Company offices at 4 Tokenhouse Buildings, Kings' Arms Yard, London.

As well as many London merchants and business men and women, some of the more notable investors in the company were the well-known photographer Samuel Bourne, who had come to prominence with his stunningly beautiful photographs of India; William Dunsmore Bohm, an avid inventor and member of the Royal Institution and Chemical Society; stationers Herbert Furze and Edmund Byrne; and Alfred Isger, who had attended the demonstration on 16 November.

The declared objectives of the Sutton Process Syndicate, were to:

"Carry on the trades or businesses of inventors, photographers, chemists, plate and metal manufacturers, printers, electrotypers, engravers, electricians, telegraphists, or photo-telegraphists, manufacturers, vendors or agents for the sale or purchase of any chemicals, photographic requisites, metals, plates, patents, licences, contrivances, inventions, or appliances what so ever in all their branches, and any other trade or business that in the opinion of the Company may be conveniently carried on in connection with such trades or business or any of them, at any place or places as may from time to time be determined."

Henry, at the time, had been congratulated by one of the leading company promoters and financiers in London upon the fact that, despite the financial circumstances of the depression, he had managed get his patents safely launched on the market.

In hindsight, Henry was late in getting his process to market; if he had left for England a couple of years earlier as he first intended, his company would have had a much easier beginning. By the time he had arrived in England,

Carl Hentschel had already established one of the biggest reproduction firms in England. An added potential setback was that Joseph Swan and George Meisenbach had both patented their half-tone processes a number of years earlier and, along with Hentschel, their companies had taken a substantial foothold on the local market. However, Henry had achieved what few had managed to accomplish: in a time of economic depression, he had successfully launched his patents on the commercial market. His process was faster and cheaper to produce and it began to make inroads into the well- established market. As the business grew, so did the company's advertisements which, at times, would consist of full page ads in photographic journals including many photographs personally taken by Henry. One particular place in England that Henry loved to photograph was Hale End and a number of these photographs were used in the company's ads. During Australia's Bicentenary in 1988, the *British Journal of Photography* published an article titled 'Victorian tales of bucolic bliss, the countryside in black and white' and used in the article was one of Henry's untouched half-tone photographs of Hale End that was taken in 1892, the same photograph that had been used in his company's advertising in 1893.

Due to the commercial success of Sutton's Process Syndicate Limited, a studio and electrotyping plant was secured at 91 Blackfriars Road, London, for the commercial workings of the company. After setting up the practical side of the business in Blackfriars Road, a number of staff where employed and Henry trained them in photography and the methods used in his process.

It was not long before his process was being used in the wider publishing market and production began to increase. Journals, magazines, newspapers, book publishers and advertisers all took advantage of the new cheaper process to enhance their publications and advertisements. Henry's role within the business was to be its promoter and to oversee and train staff, other photographers and people who had purchased licences to use the process. The long hours of work put in by Henry and his staff allowed the Sutton process to become widely used across England. As publishers began to use more and more photographs in their publications, the need for the wood cut engravings that had dominated the market for many years, steadily declined. By the mid-1890s, the production of half-tone photographs allowed for their mass use within the publishing trade, and Sutton's Process Syndicate Limited took full advantage of the now flourishing market of pictorial publications and the emerging market of photo- journalism.

While Henry was in London, he was visited by Mr H. Glenny, a well-known photographer from Ballarat. Glenny, being in the photographic business, was pleased to hear of Henry's success. He had heard nothing but praise for Henry from English scientists and those in the photographic profession alike.

Henry's time was mostly monopolised by his business which left him little time to enjoy seeing the English countryside. The only time off he had during his time in England was a trip to Hastings in the county of Sussex, his mother-in-law's home town. The family spent the time meeting the Chrismas family relatives. During his stay, Henry took a stereoscopic photograph of ships berthed in the Hastings harbour. This won him the English *Photographic Annual* gold medal award.

The trip to Hastings must have been a welcome reprieve from the noise, smog and grime of London, and Elizabeth would have revelled in meeting her aunts and uncles and cousins. When her mother, Fanny, had moved to Australia, she had left behind her two brothers, Edward and John, and her two sisters, Eliza and Ester. However, after only two weeks of relaxation by the seaside, Henry, Elizabeth and Arthur returned to London.

Even while Henry was working at his new venture, the family music business was never far from his mind. He spent a lot of time sourcing new instruments, including pianos, as well as securing agencies from various companies in England to sell their wares. He would write and keep his mother and brothers informed of those he had acquired for the music store.

For a while, all went well for Henry's photographic process business and all he had to do was sit back and wait for the royalties to come in. During this time, Henry was making plans to travel to the Columbian Exhibition in Chicago. However, by 1893, the financial crisis had gripped the world and this meant he was needed at home. Henry made the necessary arrangements to return to Australia, knowing he may have to come back to England in a few years' time.

In the end, although Henry had lived his dream of travelling to England to meet and learn from some of the world's most eminent scientists, he did not get to travel to Europe or America. His time in England had been very rewarding, providing him with exciting new avenues for experiments and new inventions, but it had also been exhausting and he was glad to be returning home. His pride in the relative success of Sutton's Process Syndicate Limited in England was expressed in an interview Henry gave to the *Ballarat Star* on 15 May 1983 after his return to Australia:

"My process has been received and highly spoken of by the highest technical authorities in the old country and evidently has a great future. I know nothing perhaps, which will better show its value than to say that when I left [England]

we had more work offered to us than we could possibly take... This consisted of two whole illustrated weeklies which meant 180 photographic blocks a week. It has been an uphill work in the face of depression in companies at home...In fact, it may be said that the only company which has been legitimately formed during the last three years to take over patents, is the one which has taken over mine... The establishment at which operations are at present conducted is not a large one, but it is capable of turning out a lot of work, and it may be improved... Before I left [England] the shares were selling at par."

Three years later, Henry was due to return to England to further the company and renew his patents. However, by that time the circumstances of the trade had changed and larger companies were slowly buying up the smaller ones and reducing the competitive market. New processes were being invented and it became harder to compete in the market. As a result, Henry reluctantly decided to sell the company. He had no trouble finding an eager buyer for his successful company and process, and on 5 May 1896, Sutton's Process Syndicate Limited ceased trading as a registered company.

The unknown company that purchased Sutton's Process Syndicate Limited still ran it under the same trading name but it was now registered under their name. Under the new ownership, Sutton's Process Syndicate Limited expanded into the American market, which in many ways was at least five years behind England in adopting half-tone photographs to print media. The company continued to thrive until 1900, when it eventually disappeared from the commercial market.

Henry's photographic process could be considered only a small success compared to others but it still has its place in the history of photography and half-tone processes.

During the 1880s, Henry was also working to produce colour photographs and by 1887 he was well on the way to inventing a process by which it could be achieved.

In 1889, Henry's colour photographic process had almost been completed and he fully intended to exhibit it at the Columbian Exhibition, along with his half-tone process. It is not known exactly when this process was first utilised, but after the long drawn-out experience of setting up his half-tone photographic company in England he was not intending to commercialise it.

The stories passed down from generation to generation within the Sutton family about the beauty and magnificent tones of Henry's coloured photographs, especially those of flowers, are legendary. Henry's son Albert stated he had seen a number of them and they were exquisite examples of colour photographs that provided graduations of colour and delicate shades not attainable with the current processes at the time.

In late 2015 it was a great surprise to find, in an old family box, a number of Henry's half-tone photographic plates from the 1890s and amongst them was one of his colour photographic plates, which miraculously was one of the legendary flower images. Taken some time in the 1890s, this photograph is thought to be one of the oldest true colour photographs in the world.

In October 1893, the excitement was building ahead of the Melbourne Cup, the 'horse race that stops the nation', which would be run on 7 November. At that time, there was also much speculation about whether it was possible to set up a camera that would capture a photo-finish of the race. However, there were several innate difficulties in trying to do so at the time. The art and skill of photography had progressed, but unfortunately not enough to be able to accurately record the finish line and determine the winner of a race if it was a very close finish. The matter had been the subject of intense debate among the public all year and it was not until Mr Robert Law, the Chief Assayer of The Royal Australian Mint, weighed into the discussion that Henry felt he needed to express his views. This began a public debate with Mr Law under the heading *"Photography on Racecourses"*, and it was one that held the readers of the *Argus* captive long after the 1893 race was won.

Although Mr Law proposed that a photograph of the finish of a horse-race would be possible, Henry was acutely aware of the limitations of the period's camera equipment and stated:

Photography on Racecourses:

"I have read with interest Mr. Law's letter on the above in The Argus of today, and whilst agreeing with him as to the accuracy and advisability of using the camera in scientific pursuits, I do not think it will come into practice in the determination of the finish on a racecourse.

The unreliability of light will be a main factor against it.

The accuracy of a photograph cannot be doubted, but the determination of what is accurate in a photograph is a most difficult matter.

I have seen some of Professor Boys' apparatus and results in scientific instantaneous photography, and I have seen Herr Anschultz's wonderful productions in the same direction. Among the latter the horses of some Uhlans are trotting in line. The camera was placed a little in front of the line; the effect produced is that the far horse is in front, when, as a matter of fact, it is in accurate line. This is purely a matter of angle of view, but it would happen on the racecourse if the photograph was taken the fraction of a second before the finish, and the reversed effect produced if taken the fraction of a second after."

Mr Law had suggested that the camera should be placed overhead on an arch, but this was accompanied by many difficulties: the camera would have to be positioned at least 100 ft above the track, and fitted with a long focus lens and an automatic shutter system, and of course the availability of light was a factor. These were just some of the problems. In the series of letters written by both men leading up to the day of the big race, many ideas and suggestions in relation to positions, cameras and technicalities were put forward of how to set up a photo-finish.

[Full Texts, see End Notes, Chapter Eight, Nos. 2, 3 & 4.]

Regardless of the lively and informed debate at the time, it took many decades, as well as improvements in photography and technology, before a camera was invented which could accurately record a photo-finish. By 1940, such cameras were being used around the world. Australia lagged behind in this as it would not be until 1946 that a photo finish camera was first used on the Flemington race track in Melbourne and not until 1948 that the first Melbourne Cup was run using a photo finish camera. Even then, the result was controversial as it was felt by many that the camera was not aligned correctly. This was much the same issue that Henry had, in fact, anticipated all those years before. Even so, he could hardly have imagined the advancement photographic technology would make in the future.

Henry Sutton at age 36 taken in England in 1891

His Imperial Majesty the German Emperor, Kaiser Wilhelm, presenting the address in the library at Guildhall, during his visit to the City of London.

Negative by HENRY SUTTON.

HALE END.

Untouched Half-tone Block by Sutton's Process.
4, Tokenhouse Buildings, London, E.C.

Photograph taken by Henry Sutton
used in Sutton's Process Syndicate Ltd Advertisement 1893

Boys at play England 1891 Photo by Henry Sutton

Arthur Sutton on the rocks at Hastings, Sussex England, March 1891
Photo by Henry Sutton.

Henry Sutton colour photograph
by the Sutton colour photographic process 1890s

CHAPTER NINE

PUBLISHING - A TOUCH OF SUCCESS

"Acquire new knowledge whilst thinking over the old, and you may become a teacher of others"

Confucius

Henry undertook two ventures into the world of print media. The first was in 1883 when he joined forces in Ballarat with Thomas Bury and George Barnes Hogg to publish a newspaper.

Thomas Bury was a passionate Dublin born Irishman who had come to Australia in 1854. A larger than life and somewhat eccentric character, he was a true man of the written word who surrounded himself with books on literature, politics, science and philosophy. Taking a step into his home was like meeting with Mark Twain and Walt Whitman only to end up finding yourself having to brush away the works of someone like Henry Wadsworth Longfellow just to find a seat. Tom was a man filled with life and passion and his showy style of journalism represented the flair of a man with too many words to say in one lifetime. He wrote under the pseudonym "Tom Touchstone" and wrote regular articles for *The Ballarat Courier* newspaper. His romantic and sometimes liberal and Utopian views would set readers tongues wagging but his aim was always to educate and bring to life an exciting literary and political view of the world.

George Barnes Hogg was an Englishman and a journalist, and he was related to the well-known American journalist, Colonel Barnes. George had travelled to Victoria as a young man to cover the gold rush at the height of its gold production and subsequently worked at a number of different newspapers around the state. He had a growing reputation as a fine journalist and had been working in Ballarat for the past eight years, specialising as a reporter in the mining sector. His anecdotes of his early days during the gold rush were at times racy and thrilling and always captured his audience, making an afternoon with Tom and George a very lively and interesting affair.

On 1 December 1883, a warm and sunny Saturday afternoon, Henry, Tom and George gathered together to discuss the idea of publishing a newspaper. The plan was to print a weekly paper with local and social content along with some current affairs and specialised regular articles which would enhance the publication. After formulating what would be in the newspaper, the dynamic trio of budding media moguls set about preparing articles to be included in the first issue. Tom would be the Editor and he and George would write the columns, as well as adding content from other sources. Henry would share his knowledge and write a weekly science page to educate the readers about all things scientific.

To attract readers, it was decided by the management that the newspaper would arrange for public essays and award prizes for the best compositions and poems submitted by readers. Caught up in the excitement of their new venture, the new newspaper was launched only one week later. By then the new newspaper had been officially registered, and a number of other Ballarat citizens, who saw value in the new venture, had become co-proprietors. These were Mr Alexander Alfred Gray, Mr Alexander Miller, Mr Lewis Pritchard, Mr Robert James McKeeman and Mr Henry Taylor.

The new Ballarat weekly newspaper was simply called *The Touchstone* and the first issue was published on Friday 7 December 1883. It was a liberal and light hearted read, more like a weekly social journal rather than a standard newspaper. The beautifully presented and easy to read format was printed by Mr Edward Ellis Campbell, a printer, stationer and book binder at 19 Lydiard Street, Ballarat. Regardless of the structure and content of the first hurried issue, sales far exceeded the three proprietors' expectations and the next issue was expanded by four pages to fulfil the required content.

It was very unusual for Henry to dive into a venture so spontaneously but the lure of writing a full page of science articles clearly appealed to the passionate scientist and inventor. Henry wrote his "Scientific Notes" page under the pseudonym "Delta" and each issue of the newspaper addressed different aspects of science covering such issues as microscopy and diseases, comets, electricity and mining. It is not known what capital was put forward for the venture but revenue was raised by some advertisements from local businesses. By the second issue, the new venture had received some reserved but very favourable reviews from other newspapers.

The Australasian Typographical Journal wrote the following review:

"The Touchstone is the title of a new weekly serial published in Ballarat. It is a genuine 'multum in parro' of a most instructive and entertaining character, and contains original articles pleasantly and ably written, upon every conceivable topic… [It is] a characteristic paper from the pen of the celebrated 'Tom Touchstone' of the Ballarat Courier, scientific notes, social statics of considerable interest to the wage-earning class.

The articles are written in the 'magazine' style and the writers are evidently imbued with a love of their respective subjects…"

This new candidate for public favour is a marvel of literary excellence… and we feel assured that it only requires to be made more widely known to secure it a circulation worthy of its merits. The Touchstone supplies a literary want in our colonial life which was much needed, not only in the 'Gold City' of Ballarat, but also in the 'Queen City of the South', where we hope soon to see an agency in connection with it established. 'The Sydney Bulletin' circulates its thousands of copies in Melbourne every week, and we can see no reason why 'The Touchstone' should not improve upon the Melbourne circulation of our Sydney contemporary, for it is more Victorian in its sympathies and aims."

Such high praise from a review of the newspaper's first issue was a lot to live up to and within a few issues, *The Touchstone* became a victim of its own success. Readers started demanding articles on the various subjects that they thought should be included and although the proprietors tried to satisfy their readers, it quickly became apparent that they could not please everyone. Other issues included finding distributors and delivery boys to service the subscribers along with the problem of raising sufficient revenue from advertisements and subscriptions to pay for the publication of the subsequent issue, for without that the newspaper could not continue. It appeared that most readers would only purchase the newspaper after it was published, which added to the burden of lack of capital to keep it in production. Therefore, after failing to gain enough commercial interest in advertising and in pre-paid subscriptions, *The Touchstone* ceased production. The last issue was published on 4 January 1884, only five weeks after its debut.

After the failed venture of the newspaper, Tom Bury kept writing for the Ballarat Courier right up to his death in 1900 and George Barnes Hogg went on to work for The Ballarat Star and The Ballarat Courier and he eventually became the Courier's Sub Editor until he died in 1896.

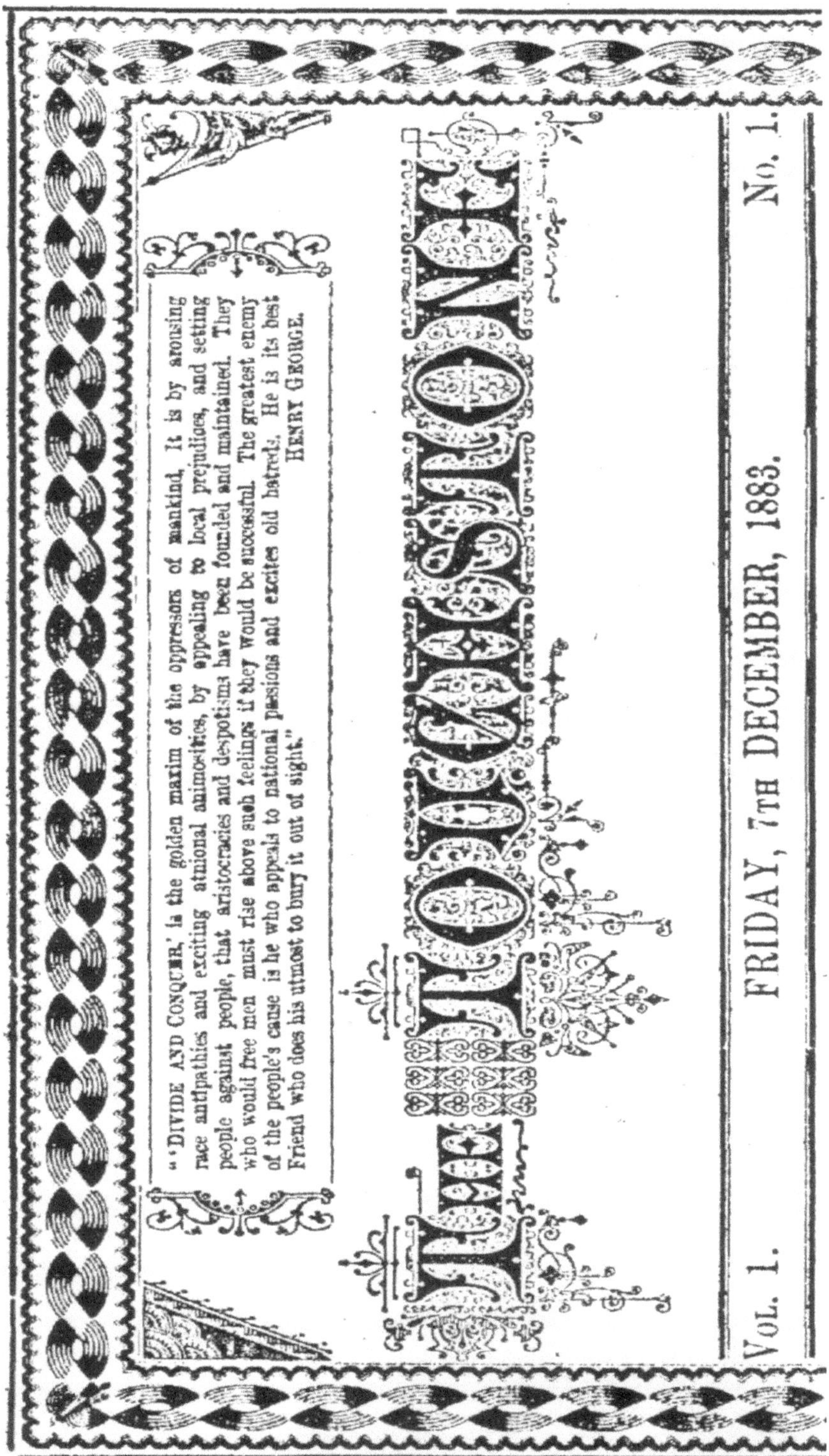

THE TOUCHSTONE

"'DIVIDE AND CONQUER,' is the golden maxim of the oppressors of mankind. It is by arousing race antipathies and exciting national animosities, by appealing to local prejudices, and setting people against people, that aristocracies and despotisms have been founded and maintained. They who would free men must rise above such feelings if they would be successful. The greatest enemy of the people's cause is he who appeals to national passions and excites old hatreds. He is its best Friend who does his utmost to bury it out of sight."

HENRY GEORGE.

VOL. 1. FRIDAY, 7TH DECEMBER, 1883. NO. 1.

Mast head of the first issue of The Touchstone Newspaper published on 7 December 1883

Henry's second foray into newspaper publishing came a decade later when travelling back to Australia from England. On 3 March 1893, the Sutton family boarded the steam ship *Australia* and headed for home. The passage was an enjoyable one but when the ship reached the Red Sea it stalled and the passengers had to find ways to amuse themselves. On board the ship, Henry had met Mr Gould, a young journalist previously connected with a London press agency, who was coming to Australia for the benefit of his health. Together, they came up with the idea of producing a manuscript illustrated newspaper to amuse the passengers. They called it *The Red Sea Scorcher*, an appropriate title given that the ship had stalled in the heat and humidity of the Red Sea.

With the permission of the ship's Captain and assistance of the ship's printing equipment, the two men set about the task of producing the newspaper. Henry had all of his photographic equipment and plates to supply the pictures and the intrepid journalist wrote the articles and stories of the passengers' travels.

The Sutton cabin was turned into a newspaper office and while the young journalist was writing the stories, Henry would be preparing the photographs. The first issue published had 32 photographs in it and all the passengers passed some time by reading about other people's adventures. More and more of the passengers were eager to share their photographs and stories and these were published. New photographs were taken as the ship continued on its journey.

Although newspapers had been published on board ships prior to 1893, it is believed that Henry's publication of the *Red Sea Scorcher* was the first newspaper on board a ship to be illustrated with half-tone photographs. It is not known how many issues were published on the ship but the often recounted story of producing of the newspaper and his journey home was a fond memory for Henry.

The pleasure and amusement on the journey home turned to distress when there was an outbreak of smallpox on board the ship. Henry, Elizabeth and young Arthur were unaffected but all of the ship's passengers were subject to quarantine restrictions. News of the outbreak on board the *Australia* travelled fast to Ballarat. The Golden City's greatest son was coming home under adverse circumstances and preparations were made. The ship *Australia* docked at Hobson's Bay at midnight on 5 May 1893 and under orders from the Board of Health, Henry, Elizabeth and Arthur travelled straight to Ballarat. Upon arrival there, the family went to the home of Henry's brother,

Frederick, at 116 Webster Street, where all the residents were quarantined until given the all clear a week or so later.

After Henry was again free to go out, he walked down to Sturt Street to see the new music store. Even though he had been sent photographs, it was the first time he was able to take in the beauty of all its architectural features, which were magnificent both inside and out.

The new purpose-built Sutton's Music Store had been constructed on the site next to the store at 31 Sturt Street while Henry and his family were in England. Mary Sutton had planned, since 1886, to build a larger store and waited patiently for the site next door to become available. Sadly though, Mary Sutton had suffered a stroke early in 1889 and although she made some recovery, she remained an invalid. This did not deter her or the Sutton family from fulfilling her wish of building Australia's first purpose built Music Emporium. Mary, by this time, was living on the first floor of the store at 31 Sturt Street and this posed a problem due to the fact that she could no longer walk up and down the stairs without aid. As the new music store was going to be much larger and Mary's new home, plans needed to be made to enable her to live on the top floor when it was built.

Otis hydraulic lift similar in design to
Henry Sutton's Music Store
lift at Sturt Street Store.

Naturally, Henry had swung into action and in conjunction with the Austral Otis Lift Company, he helped design a hydraulic lift for the yet to be constructed building. He began work on the idea just after Mary had her stroke in 1889, and completed it before he left for England and it was ready to be incorporated into the plans for the future building.

Growing up around the mines in Ballarat, Henry had firsthand knowledge of the mechanics of hydraulic lifts. Back in 1881, the

Ballarat School of Mines had considered putting a lift in their newly installed Gauge Tower and had consulted Henry about a design for one. That lift was never built due to the cost and the difficulties associated with building a lift in Ballarat. One of the main problems to solve was the design of the hydraulics system. As there was no electricity in Ballarat, a water-powered lift was the only answer and a lift would have to run off the very low pressure of the Ballarat mains water. Water hydraulic lifts usually had a steam powered backup system if water pressure was too low, but installing a steam engine would not be possible in the music store.

Similar hydraulic elevators used in that era were designed to bring water in from the mains and let it out through the sewerage system. Unfortunately Ballarat had no sewerage system at the time. The other design available required a holding tank to be placed at a higher level than the lift – such as on the roof – so that gravity increased the pressure of the water. Once again this was not possible to do in the music store. Clearly it was time for Henry to design a system which would solve the issue of boosting the water pressure from the mains.

The solution to the water pressure problem, enabled them to use one of Otis's standard passenger car designs. This type of lift employed a water-balance system of hydraulics while inside the elevator car the passenger would pull on the rope to bring the lift up or down.

The design was based on one created by Hale for which Otis had purchased the rights in 1878. It used a circulating pipe system that allowed water, after it had been used above the piston, to be returned during the succeeding stroke to the underside of the piston.

In Henry's design, a larger than normal, vertically-mounted hydraulic cylinder was used with a double piston and dual piston rods held in tension. These improvements, along with his design of a new valve system, overcame the problem of the low water pressure.

So successful was the design, that soon after the lift was built, the American firm of the Otis Company took Henry's design back to America and used it there. Austral Otis also utilised the design in Australia. Unfortunately very little remains of the lift in the music store in Ballarat but what does remain is the sheave/wheel at the top of the lift shaft, a few rope wheels in the basement and the large mains water pipes that led to the hydraulic cylinder.

However, an example of the design used in the music store can be found in Hobart, Tasmania, as a goods lift in the old Government Bond Store which is now the Tasmanian Museum and Art Gallery.

Henry's association with the Austral Otis Company continued after the lift was built and in the years to come another new venture would see their paths cross again.

As already mentioned, the land at number 33 Sturt Street, Ballarat, finally came up for sale on 20 May 1891 while Henry was in England. R. H. Sutton and Co. purchased the property through the agents Brokenshire and Coltman for £4,125. At the time of the sale this was the highest price ever paid for land in Ballarat.

There was great competition between local architects to produce the winning bid for the design of the new store. Mary Sutton and the family were very pleased with the design by Ballarat architects Gilbert and Cleggs and they were awarded the contract. Construction of the new store began in October 1891 and was completed by the end of March 1892.

When the workmen were digging out the basement to lay the foundations of the building, they found many gold pieces and these were kept by the Sutton family. Henry's son Edward inherited his father's piece of gold but sadly it is now long gone.

At the time of its construction in 1892, Sutton's Music Emporium was the highest commercial building in Ballarat and remained so for many years. The building was adorned with specially designed stained glass windows with the images of the great German composers: Beethoven, Mozart, Bach, Brahms and Mendelssohn. A magnificent motif of Minton tiles spelt out the store's name at the front door entrance and inside the ground floor spanned 80ft long and 31ft wide, the ceiling height being a very grand 21ft high. The building had three floors and a large basement. The basement stored stock; the ground floor had a magnificent display of pianos, organs and other musical instruments; the second floor was set up with professional rooms for music lessons and offices and Mary Sutton lived on the top floor.

The panelling in the lift was made of polished Kauri pine inlaid with blackwood and cedar which matched the offices' shelving and counter. The lift, which was Ballarat's first hydraulic passenger lift, remained in service long after the Sutton Store was sold in 1932 to the Ballarat Gas Company. The lift became a great novelty and many older Ballarat residents would have fond memories of travelling in the lift as young children.

Sutton's Music Emporium opened on the 8 April 1892 with a series musical events over the course of the following week. From 8pm to 10pm each evening, great orchestral concerts were given at the store for free and the huge crowds attending spilled out into Sturt Street.

The opening of Sutton's Music Emporium was a grand testament, not only to Mary Sutton, but to Richard Henry Sutton's legacy.

Heartbreakingly, Mary Sutton had another stroke at the beginning of February 1894 and by 9 March, news had broken in Ballarat that she was in critical condition and was not expected to live. Dr Pinnock had done all he could for her and Mary Sutton died on 10 March 1894, while surrounded by her family. The cause of her death was embolic hemiplegia, which in layman terms is a blood clot in the brain.

Mary's funeral service was held on the morning of the 12 March 1894 at the Lydiard Street Wesleyan Church. The family's long-time friend, Charles Morris, had acted as undertaker.

Mary's coffin had silver handles and was made of polished English oak which was a fair and wonderful tribute to her English heritage.

In addition to all of her family and many of Ballarat's prominent residents, a large number of Mary's personal friends had travelled from Melbourne to attend the funeral. It was presided over by the Reverend E. T. Cox, whose sermon reflected on the life of Mary and the sad loss sustained by her family.

The organist, Mr Charles Eyres, played the 'Dead March' in respect to the memory of Mary who was a devoted member of the church and the community. As he played the hymn, the whole congregation rose to their feet in respect and sympathy for the lady who had been a loving mother, a great friend and a champion of charitable causes throughout her life. Clearly it was not lost on anyone that Ballarat had lost one of their favourite daughters.

Mary's funeral cortege left to go to the cemetery later that day, leaving from Lewis Street at the back of the Sturt Street store.

On the day of her funeral, the *Ballarat Star* newspaper wrote a lovely tribute to Mary which read:

"Mary Sutton … was widely known and esteemed for her many excellent qualities,… as head of the firm, she always took an active interest in the business, and the successful operations of the firm were materially aided by her forethought and ability. For many years and up until her time of her death, Mrs Sutton was a warm supporter of the Lydiard Street Wesleyan Church, and her generous help was always given to aid the various movements in connection with that body. In private life the deceased lady was of a most charitable disposition, and many will deeply feel the loss of her kind and true-hearted sympathy."

Mary Sutton was buried in Ballarat's Old Cemetery alongside her husband Richard and her daughter Emilie. A large number of beautiful floral tributes were placed on her grave. Her death hit the Sutton family hard; the loving and guiding force of the family had died at the young age of 58, but the legacy of her wisdom remained with her family.

In 2013, as a tribute to Richard and Mary's memory, a number of members of the Sutton family joined forces and restored the grave and headstone of Richard, Mary and Emilie, preserving this part of the family history for the generations to come.

After arriving back from England in 1893, Henry began a new chapter in his life. He moved to Melbourne and set up home at 134 Kooyong Rd, Armadale. With the new found confidence he had gained from running his business in England, Henry threw himself into helping Alfred and Walter with the family business in Melbourne.

During his time in England, the Melbourne store had moved from 87 to 131 Elizabeth Street. However, the recession that gripped England had also hit Australia and the economic downturn was hard on the music business. Both the Melbourne and Ballarat stores, as well as the recently opened Bendigo store, were affected.

As Henry settled into life in Melbourne, he re-familiarised himself with the day to day running of the music store. With him giving it undivided attention, the business began to move forward. Soon after his arrival, new premises were found at 284-286 Bourke Street and the business began trading there in October 1893, which was a very good time of the year. The horse racing season was at its height, Melbourne was in party mode and music was the backdrop of the seasonal entertainment.

Around this time, Henry consulted with his brothers about taking on a new venture. With the business of R. H. Sutton now merged with Sutton Bros, they embraced the new idea and direction which would help the business grow.

CHAPTER TEN

SUTTON'S CYCLE AGENCY 1895-1899

"The bicycle is a curious vehicle, its passenger is its engine"

John Howard

While Henry was in England, he observed firsthand the beginnings of the boom in cycling that was starting to sweep across England and Europe. He saw the promise of this trend in the keen interest that Australians developed in bicycle racing. It was an idea for business that could be a saviour for the struggling music trade, while being a great fit with the interests of the Sutton brothers. Alfred, Walter and Frederick were all avid cyclists and Walter and Alfred had been penny farthing riders in Ballarat and both were members of cycling clubs in Melbourne. Alfred became the first captain of the South Yarra and Toorak Cycling Club, when it formed in July, 1896.

The Sutton brothers were now in charge of both R. H. Sutton & Co. and Sutton Bros., and it soon became apparent to them all that diversifying into the cycle trade might be the answer to their companies' future success. While new business preparations were underway and bicycle agencies were being sought from around the world, Henry became aware of an excellent opportunity. He had noticed in an advertisement in *The Argus* newspaper on 14 July 1894, that the property next door at 290-292 Bourke Street was up for tender and that the owner was prepared to construct a suitable building from plans approved by the successful applicant. After consulting with his brothers, they all agreed that it would be prudent to apply to erect a building there that would provide room for the new bicycle business and to their delight they were successful in their tender. Construction began immediately on the new building for the Sutton Bros. Music and Bicycle Agency and it opened on 27 September 1895. The music business operated from the ground floor and second floor of the building was allocated for the administration and display of the bicycle business. The back room and basement of the building housed the workshops for both the music and cycle businesses.

To capture the country bicycle market, room was also made within their existing Ballarat and Bendigo music stores for the bicycle business. In the Ballarat store, the bicycle display took up the whole centre aisle of the ground floor and part of the second floor and the workshop at the back. In Bendigo, as the music store was smaller, there was only a small display of bicycles along the workshop wall. The country stores catered for customers all over the state of Victoria and bicycles where shipped to customers just as pianos and musical instruments had been for many years.

Henry, who took sole charge of the bicycle venture, had learned by his experience in England that the art of promotion was one of the key factors to success and he was keen to utilise this knowledge to develop and grow the new cycle business. So, Sutton's Cycle Agency took on a number of well-known bicycle riders and engineers. Mr John Mennie became the business's chief engineer, although he also assisted Henry with some of his personal projects. Members of the Melbourne Bicycle Club and the League of Victorian Wheelmen were employed by Sutton's. These included Otto Lober, Charles B. Kellow, (Plugger) Bill Martin, Harley Tarrant and Victorian cycling champion, John (Jack) W. Parsons. In addition, Carl Augustus Adolph Schwaebsch (Adolph to his friends) and Alex D. Muirhead became managers for the business. When riders employed by Suttons won a race, their bicycles and trophies would be placed on show in the store's window to promote more sales.

Using an American idea, Sutton Bros. paid the champion Victorian rider John (Jack) Parsons to ride their *Nimrod-Zimmy* bicycles when he raced. Sutton's copyrighted four specially-produced photographs of Parsons on his bicycle and used these to promote the business. Parsons, who was trained by Joe Royal, had only begun his riding career in March 1893, but his enormous talent as a rider saw him become the Victorian champion within two years. The exclusive contracting and employment of Parsons by Sutton's was a great advertisement for the business and proved to be a financial success.

In 1893, the League of Victorian Wheelmen had been formed to promote the interests of professional riders and by 1894, the league had about five hundred members and ran many bicycle races. In the growing bicycle market, racing was a great promoter of the trade and there was an immense opportunity to capitalise on the interest of both the professional and amateur cyclists.

Many champion riders from all over the world travelled to Australia to compete in bicycle races. One of these champions was the American national champion Arthur Augustus Zimmerman, who was born in Camden, New Jersey. He was twenty six years old in 1895, and at the height of his riding career, when Sutton Bros. secured the sole agency for the *Nimrod-Zimmy* bicycles throughout Australia at a cost of £2000.

Hearing that Zimmerman, the current world cycling champion, had recently arrived in Australia to race, Frederick Sutton seized the opportunity to promote the new business, and travelled by train to Adelaide on 1 October, 1895 for the purpose of meeting him. The meeting went very well; so well, in fact that Frederick managed to convince the American to make a spur of the moment trip to Melbourne with him.

News that Zimmerman was making a visit to Melbourne spread fast and he was met at Spencer Street Station by many dignitaries and a throng of several thousand people who cheered his arrival. The Mayor of Melbourne, Sir Arthur Snowden, held a reception for the champion cyclist at Melbourne Town Hall that included Samuel H. Justin, who was Zimmerman's tour manager, members of Zimmerman's entourage as well as Henry Sutton, Carl Schwaebsch, Frederick Sutton and Mr W. J. Walford, the Managing Director of the Nimrod Company who was also visiting Australia.

Nimrod-Zimmy bicycles were made by the English firm of Nimrod Cycle Co. Bristol, England in conjunction with the company's newly formed subsidiary company Zimmerman Bicycle Manufacturers, Chicago, named after the champion cyclist. The appearance of Zimmerman was a great promotional coup for the newly opened bicycle business.

The cycling boom swept Australia and many bicycle businesses were spawned as a result. Since the introduction of the pneumatic tyre, demand for bicycles at times would outweigh supply. The Victorian Governor, Lord Brassey, had taken up cycling and his public display of bicycle riding made it socially acceptable to the community at large. As a result, many Victorians were keen to take up recreational cycling. Soon, there were so many cyclists on the roads that the Railway Department and the Tramway Company complained of the fall in revenue from the downturn in passenger numbers.

The boom continued through into 1896 as Victorians embraced the new national pastime. Riding lessons were being taken by men, women and children of all ages. More and more bicycle clubs sprang up in the suburbs of Melbourne, and all over the state, as people enjoyed the freedom and pleasure a bicycle could give them as a mode of transport and enjoyment.

On 23 May 1896, the first bicycle show in the southern hemisphere was held at the Exhibition Buildings in Melbourne. It was only a two day event but it showcased what the bicycle industry in Australia had to offer. Sutton's Cycle Agency was represented with a very large display on the ground floor, which was resplendent with many makes of bicycles and photographs of Parsons and Zimmerman. One of the exhibits in the Sutton stand was a centre-geared tricycle which had a fast frame position that proved that tricycles were not necessarily a slow machine.

By September 1896, Sutton's had trademarked the brands of bicycles *American Hero* and *Suttons Mascotte* and had obtained the agency for World bicycles. On the strength of this agency, Sutton's was able to secure the services of the specialist rider Don Walker. Walker was commissioned by Sutton's to be paid £3 for his wins. Jack Parsons, who was at the time on tour overseas, would ride for Sutton's again when he returned. Henry's association with Walker was a very successful one as he also became part of his engineering team and a great ambassador for the cycle business. Within a month of beginning the association, Don Walker broke the Australian record for a quarter mile at the Adelaide oval on a World bicycle; the time was twenty seven seconds which was two thirds of a second better than the previous record. The specially built racing bicycle which Parsons had sent over from America weighed nineteen pounds, a far cry from the light weight of a modern day racing bike, but for that time it was a remarkable improvement.

The American firm of Schwinn bicycles, who produced *World* bicycles, went on to become an American icon and World cycles in Australia sold by the thousands.

Henry's involvement in the bicycle business was reflected in his home life. His son Arthur, now fourteen years old, was a member of the Melbourne Bicycle Club and an exceptional rider who was well known amongst the cycling fraternity. Early in 1896, in the company of his uncle Alfred, Arthur had ridden from Ballarat to Melbourne in six hours on a thirty pound bike. By mid-year he had raced in the Century Run and subsequently wore the successful completion badge for his efforts. He was the youngest person in Victoria to hold such a badge of distinction. In December 1896, Arthur won the gold medal at the University High School sports track in East

Melbourne. A. E. Sutton, as he became known, proved to have a competitive nature that in the years to come would make him one of Australia's first motorcycle champions.

Early in March 1896, Elizabeth's mother Fanny took gravely ill with influenza, and Elizabeth travelled back to Ballarat to help nurse her. Robert and Fanny were now living in a cottage at 115 Lydiard Street and as Robert was now blind, he could not care for his wife. For two weeks, seventy-two year old Fanny clung to life but sadly, on the 20 March, Fanny died of influenza and pleurisy. Henry and Arthur rushed to Ballarat to comfort the naturally very distressed Elizabeth and to attend the funeral. It was a sad affair as Robert had loved Fanny deeply and she had been highly regarded both in the local community and as a member of the Sutton family. It was a tragic loss for everyone.

Robert Wyatt, who was now eighty years old, had been very dependent on Fanny for his everyday needs, so after the funeral, before they returned to Melbourne, Henry and Elizabeth made arrangements for someone to care for him on a full time basis.

Fanny's death brought Henry and Elizabeth closer together and the word "comfort" brought on a whole new meaning. Nine months later, on 26 February 1897, Elizabeth gave birth to another son, Edward. After enduring the tragic loss of two sons, the joy of Edward's birth was not lost on anyone in the Sutton family.

The year 1897 started off very well for the cycle business. Numerous agencies for different brands of bicycles were obtained, the names of which were designed to appeal to both the male and female buyer. Henry's role in running the cycle business made it a financial success and enabled the opening of a store in Geelong.

Since the beginning, Henry had been in sole charge of the importing and advertising arrangements for all the bicycle departments, as well as in charge of up to two hundred staff who worked for the bicycle business in Melbourne, and around the state. His duties were to obtain the highest quality goods from all parts of the world and to attract customers. The business had grown and company representatives in England, France, Germany and America informed Henry as soon as anything new came onto the market.

Testament to the success of this approach was that Sutton's were the first to introduce into Australia the horizontal upright framed bicycles and triplet pacing bicycles for racing. Henry ensured that country racing was not left out and put at the disposal of Ballarat's and Bendigo's cycling tracks a pair of specially built triplet pacers imported from France.

Triplet bicycles were not racing bicycles. They were designed to pace the riders at the start of races and assist riders to gain extra speed. They helped Carl Zimmerman break the world's mile record in Paris – finishing the distance in under two minutes.

Sutton's triplets were also very successful when introduced and immediately helped secure nearly every track record in Australia.

Sutton's also were the first company in Australia to introduce the flush joint bicycle, the one piece crank, the wire laced and laminated wood rim, the self-oiling hub and the Baldwin chain. Sutton's would also custom build bicycles. Customers could choose which brand or type of tyre they would like such as a Gormully & Jeffery tyre or a Dunlop pneumatic tyre. Accessories and parts and the latest patented improvements were custom fitted to new or existing bicycles.

Sutton's Cycle Agency had a very high reputation thanks to Henry's attention to detail, and as a result, they quickly grew into one of the largest cycle businesses in Australia. On the back of this success, the four brothers seriously considered disposing of the music store to concentrate solely on the bicycle business providing it continued to remain successful.

On 21 January 1897, the champion rider J. E. Snell arrived in Melbourne from Adelaide on a *World* bicycle, his journey having broken, by eight and a half hours, the existing record held by the cyclist William Kux. News that Snell was expected to arrive in Melbourne at twelve o'clock was quickly circulated from Sutton's in Bourke Street. Just before midday, a large and very enthusiastic crowd had gathered at the Post Office which was just up the road from Sutton's. While the swelling crowd waited in eager anticipation, the police found it hard to control the traffic. Throngs of people were streaming onto the middle of the road, straining to see Snell come over the rise and down Elizabeth Street from the Haymarket hill. Suddenly a large group of cyclists rode up Elizabeth Street towards Haymarket, whizzing past the hundreds of spectators and disappearing over the hill. Moments later, as the Post Office clock struck twelve, the intrepid rider Snell and the cyclists that rode up to greet him, appeared at the top of the hill and the crowd erupted in cheers, shouting: "Here he comes!". Even the Post Office employees rushed to the windows, totally forgetting their work in the excitement. The trams rang their bells and the Post Office bells rang to the tune, "*Good old Mary Ann*".

There to greet Snell was the well-known cyclist Henry (Harry) B. James, who later whisked Snell back to his home in St Kilda to rest as he had not slept for the whole journey from Adelaide. The next day, Snell arrived for a reception at the Bourke Street Sutton store, that was so packed with people that they spilled onto the street. The riding of a *World* bicycle by Snell and breaking of the record was an unprecedented advertisement for Sutton's Bicycle Agency.

It was only one month after Snell's arrival in Melbourne that Henry and Elizabeth became the proud parents of their son Edward, who was born on the 26 February 1897, at 40 Lisson Grove, Hawthorn. Henry had moved into that house late in 1896 as it was a much larger property and he had more room to work on his inventions. Over the last few years, he'd come up with a number of improvements to bicycles but although patents were taken out, copies or files on these patents have not been found. It is known, though, that Henry received royalties from these patents, so he might have eventually sold them to either, a syndicate or directly to bicycle manufacturers.

The role of Alfred and Walter Sutton in the running and promotion of the bicycle business cannot be ignored, because it was their involvement in and their personal enjoyment of cycling that was to pave the way in many areas of the sport of cycling. This was especially true in the case of Walter Sutton and his wife Mary, who both enjoyed cycling. On one occasion, in June 1896, Walter and Mary, accompanied by their childhood friend Mr John Whitburn Rail of Ballarat, rode from Ballarat to Melbourne. The cycling trio left Ballarat at 7am and after negotiating the slush and the rain they arrived in Melbourne at 5pm. Mary, who had recently ridden from Ballarat to Bendigo and back again was a very competent rider. The ride from Ballarat to Melbourne was a significant effort for a woman and did much to promote and expand the interest of female riders. Carl Schwaebsch's wife Ellen also rode bicycles and the two families spent many hours riding together.

In April 1897, Walter Sutton along with Carl and Ellen Schwaebsch; Mr E. . Stokes, who wrote for the journal *The Australian Cyclist*; Dr William Lowe as well as Dr Morton and his son, were the first people to cross the Australian Alps on bicycles.

When news of this momentous and unprecedented feat reached Melbourne, it made headlines. Upon their return, Stokes documented, in

The Australian Cyclist, the details of the touring party's history making cycle journey, for not only had they been the first group to cross the Alps by this means of transport but Mrs Schwaebsch was the first woman to do so. The success of this expedition opened up the Australian Alps to riders all over Australia, and today that tradition of crossing the Australian Alps on bicycles still continues.

Walter Sutton, along with the Mountjoy family who were pioneers of the Lorne district in Victoria, made cycling history when, in the 1890s, they opened up riding tracks for cyclists along what we now know as the Great Ocean Road in Victoria. Walter's efforts contributed not only to the promotion of the Sutton business but also to the history of cycling in Australia. In April 1898, Carl Adolph Schwaebsch retired from Sutton's to take up the management of the Austral Cycle Agency Company's depot. Before leaving Sutton's, a smoke night was arranged by Henry who, on behalf of the firm, presented Carl with a handsome silver-mounted smoking pipe which had been specially engraved.

Later that year, however, bicycle sales began to decline and by the following year the bicycle boom had busted and many smaller businesses closed. With the over-saturation of the market clearly spelling doom for many, Sutton's saw the writing on the wall and Henry began to gradually down-size their cycle business. By mid-1899, the end became inevitable for the firms that remained, including Sutton's, and they began to advertise the sale of all their stock. Sutton's had once sold so many bicycles in a two year period that Henry claimed "*If the bicycles were placed at equidistant around the earth and ridden at a 2.30 pace the whole 24,000 miles of earth's circumference would be accomplished in thirty minutes.*"

It came as a great shock to the industry that November, when Sutton's closed its cycle business to concentrate solely on the music arm of the company. By the following year only a few bicycle businesses remained.

Nevertheless, out of the dust of the bicycle craze, cycle manufacturing in Australia rose from the ashes. In 1902, Australian cyclist and 1898 Austral Wheel winner Tom Finnigan opened up a bicycle shop called Malvern Star in small premises at 58 Glenferrie Road, Malvern. The bicycle brand eventually became an Australian icon.

The bicycle era gave rise to many talented men, some of whom have already been mentioned, and those men and many others including Henry, went on to become part of not only Australia's bicycle history but also Australia's motoring history.

Sutton exhibit - The first bicycle show in the southern hemisphere.
23 May 1896, Exhibition Buildings in Melbourne.

Arthur Sutton on his bike circa. 1896

Snell's arrival at the GPO in Melbourne. 21 January 1897.

Photo by Barroni.
Dr. Lowe. Mr. Morton, Junr. Dr. Morton. Mr. E. Stokes. Mr. W. Sutton.
Mr. and Mrs. Schwaebsch.

The first touring party to cross the Australian Alps. April 1897.

CHAPTER ELEVEN

ENGINES AND AUTOMOBILES 1894-1906

"The motor industry will be the biggest industry the world has seen"

Henry Sutton 1898

Having witnessed the birth of the automobile in England, Henry arrived back in Australia with a burning desire to design and construct automobiles. He saw the huge potential of motorized transport and of automobiles as the transport of the future. He also saw the enormous potential of automobile manufacturing. *"It would be the biggest industry the world has seen."* he predicted in an interview published on 7 April 1898, in *The Australian Cyclist.*

Henry had been studying the subject of motors for many years and it was now time to put his extensive knowledge to practical use. He began his first experiments specifically with automobile engines in 1895 when he began testing to determine what type of engine would be best suited to Australian conditions. His first combustion engine was built in 1895 and was a 2 cylinder petrol engine electrically fired with a high tension coil and a spark plug.

Realising, that an economical and safe fuel was paramount to designing a successful engine and automobile, he began to work on that basis. Over the following eighteen months he built numerous different engines and conducted hundreds of experiments on different types of fuels.

In order to test the engines and the various fuels, Henry set the engines up in his workshop at the family's music and bicycle store and attached them to a bicycle wheel. One particular engine was able to run the bicycle wheel at a rate of one and quarter miles per minute. Some engines ran on petrol, others on kerosene or benzene. Initially, He had contemplated a steam engine but quickly ruled it out.

Henry went on to consider petrol fuelled engines, but at that time petrol was an extremely volatile product, much more so than today. In hot climates it was very dangerous, for if it were not stored safely, it could readily explode. Variations in temperature and humidity would also affect the performance of a petrol engine. In a country like Australia where heat can affect large areas of the continent, this was an important factor to consider.

Perhaps more significantly, petrol was not readily available in Australia and at the time it was also very expensive. If you were going to drive any sort of distance, you needed to carry your fuel with you.

After two years of experiments, Henry decided that there were too many issues with petrol as a fuel in addition to it being unsafe and uneconomical to use.

He then toyed with the idea of an electric powered vehicle but ruled that out as the weight of the batteries needed was impractical. In an article published in *The Austral Wheel*, June 1898, 'Electric Motor vs. Oil Motor', he wrote about the economy of the various types of automotive power, going into great detail in the article on the pros and cons of each form of engine and fuel and giving some insight into the technical and economic difficulties experienced by engineers and inventors in producing a successful and economical engine.

"However desirable it may be that electricity should be the mode of energy used in automotive vehicles, the time has not yet arrived for its practical application. There is only one known method of utilising electrical energy for automotive purposes, and that is through the agency of the storage battery; this latter, however, has advanced very little during the past twenty years. We still adopt some method of using oxidised lead for electrodes. The crux of the whole position being the amount of chemical, and therefore electrical energy that may be conserved per pound weight of lead; if we use the lightest known materials, the weight can never be reduced to less than one-fourth of what it is at present."

[Full Text, see End Notes, Chapter Eleven, No. 1.]

His experience and views about safe and economical fuels for engines led him into a public debate a few years later, in March 1905, in relation to the economics of constructing an electric train service from St Kilda to Brighton. The debate centred on what type of transport would be more efficient and suitable to run. Many people were in favour of running a service of omnibuses instead of the train line. In the end the subject boiled down to a matter of economics and there was a lot of debate about which fuel would be more economical for the omnibuses to run on.

Fuel for omnibuses:

"It is stated that 'the advantage in price of using cheap paraffin fuel instead of petrol far outweighed the difference theoretical efficiency between the internal combustion engine and the steam engine.' Those in the forefront of the motor movement know the great commercial problem is fuel cost. In Australia kerosene is half the price of petrol...

The internal combustion engine, as used in motor cars, has beaten the steam engine out of the field, where speed and power combined are required. Manufacturers of steam are now turning their attention to internal combustion engines, and also turning them out in their factories. Some of the most noted steam engineers have veered round, and are designing internal combustion engined omnibuses."

[Full Text, see End Notes, Chapter Eleven, No. 2.]

Henry also outlined the advantages and disadvantages of the different fuels and particularly the safety concerns when using petrol. He cited the volatility of the vapour if stored in large quantities, and brought awareness to the public of some very serious industrial accidents in England that were related to the fuel and omnibuses.

In December 1905, the Victorian Railways established the first motorized bus service from Prahran train station to the Malvern Town Hall. Each bus was powered by a kerosene-fired imported steam engine, mounted on a Chelmsford bus chassis brought in from England. They seated 24 passengers. Unfortunately, the service only lasted six months, and it would not be until November 1925, that the first petrol-powered railway buses began service between Melbourne and Geelong.

As a result of the difficulties he saw inherent in petrol, Henry eventually turned his attention to heavy oils such as kerosene which were more readily available and could be used more safely and effectively at all latitudes around the world. In Australia, people who required petrol had to import their own personal supplies for their private use at a cost of 4 shillings per gallon, whereas heavy oil such as kerosene was available at every corner store and only cost 1 shilling and 7 pence per gallon. After meticulously documenting the results of all the trials of his engines and fuels, Henry settled on perfecting a kerosene engine. At the time, kerosene was commonly used for stoves and painters' blow-lamps. Henry was certainly one of the first Australians to consider it as an engine fuel.

He calculated that one gallon of kerosene would run a tricycle a distance of 100 miles (approx. 160 kilometres). This worked out to be very economical.

The next problem to be solved was the type of battery to use to provide the ignition spark. Henry felt that a wet cell battery was too heavy, and decided to use a dry cell battery as it was simple, lightweight and had a low current.

The Sutton dry cell battery, which was calculated to last three years without any attention, would take a motor-tricycle, such as a de Dion 5,000 miles (approx. 8,000 kilometres) without replacement. The use of this battery put Henry one year in advance of the De Dion-Bouton Company.

After four long years, and hundreds of experiments using a number of different cylinders to determine which would perform the best with his engine design, Henry was finally satisfied that his design was ready to patent. On 7 December 1898, he lodged his Australian patent for *"Improvements in and relating to internal combustion engines."* and over the next year took out patents around the world.

[Full Text, see Patents, Chapter Eleven, No. 1.]

News of Henry's combustion engine received little attention in Australia as automobiles and engines had yet to gain the interest of the public. However, the details of Henry's new engine were published in many overseas journals and immediately attracted world-wide attention.

Inquiries about, and orders for, his combustion engine soon began to arrive and it was quickly apparent that there were more potential uses for his new engine than just being installed in automobiles. So as well as producing engines for that purpose, Henry began to produce stationary engines for electric lighting, farm machinery and where power was needed to run equipment.

By 1901, Henry had made significant improvements to his combustion engine and had incorporated the carburettor into the construction of the engine, producing what we know today as an automatic fuel injection system. The engine was now able to start from cold with no external vaporiser needed to heat up the engine, and in *The Autocar* on 23 March

1901, Henry stated that:

"[The] oil is not gasified at any time, and it is not vaporised until some minutes after starting. I do not heat the air. I am against gasification, as the decomposition produces tarry deposits. I am against vaporisation, as this means the application of external heat to start with- such as a blow lamp, and I am decidedly against heating up the air. I take the air in as cold as possible, always provided the temperature at compression is sufficiently high for the electric spark to fire the charge. The air is the working medium,6 and any addition of heat to it prior to admission to the cylinder is a retrograde step."

Also at this time, Henry had built a bigger combustion engine and had also produced a suitable and effective wet battery for use with his automobiles. News of his new engine was widely reported in newspapers and journals around the world and he received many inquiries, some from as far away as Africa, The Middle East and Asia.

Over the ensuing year a lot of interest was shown in Henry's engine and on 9 July 1902, he gave a public demonstration of the engine at his home in Armadale; the details of which appeared in various newspapers;

"For the last seven years Mr H. Sutton, of Sutton's Music Warehouse, Bourke Street, who has produced several clever, mechanical ideas at different times, has been at work on an oil-engine, with a view to its adaption to the use of kerosene. The result of his efforts is embodied in an engine which has been in use on a motor-car for several months past, and in which Mr Sutton is considered by practical engineers to have completely solved the problem of the utilisation of kerosene as fuel for oil-engines. This, of course, when extended, will revolutionise the use of oil-engines, particularly in relation to the motor-car."

[Full Text, see End Notes, Chapter Eleven, No. 3.]

While producing combustion engines, Henry had worked to perfect his carburettor. He used them to convert other automobiles to run on kerosene as well as for modifying existing machinery such as engines for motor boats, farm equipment, electric lighting and other industrial equipment which used engines for power.

In June 1904, a four page article titled 'The Sutton Heavy Oil Engine' was published in the *The Automotor Journal.* The working of Henry's carburettor was described, and a number of photographs of one of his more recent versions were reproduced.

Henry's carburettor did not rely on having a sufficient amount of heat to vaporise the fuel.

On a stationary engine, running at a steady speed, the amount of fuel needed is constant, and it was not hard to maintain a constant degree of heat. With automobile engines, the speed of the engine and the amount of work they are being called on to do, are constantly varying. It is difficult to prevent the heat of a vaporiser from varying between too hot, so the oil is 'cracked up', and insufficient heat so that the oil is imperfectly vaporised.

Henry perfected a method of 'spraying' the oil into the engine, along with air – mechanically forming the needed vapour. His carburettors were being commercially sold with the branding of "Sutton Heavy Oil" moulded on their side.

The advantage of what Henry called "The Sutton Heavy Oil Engine", was that it could run on kerosene, which at the time was far cheaper than petrol. The term "heavy oil engine" actually referred to Henry's carburettor system as an integral part of the Sutton combustion engine.

[Partial Text, see End Notes, Chapter Eleven, No. 4.]

When a person purchased one of Henry's engines or carburettors they were given an information and promotional brochure on the engine and carburettor detailing the success and advantages of the Sutton heavy oil engine. The brochure outlined how the engine worked and that it could be used for automobiles, motorboats and for commercial uses.

[Full Text, see End Notes, Chapter Eleven, No. 5.]

Evidence of overseas usage of Henry's carburettor comes with the purchase of one by the British inventor Edward Butler. He installed it on the Fairbanks-Morse spirit engine in his motorboat and went on to state in the British journal The Rudder in 1906, that *"The carburettor permits of a very easy start, but requires the use of spirit to take the chill off the cylinder walls when the kerosene feed may be turned on without interfering with the working of the motor. The ignition in this case being obtained from a small dynamo and wire contact capable of giving a very thick spark."*

During the constructing and testing period of designing his engines, Henry needed a vehicle in which to mount them in order to see how well the engine and fuel performed, but there was no suitable vehicle available in Australia. He had been researching new advancements and inventions in the motor vehicle industry all over the world and had been looking for a suitable vehicle for over eighteen months when, in 1897, the De Dion-Bouton Company in Coventry, England, began production of a new motorized tricycle called the De Dion Beeston. As soon as he read about it, Henry knew that the Beeston would be the vehicle to use for his tests. He quickly contacted the De Dion Company and made the necessary arrangements to purchase one of the new motorized tricycles. At the same time, he was able to secure Sutton's Cycle Agency as the new Australian agency of the De Dion-Bouton Company. The new motorized tricycle, which bore the production number 17, was priced at approximately £70, plus there was the cost of shipping to Australia, which was no small outlay.

The De Dion tricycle was the first modern motor-tricycle to be imported into Victoria, and it arrived in Melbourne in late May 1897. The tricycle was proudly displayed in the front window of the Sutton store to the envy of engineers all over Melbourne, and crowds of people streamed into the store to see the new invention. Immediately, newspapers and Australian cycling journals ran articles on the tricycle and the combustion engine that Henry had installed on it. These generated a lot of interest among the general public.

On 8 July 1897, Suttons took out a full page advertisement in T*he Australian Cyclist* journal featuring the De Dion Beeston tricycle. This advertisement is most likely the first advertisement of a motorized vehicle in Australia.

By August 1897, the tricycle fitted with Henry's engine had travelled over 250 miles (approx. 402 kilometres) on Melbourne roads and was ready for its next big test. Other modifications had been made to the tricycle such as adding his new vaporizer, and modifying the ignition. Henry rearranged one electric circuit to do the work of six and the engine now started by simply pressing a button and causing the dry cell battery to produce the ignition spark.

Henry now needed to test the economy and endurance of his engine, so he and his engineer John Mennie set about planning a long distance journey from Melbourne to Ballarat and return: a distance of approximately 150 miles (approx. 241 kilometres).

In September 1897, after being fully equipped with Henry's new engine and the new dry cell battery, the De Dion tricycle was ready for its first long-distance run. Mennie rode the tricycle from Melbourne to Keilor, where it successfully travelled up Keilor Hill. The motorized tricycle was now ready for an endurance and fuel test. The object of the trip was to demonstrate that Henry's motor could be worked on low-grade oil, whereas all other motors required high grade fuel.

The city of Melbourne was a-flutter with excitement because this journey from Melbourne to Ballarat, a distance of approximately 72 miles, would be the first long-distance journey of a motor vehicle in Australia.

At 8am on Saturday 11 September 1897, a huge crowd gathered outside the Melbourne Post Office on the corner of Elizabeth and Bourke Streets to witness the beginning of the historic journey. Also along for the trip to Ballarat were members of the Sutton family and the Melbourne Bicycle Club, who were to ride their bicycles alongside the motor-tricycle. The Sutton cyclists were Henry's son, Arthur, and his brothers Alfred and Walter, and accompanying them were Carl Schwaebsch and his wife Ellen and Mr J. Railton. A number of other cyclists were scheduled to meet up with the cycling party later: Mr B. Garnet, the Secretary of the Victorian League of Wheelmen; the Proprietor of *The Australian Cyclist* journal Mr W. G. Rae; the Italian champion cyclist, Porta; Mr Lee, Mr P. R. Tierney and Alexander Johnston.

By nine o'clock, all was ready and as the Melbourne Post Office clock struck the hour, Henry Sutton stood proudly beside his motor-tricycle as John Mennie started up the engine. The crowd roared with excitement and with exuberant cheers echoing through the city streets, the motor-tricycle headed up Elizabeth Street with the posse of cyclists. The crowd surged forward to gain the last view of the intrepid team as the motor-tricycle and cyclists disappeared over the Haymarket Hill.

The motor-tricycle travelled through Keilor and Melton, reaching Bacchus Marsh just before lunch. A crowd of over five hundred people greeted them there. All along the journey, many cyclists had joined them and throngs of spectators had lined the sides of the roads to see the new motor-tricycle and cheer the riders on.

The crowd, which had increased to over six hundred by the end of lunch, watched the motor-tricycle leave Bacchus Marsh and take the back road to Ballan.

The choice of route proved to be a grave mistake, as the unmade and unformed road was in such poor condition that it was the cause of great hardship for both the motor-tricycle and the cyclists. Consequently, the riders lost valuable time on the journey from Bacchus Marsh to Ballan and so, although the motor-tricycle worked very well and successfully negotiated the terrible terrain over the Pentland Hills, it took the riders six and a half hours to reach Ballan.

After a short rest in Ballan, they travelled onto the small town of Gordon. By this time a heavy shower of rain had fallen but it did not dampen the enthusiastic crowd of cyclists and spectators that greeted them. At around

3pm, a large posse of five hundred cyclists rode from Ballarat to Gordon to greet the entourage, after which the now very large party of riders headed off on the home stretch to Ballarat.

At intervals all day, telegrams were sent to Melbourne so that Henry could keep the public up to date of the motor-tricycle's progress. Telegrams were also sent to Sutton's Music Store in Ballarat advising them of the progress of the journey. In the afternoon a large crowd gathered outside the music store in Sturt Street eager to learn the news and to greet the motor-tricycle on its arrival. The crowd had waited patiently since 4pm and by 9pm, it had grown to such an extent that each side of Sturt Street was now crammed with people. At 9:30pm the motor cycle party finally entered the boundary of Ballarat. Such a scene had never been seen there before. As the motor-tricycle and the hundreds of cyclists that had joined the entourage travelled along the main road into Ballarat, hundreds of people cheered them on. In the city centre, the sight was one to behold as from one end of Sturt Street to the Town Hall there was a sea of eager and anxious spectators. The crowd, who had been in a heightened state of excitement since four o'clock, were now metaphorically on the edge of their seats as they jostled and fought each other to see the arrival of the motor-tricycle and cyclists.

Just before 10pm, the motor-tricycle was in sight and as it glided down Bridge Street the crowd, which now numbered thousands, surged forward in a mad rush - each person trying to push others aside in an attempt to get a better view. The crowd's roars and cheers heralded the arrival of the motoring party and echoed all up and down Sturt Street.

When the motor-tricycle had nearly reached Sutton's Music Store, John Mennie struggled to get a free run through the crowd who had been swarming him since he drove into Sturt Street. Eventually, he freed himself from the crowd and, to protect himself and the vehicle from the frenzy, he drove the motor-tricycle directly into the music store and the doors were immediately shut behind him.

Outside the store, the crowd yelled and cheered for the motor-tricycle to come out again and the police, who had been trying to hold the people back, were seriously concerned that the crowd might break down the front of the store. Frederick Sutton was requested by the police to let the engineer bring the motor-tricycle back out for the people to see. Due to the imminent danger that the crowd would wreck the premises, the vehicle was bought back outside and then driven up and down Sturt Street while the crowd jumped and skipped out of its path. Eventually the crowd was satisfied and the motor-tricycle was taken inside and the spectators slowly dispersed.

After the crowd had cleared, the original party of cyclists, the three Sutton brothers and Henry's son Arthur, John Mennie, Mr William Denholm McKee along with a number of principle Ballarat residents were entertained with a splendid supper given by the Ballarat Bicycle Club at McIntyre's Cafe. Owing to the lateness of the hour, the official welcome to the city by the Mayor of Ballarat, Cr. Charles C. Shoppee, was postponed. So, on Monday morning, the motor-tricycle was driven up to the Ballarat Town Hall where the pioneering troupe was officially welcomed by the mayor, who raised a toast to the health and future of Sutton Bros. In return, Frederick Sutton thanked the mayor and informed the guests that it was the intention of the firm to place upon the cycle racing track, a machine which would travel at an incredible rate of speed and also to expand into the automotive business.

In the afternoon, the motor-tricycle went on exhibition again and the crowds once more lined the streets. The motor-tricycle was driven up and down Sturt and Bridge Streets several times at various speeds, causing cries of joy and excitement from the crowd which were, at times, so loud they drowned out the noise of the engine.

To gauge the enormity of the spectacle that greeted the motor-tricycle in Ballarat, compare Henry's event to the Minardi Formula 1 event in Ballarat in 2011, when history was inadvertently repeated. In March of that year, the city of Ballarat hosted the Minardi Formula 1 cars which put on a similar spectacle in Sturt Street as Henry's motor-tricycle did in 1897. That demonstration was certainly a testament to how far the automobiles had advanced in 114 years.

The journey from Melbourne to Ballarat was a triumphant success, even though the travel time was long. While the trip highlighted some needed improvements, such as using an engine with a larger cylinder and two pistons instead of one, Henry was satisfied with the results. That journey was the first of its kind in the world using heavy oil as fuel in a motor vehicle and the great feat of getting over the Pentland Hills and completing the journey proved the success of both the engine and fuel. The cost of the fuel for the journey; was a mere one shilling and seven pence.

The difficulties of the journey to Ballarat were avoided by travelling back to Melbourne on the main road. The motor-tricycle arrived at the Melbourne Post Office just before 3pm where, once again, a large crowd had gathered to greet the motor-tricycle and cyclists. Henry was wholeheartedly congratulated for the firm's successful demonstration of the future possibilities of long distance road travel by motor vehicles.

Henry was now justified in feeling there was a great future for the introduction and manufacture of motorcycles and automobiles in Australia.

After attaching a new, larger 5 hp engine and installing a new Baldwin chain, the motor-tricycle made an even longer distance journey starting on 28 April 1898, and travelling through the Western District of Victoria, covering over 350 miles (approx. 563 kilometres).

The motor-tricycle, ridden again by John Mennie, was accompanied by Henry's son, Arthur, on a bicycle while other cyclists joined them along the way. The motor-tricycle comfortably outclassed a pair of riders on a tandem bicycle on the hill going into Melton.

Just as with the journey to Ballarat, the motor-tricycle was greeted by people and cyclists all through the Western District, from Melbourne through Melton, Bacchus Marsh, Ballan, Ballarat, Linton, Skipton, Lismore, Darlington, Mortlake, Warrnambool, Camperdown, Colac, Winchelsea, Geelong and back to Melbourne. Throughout the journey an average speed of 24 miles per hour was obtained and an average of 76 miles (approx. 121.6 kilometres) per day was covered. The total fuel consumption was 3½ gallons (approx 13.24 litres) at a cost of 5s/3p. This long distance journey heralded the birth of Australian motoring and cemented into history the beginning of the era of personalised motor vehicle travel in Australia.

The motor-tricycle was eventually sold for £150 to a man who lived in the Riverina Region in New South Wales

In an interview in April 1898 Henry was asked his opinion on the future of the motor industry and trade overseas he stated:

"...Millions of pounds in capital have already been put into the enterprise. In Melbourne alone during the past eighteen months something like £20,000 must have been expended in the endeavour to design a practical motor"

One [company] in England is issuing cars at a rate of four per week, average price. £300. This same company is doing a trade in engines independent of cars, say, £100,000 per annum. Then, incomplete as the little de Dion is, nearly £100,000 worth was sold in Paris alone last year, £30,000 of which was from October to December. From my own knowledge I should say £250,000 worth of Dions should be sold this year.

One company has actual orders on its books for six hundred phaetons at about £300 each that is £180,000. Another concern has orders on its books for 150 at around the same price which amounts to about £50,000. These last three

concerns in Paris alone. In the United States large sums of money are being expended to cater for the coming revolution. One factory there, to my personal knowledge, has devoted four acres of land, has an immense staff of men and thousands of pounds, in order to be in time. I shall say that the actual trade for automobiles in 1897 would be no less than £1,000,000."

Obviously, Henry was well aware of the financial rewards of being able to produce a successful vehicle and he was clearly working hard, given his humble circumstances, to be a motor car pioneer in Australia.

By November 1898, Henry had constructed two motorcycles that were wholly of his own design. They were tandem motor pacers built on a triplet bicycle frame. These machines were designed to pace racing cyclists on the track to increase their speed. These machines were an advancement in the professional sport of cycling and had begun being used overseas. The engines were built by John Mennie to Henry's blueprints. A Mr Dixon made the patterns and on both were 4hp horizontal twins that ran at 1200 rpm and which weighed 150 pounds - about equivalent to the weight of a rider. The completed engines were fitted to triplet frames designed by Henry and made by the bicycle building firm of Wallis & Sampson at 244 Latrobe Street, Melbourne. The engine took the place of the centre rider of the triplet, while the front seat was occupied by a rider who steered the vehicle and the rear seat rider was in charge of the motor control which regulated the speed. The tandems were started by simply giving the wheel a smart turn. In its fixed position the wheel was calculated to revolve at over 60 mph. There was no noise, no smell, no ignition required as an automatic arrangement immediately set the machine in motion as soon as the wheel started revolving. Unlike other motorized tandem pacers Henry's tandems had no pedals and were capable of such a speed that the driving crank became invisible showing only the vision of a blurred nebulae to the naked eye.

By late April 1899, Henry's motor pacers had been fitted with larger 5 hp engines and had been timed at one minute and thirty five seconds for the mile giving the tandem pacers a speed of 43.3 miles per hour. A private trial of one of the tandem pacers and its new engine was undertaken at the St Kilda racing track and although it did not reach the expected speeds, Henry was confident that the time could be improved.

There had been a certain amount of friendly rivalry between Henry and the Dunlop Company over the years as each had employed champion

Australian cyclists. The Sutton camp had Jack Parsons and Don Walker and the Dunlop Company had Henry B. James, Harry to his friends. Upon hearing that Dunlop was soon to receive two new De Dion-Bouton motor pacing machines, Henry publicly challenged the Dunlop Company to a friendly race between his tandem and one of theirs.

The excitement of the proposed race was anticipated by everyone. Recently in London, on a test run at The Crystal Palace, one of Dunlop's new tandem pacers ran a record time of one minute and forty seconds. Henry was confident that his tandem was capable of beating that record and as one newspaper wrote at the time, the race between the two tandems should be fun! While the Dunlop Company were waiting for the arrival of their De Dion pacers, Henry placed his motorized tandems at their disposal and they used them for record breaking attempts by cyclists as part of its corporate promotional program.

The first public appearance of the Dunlop De Dion Pacers in Victoria came on 25 November 1899. The President of the East Melbourne Bicycle Club, Mr Frank Stuart had organised an International race meeting at the St Kilda cricket ground. The feature drawcard of the tournament was Dunlop's new tandem pacers; both machines were featured in the Grand Motor Paced Tournament races of the days racing program. It is not known if there was a one-on-one match between Henry's and Dunlop's tandem pacers and if so whose tandem won, but the advent of the motorized tandem pacers bought a whole new interest to the sport of cycling. To the cycling fraternity and promoters of the sport, the Sutton and Dunlop tandem pacers were a sensational drawcard and added a new dimension to bicycle racing. The tandem motor pacer was instrumental in beginning a new era in cycling, not only in Australia but all around the world.

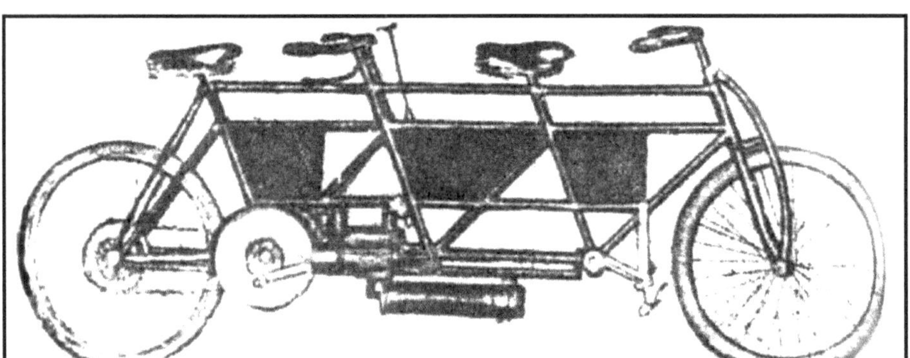

Sutton Tandem Pacer 1898

Henry Sutton heavy oil stationary combustion engine 1901

Assembled Sutton carburettor

Pieces of the Sutton carburettor

Restored Beeston tricycle on display in English museum.

Drawing of Henry Sutton
on Beeston Tricycle 1897

Suttons Cycle Agency - Advertisement 1897

Early in 1897, John Mennie and Sutton employees Don Walker, Alex Muirhead and Otto Lober all helped to construct two automobiles that Henry had designed - both were two seater Doctors' Phaetons. The Phaeton automobiles were constructed at the workshop of the Sutton Music Stores warehouse in Little Collins Street Melbourne. John Mennie was instrumental in making the chassis, body and wheels of the two vehicles. The cylinders, cranks, flywheels, axles and steering gears for the two cars were made and machined at the Joe Royal & Co foundry in Leicester Street Carlton. At the foundry, it was found that heavier flywheels were needed for the engines and in the process of attaching one of the flywheels, it accidently fell from the drilling machine table and onto the foot of one of the foundry workers. The accident happened in April 1898 and the foundry worker Mr Edward Krieger could still vividly recall the painful incident twenty years later.

The first information about 'The Sutton Autocar' was reported on 17 September 1897, when the Sutton Tricycle was making the first long distance journey to Ballarat. The information reported in the *Geelong Advertiser* mistakenly referred to the car as a Brougham, mainly due to the fact that the reporter had not actually seen the car. The article read:

"Sutton Brothers are now manufacturing two more motors which they intend to attach to racing triplets, in order to be used as pacing machines for the breaking of records. Confident that ultimately the majority of carriages will run without horses, the firm are also manufacturing a smart brougham capable of seating half-a-dozen, which will be a horseless carriage, and is expected to be capable of a speed of from 15 to 20 miles an hour. The racing machines are expected to be completed in ten weeks and the brougham in three or four months."

The new automobile was completed by early 1898 and by July of that year, the Sutton Autocar, which was named after the English motoring journal *The Autocar*, was ready to be tested with a journey of its own from Melbourne to Bendigo and back. For the test run to Bendigo, Henry's automobile was fitted with a twin cylinder, air-cooled engine, both cylinders having been cast en-bloc at Joe Royal's foundry. The secret test run to Bendigo and back was intentional on Henry's part as he did not want the public to know too much about his new automobile until it was fully tested. In a letter to the English automobile journal *The Autocar,* Henry noted that, *"I do not say much about the car in Australia as I prefer to let the information come through The Autocar."* Henry's statement in the letter confirms that it was always his intention to keep his car under wraps until the car was ready.

The success of both the engine and automobile convinced Henry that now was the time to patent his combustion engine, which he did in December 1898. Due to the downturn in the bicycle trade in 1899, Henry was slowly closing down the cycle business and, until that was completed, his announcement of the automobile would have to wait. Undoubtedly he was clearing the decks for his next venture, which was entering the automobile industry. Since Henry had begun construction of his engines and automobiles, he had been sending information about them to *The Autocar* and even before he had finished his first 'Doctor's Phaeton', he was busy constructing a second car of the same design. This was finished by 1899 on a stronger frame to carry the extra weight of a larger engine. During the latter part of 1899, Henry sent *The Autocar* the information about the car and engine, which he was now ready to announce to the public. On 13 January 1900, Henry finally gave the journal permission to publish the details of his new car, although the journal had published the early photographs taken of the first car in 1898 with Henry and his son Arthur seated in it. The Phaeton seated two people and could house a box seat at the front of the car which sat over the motor for another passenger or baggage. There was a simple elegance to the design and the description of Henry's Phaeton was as follows:

"The car is fitted with a 3½ inch diameter and 4½ inch stroke single cylinder vertically mounted engine and is air cooled by radiators. The flywheel is encased in the oil drum and the weight of the motor is about 150 pounds. It has an electric ignition which can be advanced or retarded and dry cell batteries are used. The oil feed is automatic, there are no floats or pumps in use and there is a perfect regulation of supply and perfect combustion and ignition. The engine is started from the seat and the driving gear consists of a sprocket on motor axle which transmits power by chain to the large sprocket on the countershaft. The countershaft carries a double friction clutch and two chains one for high and low speeds are connected to the differential shaft and a chain drive from differential to the front wheels. The car has front driving and rear steering wheels, steering is done by a lever rising from the driver's side of the car and the clutch is worked by a handle lever on a pillar rising up from the footboard. The advance ignition lever is attached to the clutch lever which is in front of the driver's left hand. The oil tank is mounted on the dashboard and the starting lever is placed on the footboard, the car is fitted with pneumatic tyres. To start the motor the driver turns on the oil at the tank and raises the starting lever two or three times where the motor starts immediately. The clutch is pressed to the right and low speed clutches in very gently and the car starts. If high speed is required the clutch lever

is pressed to the left, any further acceleration or retardation of speed is obtained by advancing or retarding ignition. The motor responds at once to the ignition and on the present gear the speed on level is about fourteen miles an hour. The two frames were pivoted front and rear to allow each axle to tilt independently which allowed each driving wheel to obtain an equal grip on the road."

The Sutton Autocar is thought to be the first front wheel drive car in the world and one of the first with four wheel steering, which was very advanced for its time. On 31 March 1900, *The Autocar* published another article on Henry's car and this time the photographs were of the second Phaeton he had built. The only problems Henry had with the car were with the pneumatic tyres and the springs, each were too small. In the case of the tyres, good pneumatic tyres were not yet available in Australia and the ones that were available, were too light.

Early in 1900, Henry attempted to put his Autocar into production and approached the members of the about to be formed Thomson Motor-car Syndicate. By March, the rights to the Sutton Autocar were under offer to the new company, which was incorporated on 18 June 1900. The new Thomson Motorcar Company Ltd was started by Mr. Edward L. Holmes, cousin to Herbert Thomson who is best known as the inventor of the first steam car in Australia. The newly formed company had purchased his patents as well, as it was formed to procure patents from around the world and produce motorcars, with a view of establishing large manufacturing companies throughout Australasia. Although the company apparently never took up the offer of Henry's car, he seems to have taken it in his stride and kept on working with two new automobile designs. By May 1900, one was nearly complete.

As the cycling business slowly ground to a halt in the latter part of 1899, Henry and his family spent much more of their time travelling between Melbourne and Ballarat. By the turn of the century Henry had started building a much smaller automobile in Ballarat. This voiturette design with pedals seated two people. The car, which by contrast was somewhat crudely made, had parts sourced from leftover parts from the bicycle business and other places and was powered by one of Henry's early engines and carburettors. By May, this little Ballarat car was nearing completion and in June 1900 it was ready for its first run. A photograph of Henry and his wife Elizabeth sitting in the car in Ballarat was taken at the time of completion and a copy of this photograph was sent to *The Autocar* in England, but the photograph was not published. In the years to come it would contribute to over a century of great mystery relating to the origins of this previously unknown Ballarat car. (More about this car's later appearance is at the end of this chapter.)

Following the Voiturette, Henry began working on the design and construction of the Sutton Racine. From early in 1900, it took nearly a year to successfully design and construct the new larger 6 hp automobile. In January 1901, it was announced publicly that Henry had built a new 6 hp motorcar to the order of a Melbourne gentleman. Henry, and its owner who is yet to be identified, took the car on a trial run through the Western District of Victoria and the vehicle performed excellently. Henry then set about the task of building another one for his own personal use and this was completed by June 1901. The Sutton Racine was designed along American lines and resembled a Tonneau, the description at the time was as follows:

"The body of the car is suspended above the front axle by semi-elliptic side springs, the front ends of which are carried by a traverse spring sloping downward and forward centrally. Semi-elliptic springs are also used in the rear, the back ends being connected together by a traverse spring. The car uses a jump start ignition and has a 6 hp kerosene burning engine and is fixed in a vertical position beneath the driver's seat, the power is taken from the engine through either of the two chains to a countershaft and from the countershaft by a single chain to the differential gear on the live rear axle. The motor is water cooled, and a radiator is placed beneath the body behind the cross-spring in front. Two speeds only are provided, giving a gear ratio between the motor and the driving wheels of one to five and one to nine respectively. Either speed is bought into play by means of a clutch. The form of the body is clearly on American lines, the steering pillar is vertical and that the Ackerman principle of steering has been adopted, a side lever controls the rear brakes and it is fitted with 2½ inch Dunlop pneumatic tyres. The car is fitted with an exhaust valve lifter and the method of controlling the quantity of charge admitted to the cylinder is by varying the compression of the inlet valve spring. The Sutton Racine's engine had a bore of 4½ inch (114.3mm) x a stroke of 6in (152.4mm) 1565cc engine which had external flywheels which weighed 2½ cwt. The engine ran at 640 rpm for a weight penalty of 280 pounds. With four passengers the car weighed just over a ton and was capable of getting up to 20 miles (32.187 kilometres) to half a gallon (2.2½ litres) of kerosene, the cost of the car was between £350 - £400."

On the success of his new Racine automobile, Henry was now ready to go into automobile production himself and so he contacted the Austral Otis Lift Company again, having had a good working relationship with them during the design of the lift in the Ballarat music Store. Henry had formed a great friendship with Mr Edward Joseph Rigby, a manager of Austral Otis, and his approach to the firm was initially met with favourable interest. Austral Otis had a large engineering works in Grant Street, South Melbourne, which had the capabilities and facilities to manufacture Henry's automobiles and the new and promising era of automobile production posed a possible means of expansion to the company's business. At the time Henry first entered into negotiations with Austral Otis, they were very interested in the business venture and seriously considered putting Henry's car into production. However, after many months of negotiation and careful consideration, Austral Otis decided not to proceed with the manufacture of automobiles. It was estimated that each car would cost around £400 to produce, whereas imported cars were selling as cheaply as £350 each, so that the cost of producing Henry's car was not financially viable.

In January 1903, Henry's Racine Tonneau graced the front cover of the English car journal *The Automotor.* In the same year, Henry had also constructed a new Voiturette called The Sutton Australian Voiturette. This automobile, seen pictured for the first time in the *Melbourne Punch* newspaper, has never before been recorded in Australian automobile history and publication of the photograph dates its provenance as February 5 1903.

Regardless of the failed bid with Austral Otis, and owing to the success and interest of the Sutton Racine and kerosene engines, Henry had received many enquiries from such countries as India, Burma and Egypt. He also had a number of orders from friends and acquaintances to build new cars, which undoubtedly kept him very busy over the next few years. Beauchamp Motor Company were instrumental in manufacturing and selling Henry's carburettors and they converted a number of vehicles for customers using the Sutton patents, including a lady's Victoria and an 8hp doctors Brougham.

Henry Sutton was one of the first automobile manufacturers in Australia and once stated; *"The motoring industry will be the greatest industry the world has ever seen"* and with that statement a new era had begun in Australia. Henry and Harry B. James from the Dunlop Tyre Company were both keen to see the expansion and promotion of automobiles and motoring in Australia and to grow the trade and industry. In June 1901, just after Henry had completed building his second Racine car, Harry James discussed with Mr

W. J. Proctor the possibility of an annual Australasian automobile race being held. The race, he thought, could be similar to the great European international event known as the Gordon Bennett Cup.

One night after a long country drive over the Black Spur country, and with the idea of an annual Australasian race playing on his mind, Harry James dozed off in front of a cosy winter fire and dreamt that it was November 1902, and he had taken part in a big Australian motor race called 'The Federal Cup'. This dream was so vivid that upon awakening, he immediately wrote down the details, and these were subsequently published in a number of newspapers.

The competitors in the dream race were the who's who of Australian motoring: there was himself and Jack Proctor in a car modelled on the lines of a Panhard as well as Henry Sutton in his Racine with John Mennie. Other competitors were Charles. B. Kellow in a Voiturette, Herbert Thomson and E. L. Holmes in a Thomson steam car, Vivian Lewis and T. O'Grady in a car they brought over from Adelaide, W. H. Lewis in a Tarrant, Sydney Day in a De Dion and numerous other well-known motoring men from all over Australia in their imported cars. Harry James's account of the dream of the race was very detailed and exciting, as shown in the following exerpt:

"We are now close up behind the leading cars, but finding it unpleasant getting their dust, we decide to 'jump' them. The first car we reach is Lewis' and with a brief 'how's things' we range up close behind awaiting a chance to slip by. At last the road being favourable the high gear is thrown in, and we gradually draw up. We can hear O'Grady call out 'Look out they are passing us'. Vivian Lewis does his utmost but as our electric ignition is slowly advanced we gradually gain until we cross over into the centre of the road ahead of him. We now ascertain from Lewis that Elliot's Daimler is only a couple of hundred yards ahead in company with the Tarrant and Racine. The Tarrant seems to have some trouble as Hunter is noticed several times trying to fix something in the front of the car. We therefore decide to push forward and after covering some two miles are right up behind the Daimler and the two others. As we pass Elliot both the Tarrant and Racine follow us. The road now being perfect for our car is put at its top speed and mile after mile is left behind."

Of course, being Harry's dream, by the finish his car had won the race. There is no doubt that the race would have been a very exciting event had it actually happened. The dream was an idea of the future for automobiles and on the back of that dream more humble events were to start the rise of the automobile in Australia. One such event was arranged by The Dunlop Company and Harry James a few months later.

On the occasion of the upcoming Warrnambool to Melbourne bicycle race organised by Harry James and the Dunlop Company, a number of motorists joined together to accompany the riders to Melbourne and make observations of the race.

Four cars and their owners actually took part in the motor vehicle demonstration. Although it was not an official car race, the cars were competing against each other in a friendly manner to show the endurance of their cars and for the glory of being first to arrive in Melbourne. So it could be said that it was the first time vehicles competed against each other in Australia. The line-up consisted of Henry Sutton in his 4 seater, 6 hp Sutton Racine; Harry B. James in the Dunlop 3 seater, 3½ hp De Dion Voiturette, accompanied by the Managing Director of Dunlop Tyres, J. W. Proctor; the well-known cyclist Charles B. Kellow in his newly imported 4 seater, 6½ hp Darracq Voiturette and a Frenchman from Brisbane by the name of Arthur Constant Aucher in his 4 seater, 8 hp De Dietrich car. Travelling with Aucher was the former champion cyclist Con Dwyer and E. H. Lenne who was the officer in charge of the cyclists of the Melbourne General Post Office. Mr Aucher was currently in the process of demonstrating the use of automobiles for the possible future use by the post office.

The four automobiles left Melbourne on 9 October 1901, and stopped over-night in Geelong before continuing onto Warrnambool. Henry's car used kerosene which was available at any store along the road but the other drivers carried their own fuel for the trip to Warrnambool and had to forward their petrol for the return journey to Warrnambool by train prior to their arrival. The rules for the riders in relation to the cars participation were as follows: the cars would leave Warrnambool two minutes after the scratch men had started the race and the cars would not pass the riders until they had reached Geelong. Strict injunctions were issued to riders not to attempt to take pace from the cars under penalty of disqualification. It had been arranged that members of the Press would ride in the spare seats of the cars and report on the race as it progressed and then to beat the riders into Melbourne and report on the finish of the race. Well that was the plan!

All week prior to the race, cycle riders arrived in twos and threes from all points of the compass but the highpoint for the locals was reached when the motor cars arrived in Warrnambool. People came from all over the district and were never tired of seeing the strange vehicles dashing through the streets.

The Warrnambool to Melbourne Road Race began outside the Warrnambool Post Office on Saturday 12 October 1901. By early that morning the crowd had swelled to nearly 300 people and by 4.55am, all the riders were ready. Out of seventy acceptances, fifty eight riders presented and at 6am, just as a fine mist of rain began to fall over Warrnambool, the local Mayor, Cr. William Ferguson, started the race. He first flagged off the limit man, J. R. McDonald (the man with the greatest advantage in a handicap race), followed five minutes later by C. F. Taylor, J. Oakes, A. Hall, J. Jordan and T. Fallon. As each rider started the race the excitement of the crowd grew.

The riders made a quick start on their journey to Melbourne. The course was from Warrnambool to Terang (29 mile mark), then onto Camperdown (43 mile mark), Colac (72 mile mark), Winchelsea (95 mile mark), Geelong (119 mile mark), Werribee (145 mile mark) and then through Footscray, past the racecourse and onto Flemington Road close to the North Melbourne tram sheds, and then onto Haymarket - a total of 165 miles (265.54 kilometres).

The four cars started off well but over the course of the race they all came to grief. The Kellow car was the first to strike trouble and had to bow out. In Colac, discovering that his oil had been taken by another driver, the Frenchman Mr Aucher was incensed and bellowed many loud expletives in French, German and all other languages of Europe. Mr Lenne and Mr Dwyer helped him with Colonial and Irish tirades and afterwards Mr Lenne frantically sent telegrams in all directions seeking assistance with getting more oil. As to which driver stole Mr Aucher's oil, that remains a mystery; only the other three drivers could answer that question. Ironically, though, the stolen oil did not assist anyone as all the cars failed to beat the riders to Melbourne. The Dunlop car failed when one of the De Dion's electric wires short-circuited about seven miles from Winchelsea. This was due to excessive speed and heavy rain. Henry managed to use his Racine car to tow the Dunlop car into Winchelsea, where repairs were eventually made. Henry's Racine car handled the distance better but in the end was held up due to the atrocious conditions. At the start of the race, these were favourable but in the latter part of the race, the weather turned horrendous and the road had become almost impassable, both for the cars and the cycle riders, but in true grit and spirit both the drivers and riders pushed forward.

The terrible state of the roads were described at the time as like riding through porridge, such was the slush and mud. News of the progress of the race was being passed onto Melbourne through telegrams and phone calls. Due to the conditions, the race could be won by anyone and the excitement began to swell. A crowd of 4000 people assembled at Haymarket at the north end of Elizabeth Street. By 2.30pm the crowd had swelled so much that Charles Proctor of the Dunlop Company arranged to have forty mounted and foot police to help control the crowd as the winner, Albert Nioa, crossed the finish line at 2.45pm.

Although the cars did not fare as well as hoped in the conditions on the day, their participation in the race helped to highlight the use of motor vehicles and brought a new awareness of the value and utility of the motor car to the public and this helped move Australian motoring forward. The suitability and affordability of motor cars for travelling over the rough roads of Australia was established as a result of this race day as eventually all the cars made it back to Melbourne. Although the Dunlop De Dion took just over nine and half hours for the trip, it consumed five gallons (just over twenty two litres) of petrol, at four shillings per gallon, and a quart of lubricating oil. On the other hand, Henry's car consumed approximately four gallons of kerosene, a little over eighteen litres, at a cost of one shilling and seven pence per gallon, demonstrating that the economy of Henry's engine and automobile far outweighed the petrol driven cars. The participation of the motor cars in the friendly race, had cemented the future of automobile travel in Australia and the names Henry Sutton, Harry B. James and J.W. Proctor would soon become major influences in the new era of Australian motoring.

Henry arrived back from the Warrnambool to Melbourne race to the shock news that his wife, had been taken gravely ill on 11 October, the day before the start of the race. Elizabeth's life with Henry had, at times, been very lonely and testing; he'd spent many long hours, day and night, year after year, working on his inventions and this had not been easy for her. It made for a lonely and somewhat neglected life. In the 1880s when Henry and Elizabeth had first married, life with Henry would have been exciting for a young wife but the emotional and social distance grew over the years. The deaths of two of their children, infants Reginald and Percy, created a great void in Elizabeth's life and left her facing years of grief at the loss. Henry, although loving and caring and also feeling the loss of their two boys, retreated back to his work, but Elizabeth was left idle for many years and consequently drifted into depression. It is not clear at what point Elizabeth started drinking, but due to the illness that clouded her life at the turn of the century, it must have been for many years. Henry clearly would have known about Elizabeth's drinking, but although he probably tried to help her, his work consumed much of his time. If Henry had any failings, they were the neglect of his long-suffering wife.

On 24 October 1901, Elizabeth died of chronic alcoholism, exhaustion and organ failure at their home, "Rupetstrue", at 134 Boundary Road (now Kooyong Road), Armadale. It was a sad end for the loving and caring woman that Henry had married twenty years before. Their son Arthur Ernest was now a young man, but Elizabeth's death left their four year-old son Edward (young Teddy) without a mother. Elizabeth was buried on 27 October, in the Ballarat New Cemetery alongside Reginald and Percy.

The void Elizabeth left was huge. Henry now needed to take care of Edward on his own, but clearly he was not equipped to do that. Just as he was sorting out what to do, another cruel blow struck the family. On 24 December, Elizabeth's father, Robert Wyatt, died in Ballarat. The last surviving member of his family in Australia, Robert had been blind and frail for many years, and the death of his last surviving child had taken its toll. With no more will to live, he died of old age at his home in Lydiard Street, Ballarat. Henry once again travelled to Ballarat, this time to bury his father in-law Robert alongside his wife Fanny and his son William. Elizabeth and the two children were buried close by. Robert Wyatt's will, dated 29 October 1901, reveals that he must have revised it after Elizabeth's death in order to bequeath his grandson, Arthur, his shares in the Ballarat Banking Company Limited. He left in trust for Edward, his shares in the Ballarat Gas Company Limited. Robert's cottage property and his freehold property in Lydiard St, along with the furniture, were to be sold and the money put in trust for both his grandsons in equal shares until they were 21 years old.

It was a sad Christmas for Henry, Arthur and little Teddy, but the Sutton family rallied around them. In the new-year, Henry advertised for a live-in housemaid/nanny to care for his son and look after the house. A very lovely young lady by the name of Annie May Tatti was eventually employed in the position. Annie May was born in Daylesford and was the fourth daughter of Martino (Martin) and Ann Tatti (*née* Jackson). Previously, Annie May had been a waitress in a cafe and she was only eighteen years old when she began living in the Sutton residence to care for young Edward. At the time, there was a lot of criticism about a young single woman living in the same house as a recently bereaved forty-six year old man.

Henry was a handsome and wealthy man and tongues began to wag about them both. Many local townspeople and members of the Sutton family voiced their disapproval of the new living arrangements. So whether it was the pressure from society, family and friends or just circumstance that bought Henry and Annie May together as a couple, no one really knows, but love won out in the end. On 17 September 1902, surrounded by family and friends, Henry Sutton and Annie May Tatti were married by the Reverend F. Mason at "Rupertsrue".

It was not long before Annie May became pregnant and on 2 November 1903, Henry and Annie May became the very proud parents of a son whom they named Albert Martin Sutton.

The marriage to Annie May marked a new era in Henry's life; he became a more devoted father and attentive husband who was often seen out and about in his new Racine automobile with his very attractive young wife.

Left: Wedding photo of
Annie May Sutton, née Tatti, 1902

Having made great inroads in the motor vehicle industry in Australia. Henry felt it was time to expand into the world market with his patents. This led him to construct yet another vehicle which would be the show piece and demonstration vehicle for his patents and would be an affordable Australian made vehicle for the Australian and overseas market. He called this car the Sutton Austral Car.

After Austral Otis decided that his Racine Tonneau was not financially viable, Henry came up with an automobile design which was cheaper to produce. This time he successfully teamed up with the Austral Otis Engineering Company in South Melbourne to produce automobiles based on his kerosene combustion engine. Sometime in 1903, Austral Otis began construction of the new car and by early 1904 the first was completed and ready for the automobile market both in Australia and overseas. The initial Sutton Austral Car was built in time to be exhibited at the 1904 St Louis Exposition. Sadly, no details of the design or construction of the Austral car are available, nor are there any photographs. Future research may yet uncover more information about this car.

On a commercial scale, Henry Sutton's engines and carburettors were successful in the small market place in Australia and to industry and private individuals around the world, but Henry had bigger plans afoot. By 1904 , he was ready to sell his Combustion Engine Patent to automobile manufacturers around the world. The decision to sell his patents was a calculated one that he had spent the previous year planning with the aid of his great friend Dr Alfred Perkins Merrill, who was the American Vice Consul to Australia, his engineer Mr John Mennie and past employee, the cyclist Don Walker.

The plan was big, on a scale never attempted by an Australian inventor associated with the automobile trade. It was a bold move and financially risky but it is clear that Henry saw the potential of the world's growing automobile industry. He had amassed a very good reputation for his cars, engines and carburettors and automobile production was clearly the next step to cementing a successful future in the industry. On the strength of his reputation alone, Henry set about arranging to exhibit his engine, carburettor system and his 'Sutton Austral Car' at the 1904 St Louis Exposition. Unfortunately it turned out that the cost of sending the Austral Car over to the exhibition in St Louis was enormous and consequently much to Henry's disappointment the decision was made not to ship the automobile over. Instead Henry arranged to have an automobile in St Louis fitted with his engine and carburettor to show the adaptability of his invention.

As it was impossible for Henry to attend the Exposition himself due to his health and work as well as family commitments in the form of a new wife and baby, he arranged for John Mennie to travel to St Louis with the engine and carburettor and to set up the exhibit that would be on display.

The 1904 St. Louis Exposition ran from 30 April to 1 December and showcased some of the world's newest technologies and the industrial research around them, and it would be the first large scale exhibition of automobile manufacturers in America with no less than 160 automobiles exhibited along with many automobile-related inventions. They would all be showcased in the Palace of Transportation, a building 1,300 feet long (approx. 396 metres) and 525 feet wide (approx. 160 metres) with a total floor area of 620,400 square feet (approx. 57,637 square metres).

In the Palace of Transportation, that was full of the old and the modern wonders of the new transportation age, Henry's exhibit was situated under the auspice of Great Britain in the Australian section, which was in the western part of the palace and came under the heading, Carriages and Wheelwrights' Work, Automobiles and Cycles, Department G, Group 72.

The converted car and Henry's patents were listed under Class 455, 'Vehicles driven by mechanical motors, Motorcycles and Automobiles' and his engine and carburettor came under Class 457, 'Detached parts, material and inventions pertaining to carriage building, wheelwright's work, automobiles or cycles'.

It so happened that another relative new-comer to the automobile industry was also exhibiting his new automobile at the Exposition and his company was in the market for new and improved automobile inventions and patents. The name of that company was The Ford Motor Company which Mr Henry Ford, who spent most of his time at the St Louis Exposition that year, had started only one year before in Detroit, Michigan.

The prior publicity of Henry's carburettor system, which had been published in the various journals and especially the American *The Automotor Journal*, brought awareness and a keen interest in his system from all sectors of the automotive industry and many of the automobile companies at the Exposition. Henry had organised for Dr Merrill and Don Walker, who was scheduled to ride at the Exposition, to assist John Mennie in promoting his car, engine and carburettor system. Timing was crucial, so to coincide with the opening of the Exposition Henry sent information on his system to a number of English car journals and to *The Automotor Journal,* in which a large descriptive article and photograph's of Henry's carburettor were published on 4 June 1904. Included in the article, which spanned four full pages, was a photograph of Dr Merrill's Winton car that Henry had modified late in 1903, as well as photographs of one of Henry's carburettors. The strategy worked, and the flow-on effect was that Henry and Dr Merrill received many inquiries on the carburettor system and patents.

Henry's American dentist friend, Dr Merrill, was a character who, before coming to Australia, had led a very interesting life. As a young man, he had travelled to the West Indies and many of the South American countries. He was once ship-wrecked near Barbados on his way to Havana, Cuba, and on another occasion his ship ran aground sixty miles from Valparasio in Chile. Merrill, who was a great advocate of Henry and his engines and carburettors, was taking a trip to America and Europe and, as the well-known Vice Consul to Australia, he had a lot of weight behind him in the American market. Henry could not have wished for a better promoter and Dr Merrill was all too happy to assist his friend in his endeavour to make inroads into the automobile industry. Henry also enlisted his past employee and friend, Don Walker, who was travelling with his close friend and fellow cyclist the American champion Major Taylor to America. Both were scheduled to ride at the St Louis Exposition in July.

Mennie and Dr Merrill left Australia for America on the 14 March 1904, on the ship the *Ventura* out of Sydney. After arriving in America, Mennie travelled straight to St Louis with Henry's car, which had a Sutton engine and carburettor. He immediately set about putting the car and carburettor on display and doing some initial promotion and talking to companies in relation to Henry's patents. Soon after that, he left contact details at the Exposition and travelled onto England to seek British and European buyers of what was now called the 'Sutton's Patent Vaporiser'.

Meanwhile as a representative of the newly formed Automobile Club of Victoria, Dr Merrill was not only travelling to America but also going onto England and Europe to see the Gordon Bennett Cup motor car race, which was being held on 17 June 1904, at the Homberg Circuit in Germany. When Dr Merrill arrived back in America after the Gordon Bennett Cup, he purchased the Light Spaulding Runabout which was then converted to Henry's engine and carburettor system and put it on display at the exposition. While in New York, Dr Merrill busied himself with promoting Henry's new carburettor system and seeking out possible avenues of the sale of his patents. He had also promoted and sought potential buyers of Henry's patent while he was in England and Europe. John Mennie was going to follow up the contacts Dr Merrill had made.

After a short stay at the Exposition to see the sights and promote the Sutton carburettor system, Dr Merrill returned to Australia on the *Sonoma* on 22 August 1904.

In an interview given a few months after his arrival back in Australia Dr

Merrill stated:

"Within the last few years a perfect kerosene carburetting system for motors has been introduced by Mr Henry Sutton. In my opinion the motor world will see this Australian invention universally adopted. Having just returned from a trip to America in which places I received many enquires to Mr Sutton's carburettor, I feel that for simplicity, economy and utility it has few equals."

Don Walker, who was born in Yamachiche in the Province of Quebec in Canada, had come to Australia to live when he was seven years old. He left Australia with Major Taylor on the ship the *Sierra* out of Sydney on the 15 June 1904. After arriving in San Francisco, they both travelled to Major Taylor's home in Worcester, Massachusetts.

After a few days rest in Worcester, Don Walker travelled to Yamachiche to visit family and then went on to Toronto for a few days to visit the Massey-Harris Factory (now known as Massey Ferguson) and The Canada Cycle and Motor Company both of which were based there. With the very small market in Australia, Henry had decided to sell his combustion engine patents to overseas companies, which was the next obvious step to putting his carburetting system in mass-produced motor vehicles. He picked a ripe time in the Canadian and American automobile markets to do this as many new motor companies were looking for the latest advances in motor engineering to build their new motor cycles, automobiles, trucks and farming equipment. There was also a market in various industries throughout the world.

Henry had arranged with Don Walker to take a number of his carburettors with him to America and Canada to demonstrate them to the Massey-Harris Company and The Canada Cycle and Motor Company. At the time of Don's visit, the Massey-Harris Company was the biggest farm machinery producers in the British Empire. The Canada Cycle and Motor Company, who were Don Walker's sponsors, manufactured bicycles and motorcycles and were about to start producing motorcars. Don not only went to Toronto to look at their bicycles but he was also there as Henry's representative and demonstrated the carburettor system to the two companies' engineers. Don, who was due to ride at the Exposition, arrived in St Louis in July as scheduled.

Henry's brother, Alfred Sutton, and his wife Eliza, who were on a world tour and visiting various companies in the US and Europe in relation to the family's music business, also travelled to the St Louis Exposition in July 1904 to assist in Henry's venture.

After St Louis, Alfred and Eliza travelled onto Elkhart Indiana to the famous

American musical instrument makers C. G. Conn to secure the agency for their musical instruments for the family music store. After securing their agency and others in America, they travelled onto England and Europe for a holiday. Alfred and Eliza returned to Australia in December 1904, and upon their return a welcome home smoke night was held for all the staff of Sutton's at the Bull and Mouth Hotel in Bourke Street Melbourne. Everyone, including Henry, was eager to hear the news about their travels and about the St Louis Exposition.

John Mennie arrived in England and immediately took up residence at the Star Hotel in Blackfriars Road, London and set about following up on the contacts he and Dr Merrill made both in the US and England. Arrangements were soon made with the English company Thornycroft & Company. Representatives of Fairbanks Morse and Company, a Chicago engine manufacturer, and the French Renault Car Company were also interested in Henry's patent. In August, Mennie gave a demonstration using one of Henry's carburettors on one of Fairbanks-Morse's engines. He immediately wrote to Henry and told him about the success of the demonstration and that the company was now very keen interest in purchasing the rights to Henry's patent. On the 20 September 1904, Henry wrote back to Mennie congratulating him, as the results of the test were better than they had managed at Otis in Australia and was the best result per horse power and economy yet seen in England. Henry immediately wrote to his brother Alfred, who was now travelling around England, and with his help they started brokering the terms of the purchase of a licensing agreement for the patent. Alfred acting as Henry's attorney in the matter set out the terms in a letter to Mennie which is as follows:

"The Amount required in Royalties would be;

US $1.25 On Atomisers to be attached to Cycle Engines up to and not exceeding three brake horse power, 3bhp.

US $2.50 On Atomisers to be attached to Motors or Stationary Engines up to and not exceeding nine brake horse power, 9 bhp.

US $5.00 On Atomisers to be attached to Motors Stationary Engines above or exceeding nine brake horse power 9bhp.

In exclusive Royalty of System for all countries covered by Henry Sutton's patents a payment of US $10,000 would be required.

Any variations to above terms to be submitted to and approved by Henry Sutton."

The conditions set out to Fairbank Morse by Henry also stated that the terms eventually agreed to would require to be signed by the 3 January 1905.

By November, Alfred Sutton was in London and on the 3 November

1904, a legal agreement was drawn up with Fairbanks Morse's representative in London, Mr Orin S Wilson. It is unclear if Fairbanks Morse agreed to the royalty terms and signed the agreement as negotiated or if they agreed to pay the $10,000 for exclusive rights to Henry's patent in the United States. The legal agency agreement, which was drawn up by Fairbanks-Morse's representative in England, states that it will need to be ratified and signed by Fairbanks Morse before 3 January 1905. If the agreement was not signed by Fairbanks-Morse then Mr Orin S Wilson would become Henry's agent at a commission of 5% upon all money's received by him or payable under or virtue of any licence or licences which had been issued.

Another interested party was the British Company Thornycroft but although Henry's carburettor was demonstrated to the owner, Sir John Isaac Thornycroft, it is unclear if the company took up Henry's patent. In France the Renault car company did purchase the licence for Henry's patent but they immediately shelved it, which must have angered both Henry and his brother Alfred. Companies tended to do this just to be rid of any competition and this was most likely Renault's reason for shelving the patent. Further research may uncover more facts around the purchase of Henry's carburettor patent by these various companies as well as other manufacturers who were interested in using it commercially.

With business complete, Mennie left England on the 19 November 1904 and travelled to St Louis to pack up the Sutton exhibit when it finished on 1 December. He returned to Australia on the *Ventura* on 16 January 1905. Sadly though, when he returned to Australia, Henry had a falling out with him over his expenses while he was in England. He felt that Mennie had taken advantage of the open letter of credit he had been provided with.

This falling out immediately ended his employment as Henry's engineer and was a sad end to what had been a wonderful and long personal and professional relationship.

Author's addendum:

The Sutton Voiturette Pedal Car built by Henry in Ballarat and completed in mid-1900, was to become the centre of a century of great mystery.

Around 1952, Peter Menere of Peter Menere Motors in Brighton, Victoria, first heard from a friend the story about a mysterious car that was built in Ballarat. In fact, suggested the friend, it might be possible to buy what could be the oldest car in Australia. At the time, Menere did not believe the story but he was nevertheless intrigued, so he agreed to go and look at the car. Long abandoned on a rural Sunshine property, the car was eventually discovered in a shed owned by Mr Reilly who had moved to Sunshine from Ballarat in 1907. When Menere first saw the car it was tangled up with chicken wire and chickens were perched on top. The engine was found under years of rubbish, the horn was found in the ceiling of the old farm house and the seat frame was upside down in the middle of a paddock covered in plants with a duck and her eggs for company. Menere had to wait a fortnight until the eggs hatched before he could collect the seat. He never revealed how much he paid for the car, but the journey of finding the origins of the car would take him all the way to England to try and find the answers.

Reputedly the first car in Ballarat, the find had commonly been referred to as a Benz, but vintage experts quickly discounted that, although for some time no-one could actually decide exactly what it really was. Some thought it might be a New Orleans or a Vivinus but these were also discounted.

On a trip to England, Menere sought the expertise of two library assistants at Lord Montagu's National Motor Museum in Beaulieu, New Forest, Hampshire. Menere spent many long hours trawling through endless journals, books, papers and magazines to find any information on the car, but had no success. In desperation, an article was published in *The British Veteran Car Club Gazette* in 1959 along with the photograph of the original owner sitting in the car, but the article failed to elicit any information. A few years later a newspaper article was published in Australia about the mystery of the car but this too failed to shed light on the origin or maker of the car.

Eventually Menere sold the car to Tim Clemons in Tasmania and he and his brother Chris Clemons set about the task of restoring the Ballarat car. Even so, the mystery of who made the car remained. After the car was restored, it went on display in The Whitehouse Museum in Westbury, Tasmania, which was run by the Clemons brothers.

The mystery continued until one night in 2009 when Henry Sutton's car featured on the ABC show, *The Collectors.*

I was right in the middle of researching my great grandfather, Henry Sutton, when by chance, I turned on the television as the program presenter, Gordon Brown, stated that he was going to talk about veteran cars. This immediately caught my attention as I knew my great grandfather had built cars, including some that were yet to be found. I thought I might learn something and so kept watching the program.

At the end of the piece, Brown sat in a little green car and announced that this car had been made in Ballarat back in about 1900 but no-one knew who exactly had made it. I immediately thought that this could be one of Henry's cars.

The next morning, I contacted The Whitehouse Museum and explained who I was and who Henry was and asked them for any information they had on the car. The museum emailed the information and a photograph that was taken around 1900 of the car with its original owner sitting in it. You could imagine my surprise when the photograph arrived as it was a photograph of Henry and his wife Elizabeth sitting in the car. Overwhelmed with the sheer joy and excitement of the discovery, the information was quickly relayed back to the Museum's owners. Finally the mystery was solved, but the missing link had to be found in the form of information on Henry building the car to claim provenance on it.

This endeavour would take another three years and a trip to England to solve. In July 2012, portions of two letters written by Henry, one published in May and one in June of 1900 in the English journal *The Autocar*, where found. In them, Henry talks about building the car which he called a Voiturette. He completed the Voiturette in July 1900, in Ballarat. Although he'd sent the photo of himself and his wife sitting in the car to *The Autocar* magazine, it was never published. If it had, then the mystery might have been solved much earlier. Finally, there was provenance on a previously unknown Australian car, built by Henry Sutton, and the car can now be added to the annals of Australian motor history. The irony was that the information was in Lord Montagu's library all along.

The car was not in original condition when Peter Menere found it in 1952 - a Longuemane carburettor had been installed and other small modifications had been done. When restored by the Clemons brothers, a few more modifications had been made, but the engine was original and was clearly one of Henry's early prototypes. The car had a one cylinder engine and a wick-type carburettor, and there was no accelerator as the speed was regulated by Henry's advance and retard method. The clutch was on the steering column and the car had a one wheel brake.

In May 2011, the Sutton Voiturette Pedal Car made a trip back to Ballarat and was the feature exhibit of the Henry Sutton display in Mining Exchange Building at the Ballarat Heritage Weekend. It was the first time the Sutton Voiturette had made a public appearance in Ballarat in over a century, and drew a lot of attention from all over Australia and overseas. Today the Sutton Voiturette is the oldest privately owned Australian made car in Australia.

According to Henry's son Arthur, the pedal car cost Henry £50 to make and he sold it to Mr Reilly for £70.

Added note: At the time of Henry's death in 1912, he had been working on another new engine.

First Sutton Autocar circa. 1898
Henry and Arthur Sutton pictured in car.

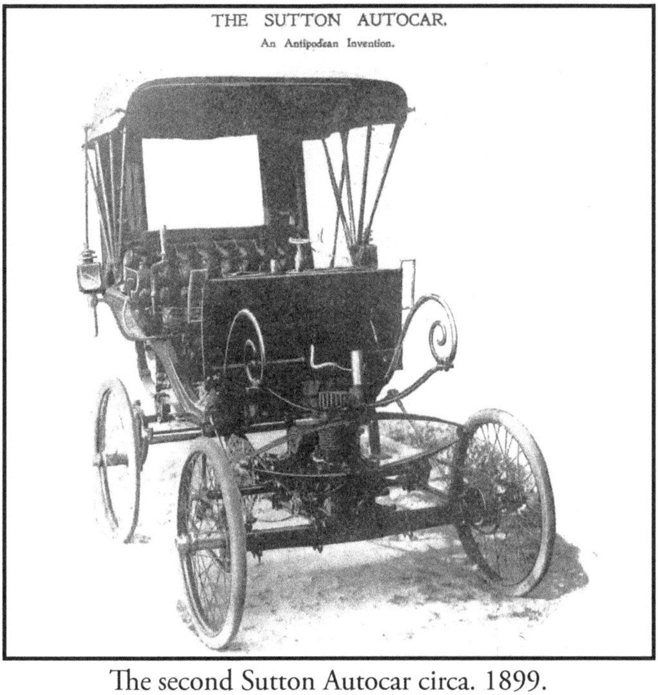

The second Sutton Autocar circa. 1899.

The Sutton Voiturette Pedal Car circa. 1900

The Sutton Voiturette Pedal Car
- currently restored.
Photograph courtesy of Tim and Chris Clemons.

Sutton Racine Tonneau
Henry Sutton and wife Annie in back seat. Arthur and Edward Sutton in the front seat.
Photo taken in 1902.

1903 Sutton Voiturette
Henry Sutton with wife Annie

CHAPTER TWELVE

THE AUTOMOBILE CLUB OF VICTORIA 1903-1908

On 26 September, 1903, a party of fifteen motorcyclists took part in a ride to Tooradin on Westernport Bay. Among the riders that day were Harry B. James, Sydney Day and James G. Colman, who would later become known as the Tooradin Trio.

Harry James had decided that the time was right to form an automobile club and had joined forces with other like-minded motoring men. James had many discussions with Colman and Day about the idea and on that trip to Tooradin, they decided they would form the club. Upon their return that day, Harry James sent out invitations to every interested motorist and motorcyclist in Victoria to join a run to Mordialloc in the aim of forming an Automobile Club. At the time, there were approximately 30 car owners and 60 motor cyclists in the state.

Bad weather delayed the run until 6 December, but on that day, almost a dozen automobiles and thirty motorcycles departed from outside the Melbourne Town Hall at 10.45am in perfect weather. On the day, Henry was driving his Racine car and his son Arthur, rode his motorcycle. It was fortunate that the cavalcade of cars and motorcycles did not travel too closely together, as the dust they churned up blanketed motorists, motorcyclists and the crowd of onlookers alike. Other members of the public joined in on the procession as it made its way around the eastern side of Port Philip Bay, and when they reached Mordialloc, the run was deemed very successful.

To celebrate the occasion, the motorists sat down to a hearty dinner at the local Pier Hotel where James addressed everyone on the need to form an automobile club that would promote motor transport and the new era of motoring. Right there, Henry Sutton drafted the motion that founded the Automobile Club of Victoria. The motion he put forward on that day had only one adjustment, it had been agreed by all the members present that the word "pleasure," be removed from the motion.

After some discussion, they decided to hold another meeting in three days' time at the Port Phillip Club Hotel in Flinders Street, Melbourne, to officially form the Automobile Club. Along with Harry James, James Coleman and Sydney Day, Henry Sutton had been working behind the scenes in preparation for this evening. All the prominent motorists of the day attended and apologies were received from a number of other motorists who could not attend. Instead, they wrote letters stating they would accept membership under whatever conditions the meeting agreed on. In fact, Harry James had received forty applications for membership before the meeting began. In all, some 56 motorists, including Henry and his son Arthur, came along and Henry took the principle role in putting forward the motion for the objectives of the club which would be:

"The promotion of a social organisation and club composed mainly of persons owning self propelled vehicles or motorcycles. To afford a means of recording the experiences of members, and others using motor cars and motorcycles. To promote investigation in their development. To co-operate in securing rational legislation; and the formation of proper rules and regulations governing the use of motorcars and motorcycles in cities, towns and country districts. To maintain the lawful rights and privileges and protect the interests of owners and users of all forms of self-propelled vehicles whenever and wherever such interests, rights and privileges are menaced. To promote and encourage the improvement, construction and maintenance of roads and highways, and the development generally in the State of motoring, and to maintain a club devoted to the interests and advancement of automobilism."

[Full Minutes of Meeting, see End Notes, Chapter Twelve, No. 1.]

During the first official meeting, a Mr Holmes explained to the members present why the word "pleasure" had been omitted from the motion, he stated that eventually there would be many trade vehicles on the roads, as influential firms adopted motor vehicles for use in their commercial businesses and these did not come under the heading of "pleasure" vehicles and so their drivers would be debarred from joining the club. The word "pleasure" existed in the rules of the large associations in England, and if the offending word was removed, a by-law could be drawn up regulating what class of vehicle - trade or otherwise - would be allowed to participate in club runs. A Mr Thornely then asked whether it was intended to admit to the membership of the club any gentlemen not owning a motor car or motor cycle. The chairman stated that the question was covered by the word "mainly". Finally, all members present agreed to the omission of the word "pleasure," and the motion was officially put forward by Henry, then, carried unanimously as amended.

At the meeting, a Provisional Committee was formed and Henry was elected with a view to drafting the rules and by-laws. These were to be discussed at a further meeting to be held on 15 January, 1904, when the entrance and subscription fees would also be determined. The first meeting of the newly formed Automobile Club of Victoria ended after three hours, the members leaving the room with a new purpose and vision for building the future of Australian motoring.

When the members of the Automobile Club next met, considerable discussion ensued regarding the proposed rules and by-laws. By the time that meeting closed, one hundred names had been lodged as foundation members and the entrance and acceptance fees for the club had been fixed at £2/2s, in each case to be payable in advance. In order to ensure a large membership at the outset, the meeting decided that any gentlemen nominated before 1 March, would be exempt from paying an entrance fee, while those joining before 1 June, would have a reduced fee of £1/1s. In the case of country members, the fees would be half of each amount agreed upon. The club's Constitution, rules and by-laws, formulated by Henry and the other committee members, were passed by about seventy members. Types of membership were also decided and the meeting passed a motion that membership of the club would consist of three classes: Club Members, Country Members and Honorary Members, each of whom must be over eighteen years of age. The meeting then adjourned until 29 January, when the election of office-bearers and committee would be held.

At the subsequent meeting, the first official elections of The Automobile Club of Victoria were held. The chairman of the provisional committee, Mr Harry J. J. Maddox, presided and the first agenda item was the formalisation of the new by-law relating to the qualifications of committeemen. The proposal put forward by Mr Coleman was that no person connected with the automobile trade should be eligible for election. After some discussion by members, this was carried by 28 to 21 votes. The effect of the new club law was that Harry James could not be elected to any office of the very club he had worked so hard to form. It was a great disappointment to many members, but Harry was in agreement with the new law and a new secretary was later elected. The acting secretary then read a letter from His Excellency the Lieutenant Governor and Chief Justice, Sir John Madden, who in reply to an invitation, said he would be pleased to accept the position of President of the club. Dr Merrill, was elected Vice President, and Mr Sydney Stott was elected as the Honorary Secretary. The following gentlemen were elected to the committee: Mr H. C. Bagot, James R. Crooke, J. H. Craven, Sydney Day, Harry J. J. Maddox, J. W. Moffat, Walter Rose, Dr D. M. Officer and Dr Charles Ryan.

During the first few meetings of the club, an inaugural automobile run was organised to publicise the formation of the new Automobile Club of Victoria. This run was held on 20 February 1904. A lot of planning and organisation went into it and the owner of Aspendale Park racing track, Mr J. R. Crooke, generously made his venue available to the club for its inaugural run. Large numbers of excited people turned out to witness the start of the run at the Princess Bridge, and along the left-hand avenue of St Kilda Road, Melbourne. The crowd were amazed by the spectacle of the cars being marshalled into position for the start, one behind the other. It was an imposing sight as the convoy extended along the road for over a quarter of a mile. Ahead of the cars, twenty motorcyclists who included Arthur Sutton were under the direction of Mr Joshua and Mr Bagot.

The motorcycles left at 2.15pm, ahead of the cars so as to avoid the dust that would be kicked up by all the vehicles. There was a large contingent of invited guests seated in the automobiles, among whom were Mayors and councillors from the surrounding suburbs. The long line of cars was headed by Dr A. P. Merrill and Henry Sutton who were both in Merrill's Winton car. Just prior to the run, Henry had converted the car to his carburettor system which ran on kerosene, and so instead of driving himself, he travelled in Merrill's car to assist him if needed. In the car, behind Henry and Merrill, sat Sir Samuel Gillott and the Clerk of the State of Victoria, Mr George Henry Jenkins.

Shortly after 2:30pm, the cars headed off and the large crowd that lined the sides of the road cheered and waved as the cars proceeded along St Kilda Road, down to the Esplanade. The cars then travelled along Point Nepean Road to East Brighton, Cheltenham and then on to Mordialloc and Aspendale. Throngs of spectators lined the procession route and cheered in excitement, becoming totally covered in dust from the unsealed road as the motorcycles and cars passed them one after the other. All the vehicles performed well, especially the steam cars, although there was one mishap with Herbert Thomson's 8 hp car when the fuel suddenly caught fire. Needless to say, all the passengers in the car made a quick exit from the vehicle and the fire was quickly put out. Luckily the car only suffered some blistered paint and, with some loss of pride by the owner, it was able to finish the run to Aspendale. There, the entourage stopped for refreshments at Ferrero's Chalet for about an hour. Dr Merrill addressed the crowd and thanked them for their attendance, saying that he hoped the day would help the official guests when they were framing the new regulations necessary to cope with the increasing motor traffic.

After refreshments, all the motorists and guests proceeded to the Aspendale track where many local residents had come to witness the spectacle. Several of the car drivers and motorcyclists drove around the track to assess its suitability for motor racing. The verdict was that the turf was too heavy for any vehicles to get up to high speeds. Mr Crooke made immediate plans for laying down a motor speedway track. It was estimated that over £9,000 worth of motor vehicles participated in the club's inaugural run and at the end of the day the motorists' convoy headed back to Melbourne, arriving just after 6pm, concluding a very successful day.

From the first official meeting of the club, the provisional committee was put in charge of finding club rooms for the organisation and, after many attempts at securing suitable premises, a home was finally found at the Reform Club on the third floor of 243 Collins Street, Melbourne.

Another outing was organised for 3 March 1904, to give the Automobile Club Committee an opportunity of deciding what may be done in the way of promotion and activities such as car racing and automobile runs. The Melbourne Commercial Travellers Association's annual picnic, which was to be held on 12 March, 1904, at Sandown Park was a great opportunity to promote the new club, and also bring awareness to the public the future possibilities of automobiles in Australia. Members of the Automobile Club drove Mr J. Davies, the Secretary of The Commercial Travellers Association, where they were entertained by the Scottish singer Mr Harry Skinner. While at the Park, the club members undertook a number of trials to test the safety and speed of the track, both with cars and motorcycles, demonstrating that the track was perfectly safe to race on. The times recorded by many of the vehicles eclipsed horse records on the track.

The Automobile Club of Victoria seized the opportunity and scheduled a number of car races for the upcoming C.T.A. annual picnic. The first race would be a Voiturette Car Race for cars up to 6 hp, set over a course of one and a half miles consisting of one lap and 125 yards. The second race would be a Roadster Motorcycle Race over one and a half miles, and the third race would be the Heavy Car Race over a three mile course. The final event of the day would be a parade of some thirty motorcars and thirty motorcycles in or on which the lady visitors of the C.T.A. would be driven around the course.

On the day of the picnic, over sixty motorists from the Automobile Club made the run to Sandown Park - the motor vehicles leaving the city of Melbourne to travel via St Kilda, Caulfield, Oakleigh and Springvale to the track. Most members of the C.T.A. though, travelled to the picnic in two special trains that were packed to overflowing, while some of the others did go by car or cycled to the track. Fascinated spectators lined the route of the vehicle run, as they flashed by at an average speed of about 15 mph (approx 24 km/h). After the convoy had left Oakleigh behind, the rural population turned out in their droves. Various remarks from excited locals, and loud verbal complaints from sleepy-eyed market gardeners were heard as the cars passed by. These caused great amusement to the occupants of the cars. On arrival at Sandown Park, the Automobile Club members were greeted by the committee of the C.T.A. and escorted to the restrooms where they brushed off the generous layer of dust that covered them. The scene at the park was very animated; the fifteen thousand people in the crowd had been waiting in anticipation of the club members' arrival and were so thrilled by the event that they dubbed the club members and their vehicles the "Knights of the Road".

At 5pm on 12 March 1904, Australia's first car race began. Watched by thousands of enthusiastic spectators, the Voiturette race got underway after a short delay owing to a false start. The first across the finish line, the winner of the first car race in Australia, was James Crooke in a 4½ hp Locomobile. He finished the course in three minutes and 55 sec. In second place was Charles Kellow in his 5 hp Humberette, who took 4 min. and 0.025 sec.

The second event was the very exciting and close Roadster motorcycle race, and Charles Mayman won this by about 150 yards in a time of 3 min. 42.5 sec. Harry James came second, just beating Arthur Sutton by a length. The third event, the Heavy Car Race, was a very good contest as the automobiles battled for lead position. It resulted in a win for Harley Tarrant in a 10 hp Argyll car in 6 min. 55.15 sec. Thomas Rand came second in a 16 hp Decauville car and C. C. Campbell came third. The last event for the day was the parade for the women spectators who showed a lot of enthusiasm for the spectacle and when the races were over many of the ladies rushed the cars, eager to have a ride around the track which kept the drivers busy for quite a while.

This group of races organised by the Automobile Club of Victoria were the first official car races held in Australia, and the result proved beyond a shadow of a doubt that this new sport of car racing had a big future before it and that Sandown was destined to be a great car racing track. Later, at a smoke night held on 25 March by the Automobile Club, Mr Crooke was presented with the Club's colours and badge for his win in his Locomobile. A few months after these first car races, Mr Crooke built the world's first purpose built motor racing track at Aspendale Park.

Further motor racing events were organised by the Automobile Club of Victoria. The next one was the Club's inaugural Gymkhana at the Maribyrnong race course on 30 April, 1904, which was attended this time by over 5,000 spectators. This event was really the first commercialisation of motor sport in Australia, because prior to the event the Automobile Club had issued 5,000 complimentary tickets to State and civic leaders throughout Victoria, thus ensuring a very high profile event. Other spectators had to pay entrance fees of 2/6 and 1/- a head, and in fact more income was derived from the sale of programs both in the city and at the track for six pennies each. The turnout of cars for the race meet totalled 34, which at the time was the largest gathering of cars in the Southern Hemisphere.

It was at this race meeting that Arthur Sutton officially won his first motorcycle race and a few months later, in June, he won the Club's inaugural hill climb event which was held in Burgundy Street in Heidelberg. Arthur Sutton, or A. E. Sutton as he was professionally known, became one of

Australia's first motorcycle champions and had a long and successful career in motorcycle racing. Years later, Arthur's little brother Edward, eventually joined him on the race track by racing with him in his sidecar races.

In 1904 as both motorcar and motorcycle racing began to flourish, so too did the schedule of race meetings around the state of Victoria. Arthur Sutton and close friend, Charles Mayman, were both talented engineers and shared the love of motorcycles and racing. Charles worked for the Beauchamp Cycle Company in Prahran, Melbourne and had built a number of motorcycles, all of which used Henry's carburettor system. He also built two motorcycles for the Dunlop Company, one of which set an Australian speed-distance record in 1904, when driven by Harry James. This same year, Charles had also constructed an automobile for the Beauchamp Company; the car was built using drawings that were published in a British journal and was powered by a De Dion engine. On Christmas Eve, 1904, not long after the car was built, Charles and Arthur travelled together to Bendigo in the Beauchamp car having previously sent both their motorcycles ahead by train in order to race them at the A.N.A. cycling carnival at Canterbury Park.

On the morning of 26 December 1904, both Arthur and Charles arrived at the track in Bendigo to test their machines as they sometimes could be handled roughly while they were in transit. Arthur mounted his motorcycle first and went around the track two or three times before Charles started his laps. Arthur passed Mayman on one lap and was travelling at a fair rate of speed when he saw people running across the track. At that moment Arthur heard someone yell at him to stop his motorcycle and he just managed to do so before he reached the spot where he saw Charles surrounded by people. His friend had crashed his Brassey 2¾ hp motorcycle into the track's wooden picket fence just after Arthur had overtaken him. He had been thrown metres down the track, landing on his head on the asphalt track and fracturing his skull so badly that his brain was exposed.

As Arthur approached Charles, he could see that he was badly injured and immediately ran to get the doctor. Apart from the fractured skull, Charles had sustained numerous other injuries and was carried from the track and rushed to the hospital, but he died just after arriving. The motorcycle race that day was to be Charles' last and final race as he had decided to give up racing. His death was a sad and tragic end to a talented young man's life.

The funeral for Charles was held on 29 December 1904, and was attended by family, friends, work colleges, members of the Automobile Club of Victoria and the Sutton family. Charles' coffin was covered in wreaths and floral tributes and the pall bearers were Mr W. J. Dunham, Mr W. A. Dunham, Mr Sydney Day, Mr W. Barrenger and Mr George Hare. For Arthur, the loss of his best friend was a devastating blow, one that he never really recovered from. As the years passed, he never forgot his best friend and would regularly visit Charlie's family for years after the accident. The family did not have the heart to touch anything in Charles room. So, for many years his possessions remained there undisturbed, amongst them a casting of one of Henry's carburettors, that Charles had used on his motorcycles. Along with the Sutton family, the motoring and cycling world mourned the tragic loss of Charles Mayman who was described as a very talented young engineer. The Automobile Club of Victoria paid tribute to Charles by laying a tombstone over his resting place in the Boroondara Cemetery in Kew.

The photographer, Algernon Darge, who had photographed Charles Mayman sitting in his car the morning he left to pick up Arthur Sutton and travel to Bendigo, made copies of all four photographs of Charles available to the public. After the funeral copies could be obtained from Mr. E. Beauchamp at his cycle and motor works in Chapel Street, Prahran, or from the Darge Photographic Company directly.

Henry Sutton served on various committees of The Automobile Club of Victoria from 1903-1908. By 1905, Henry was no longer in the automobile trade and was free to be elected to office, he worked tirelessly behind the scenes in various committees over the ensuing years to promote motoring not only in Victoria but in Australia. During his membership at the club Henry, along with other members, helped draft up the Club's constitution, rules and by-laws, he helped behind the scenes with the organisation of the various car events including car and motorcycle races and car runs throughout Victoria and Australia. Henry helped the Club, Government and local Councils draft up road rules and legislation, including the introduction of driver's licences and signage on roads. He helped the Club lobby for better roads in Victoria and lobbied for regulated speed laws and car registration.

Many decades later, in the 1960s, the RACV were in favour of seat belt legislation but it was Henry Sutton's granddaughter, Betty Milne–Ward (*née* Sutton), who was instrumental in bringing into law the legal requirement to wear seat belts in cars in Victoria. On 22 December 1970, the Victorian Government passed the Motor Cars (Safety) Act 1970, requiring, motorists and their passengers to wear seat belts in their cars. Victoria was the first place in the world to make the use of lap-sash belts compulsory in the front seats and by the end of 1971 all the other states and territories of Australia followed suit. Seat belts for back seat passengers and children followed much later. The passing of these laws made Australia the world leaders in vehicle passenger safety, and other countries took Australia's example and passed their own laws regarding the wearing of seat belts. The passing of the laws in Australia has saved countless thousands of lives and many thousands of others from serious injury, and a debt is owed by all to not only Betty but also her dedicated band of lobbyists that fought to have these laws passed.

On 29 June 1906, Henry was elected as one of the first Trustees of the Automobile Club along with Mr W. H. Felstead and Mr Thomas Rand. The Trustees role was set out in the Clubs annual Rules and By-Laws:

The Automobile Club of Victoria Rules, By-Laws and Year Book 1906-1907
THE TRUSTEES
XVI
"Mr W. H. Felstead of Melbourne, Thomas Rand of Melbourne and Henry Sutton of Melbourne, are and shall be the first Trustees and each of them and of their successors shall hold office until his death or retirement or discharge from office. Any Trustee may at any time by writing under his hand retire from his office and any Trustee refusing or failing to act, or becoming incapable of acting may be discharged from his office at any meeting of the Club convened and held for the purpose. Any vacancy in the office of Trustees may be filled by the election of a new Trustee, or new Trustees, by the members at a meeting convened and held for the purpose."

XXVII
" All Challenge Cups for competition shall be delivered over to the Trustees, together with all such other property and funds as the members shall from time to time determine: and all Cups and other property and funds so delivered shall be, and be deemed to be, their property as Trustees for the Club, and all property shall be held by them, and investments made in their names (save in the case of Cups, except during the time winners of any Challenge Cup shall be entitled to hold the same. No such Cup shall be delivered to any such winner until a guarantee for the due return thereof, by a date to be named therein, in

such form and executed by such guarantors, as may be acceptable to the Trustees, shall be delivered to them."

In 1916, the Club was granted the use of 'Royal' in its title in honour of its efforts during the First World War, becoming the Royal Automobile Club of Victoria. Within the first few months of the Club's formation it had just over 100 members. Today, the Club has over two million members, all of whom benefit every year from their membership, and it has grown and branched out into other areas. But, the philosophy of the club remains close to the founding members' original conception, and the objectives of the club and the motion that Henry wrote back in 1903 remain as a pillar and testament to the Club's longevity. From such humble beginnings and shared ideals, Henry and the other fifty-five founding members left their mark on not only Victoria's motoring history but also on Australia's motoring history.

Start of Automobile Club of Victoria's first car run, Princess Bridge, Melbourne.
Dr A.P. Merrill (driving) and Henry Sutton seated in front seat.
20th February 1904.

The automobile meet at Alexandra Avenue, bound for Aspendale.
20th February 1904.

Motorcycle riders ACV Inaugural Run 20 February 1904

Automobile Club of Victoria's Inaugural Run 20th February 1904
Dr Merrill and Henry Sutton in lead car.

The Victoria A. C. held a race on the Maribyrnong racecourse, where a large number of spectators gathered to witness the contests, which were very successfully carried out. The above is a view from the lawn.

A.E.Sutton,
3hp Beauchamp, winner of the first VMCC 100 mile road race in 1906 .

Charles Mayman in his newly built Mayman car
on the day he left for Bendigo in 1904.
Image courtesy of the State Library of Victoria

CHAPTER THIRTEEN

SEEING AT A DISTANCE – TELEPHOTOGRAPHY 1870 - 1912

"Ideas do not reach perfection in a day,
no matter how much study is put upon them"

Alexander Graham Bell

Much has been written about Henry Sutton and his invention of the Telephane. Many people over the years have not only surmised, but added to, the story until the truth was hard to find. This chapter aims to clarify what actually transpired using new information.

According to popular legend, Henry first began thinking about the Telephane in 1885 so that the citizens of Ballarat could see the running of the Melbourne Cup. In actual fact, the story about Henry, the Melbourne Cup and his Telephane dates right back to 1870 when he was only fifteen years old. This was the year that an excited young inventor thought to himself how wonderful it would be if the people of Ballarat could actually see the race run instead of anxiously waiting for the results via telegraph and the newspapers. It was in this year that Henry first described his idea of his Telephane to his brother Alfred who, years later in 1927, recalled what Henry told him about his idea of transmitting the race:

"There were many things in which Henry interested himself which have only recently been developed in a practical manner, one of which I would very much like to mention as it is becoming of great interest at the present moment - that is Television. When Henry was a lad of 15 years he told me that he had invented a method so that any big event in Melbourne could be seen in Ballarat by the medium of telegraph. The details are as follows: A building of the Camera Obscura pattern, erected near say the Flemington racecourse, so that the rays could be reflected by the camera obscura on the table below. The table would

consist of a plate of many points and broken up by a screen. The table would revolve and as the points came into certain contact with the wire in which there was an electric current, these contacts would run along the wire and be repeated on the table on the other end. In revolving the table, all these points would come in contact with the wire and the picture would be repeated on the revolution of the table at the other end. Henry at the time was so sure of this that he wrote the particulars to Mr. Robert Lewis John Ellery who was then the Government Astronomer for Victoria so the invention could be in the hands of someone capable of stating his claim of being first in this direction."

It should be noted that prior to uncovering Henry's description of his 1870 television device, the first known people working on the concept of television were Charles William Siemens, Constantin Senlecq, Carlo Perosino and Adriano de Paiva, who did not begin their various researches into methods of transmitting images until 1876 and 1877 respectively.

The description Alfred gave of Henry's device is similar to Alexander Bain's device invented in 1843 but, although Henry's idea at the time was creative and did in many respects pose an idea of a way to transmit a moving image, it was not a solution to the problem of being able to successfully do so. The idea stayed in the back of Henry's mind while he moved onto other things, until 1884 when a young German inventor by the name of Paul Julius Gottlieb Nipkow invented and lodged a patent application for an electric telescope. This was subsequently accepted as the first patent related to television.

The apparatus was a scanning disc with a series of holes arranged into a spiral, which, when spun in front of the image, broke the image into a sequence of scanned lines. This device is known today as the Nipkow Disc (or disk), and it became a fundamental component of early mechanical television. This was the break-through that Henry had been waiting for, and he immediately went to work on improving his original idea.

Henry utilised the Kerr Cell, which operated on a principle discovered by the Scottish physicist John Kerr in 1875. This cell produced the phenomenon in which the refractive index of a material changes because of an applied electrical field, proportional to the square of the magnitude of applied electric field. In employing the Nipkow Disc and the Kerr Cell, Henry had devised his Telephane system by 1885, and in doing so he not only came up with the first practical use of the Kerr Cell but, he had devised what is recognised today as the first feasible television system.

In August 1988, on the centenary of the birth of John Logie Baird, the late Dr Clive Coogan, who was at the time the senior scientist with the CSIRO, wrote about Henry and gave the following explanation of how Henry's Telephane system worked.

"Sutton's Telephane consisted of a transmitter and a receiver linked by telephone or telegraph lines. Light coming in to the receiver was focused into an image that was rastered (scanned line-by-line) and frame by frame, by a device called a Nipkow disc, invented only one year earlier by a German, Paul Nipkow. Light from a portion of the image passing through the hole in the disc fell on a selenium photo cell and the resistance of the cell diminished with increasing light intensity. Thus Sutton's selenium cell modulated the electric current passing along the telephone line, according to the intensity of the light received from the part of the image being transmitted. At the other end, the receiver consisted of a device to convert the electric current into a light spot, whose intensity varied with the current. It used a second Nipkow disc, working in phase with the transmitter disc by what Sutton called synchronous motors, ingeniously governed by tuning forks. A Kerr cell modulated the light according to the incoming current. Kerr had only invented his cell in 1875 and Sutton's use of it was novel. The cell consisted of two electrodes with carbon disulphide between them. When polarised light was passed through the cell, its plane of polarisation could be rotated by applying a high voltage across the electrodes. Sutton used the most powerful light source at his disposal- a simple kerosene lamp- passing its light through a plane polariser and then through the Kerr cell to a second polariser, whose plane of rotation was perpendicular to the first. The point of such an arrangement was that if there was no high voltage across the Kerr cell, no light could pass through. But at higher voltages an increasing amount of light got through. There was a small difficulty neatly solved by Sutton. Electronics had not even been invented at the time, J. J. Thomson would not discover the electron for another 12 years, so there were no amplifiers of current or voltage. The Kerr cell demanded high voltages but the telephone line operated on a low voltage. The only kind of amplifier was a transformer, which could only amplify changes in current and voltage. Sutton used the transformer's changing output to modulate the higher voltage of the Kerr cell. He could thus only see the light allowed through his system as a consequence of change in the light intensity of the original- in other words, he would see a line where black changed to white, but only black where the light's intensity remained uniformed. The system was imperfect, but he should at least have seen outlines corresponding to the main features of the image. The viewer of the "Telephane" sat peeping into an eyepiece and saw the image in the plane of the Nipkow disc, which systematically scanned the image in synchrony with

the transmitting Nipkow disc at the other end. Sutton's breakthroughs were very important, even if there is no information about the quality of the transmission image. His was the first feasible TV system - not just an opinion but that of the TV technology historian George Shiers."

From 1885 to 1887, Henry was also working on other inventions which took up a lot of his time and so the advancement of his Telephane system was put on hold for a few years. In 1887, William Bramwell Withers wrote about Henry's Telephane in his book, *The History of Ballarat,* which gave Australia one of the first insights into his ideas and work on the transmission of images. Although Henry continued working on his idea of the Telephane, he had not as yet constructed it, but after he had patented and announced his photographic process late in 1887 he set about working on his Telephane system once again. This time however, he had in mind some different uses for it.

On 30 May 1888, at his home in Doveton Street, Ballarat, where Henry had previously demonstrated his half-tone photographic process to a number of scientific and newspaper men, he gave a very interesting talk and demonstration of the other processes he was working on. One of these was his Telephane system and the other was his Facsimile system. The two ideas were closely linked but differed in purpose and procedure; one involved seeing moving images at a distance and the other was sending and receiving still images over a distance. Both processes used telegraph or telephone lines for transmission as that was the only method available at the time, but this is the point in history where seeing at a distance began to diverge, eventually following different paths: one would become known as television and the other the facsimile machine.

By May 1888, Henry had constructed working models of his Telephane and Facsimile systems and felt it was time to demonstrate how they worked to Mr Ellery. At that very private demonstration, an image was transmitted but the quality of the transmission is not known. It is interesting to note, though, as stated by Thomas Bury who had been present at the demonstration, that the image took forty minutes to transmit. This is the same amount of time it took Arthur Korn, many years later, to transmit his first photograph. A short time later, on 30 May, at a demonstration of his photographic process, Henry went on to show his scientific friends and the press his Telephane and Facsimile transmission systems. He talked about his aim to be able to transmit a photograph and within hours publish it in newspapers. However, at this point, he was not ready to divulge how his two systems worked as they still needed perfecting so they could transmit clearer images.

By late in 1889, Henry no longer saw any real practical use for his Telephane system, commenting that, "he thought very little of the utility of merely being able to look into a sort of disc and see what is happening in a certain spot miles away." Although it might seem rather incredible from today's point of view, Henry was in fact quite right. Limited by the technology of the time, his Telephane image was very small and there was no such thing as a camera that could pan or change focus. One had to peer through the lens to see a tiny, flickering scene about the size of a matchbox label, and to Henry this was not useful at all. There was no way of moving the transmitter or receiver to pan the landscape as the two devices needed to be in total synchronisation to work, nor were the telegraph lines powerful enough to transmit the motion, only a static image. So, given all the limitations, Henry felt that it would be more useful to be able to transmit photographs via the telegraph lines to be reproduced at the other end and published in newspapers.

Scientific men around the world had been trying for years to invent a method of transmitting images via telegraphic lines. It first started in 1843 when the Scottish inventor and engineer Alexander Bain first contemplated the transmission of printed letters, drawings and pictures over telegraph lines and took out a patent. Bain who had invented and patented the first electric clock, also invented an experimental facsimile machine using a clock to synchronise the movement of two pendulums for line-by-line scanning of a message. His method of transmission consisted of metal pins that were arranged on a cylinder made of some insulating material, and an electric probe that transmitted on-off pulses then scanned the pins. A message would then be reproduced at the receiving station on electrochemically sensitive paper impregnated with a chemical solution. A few years later, in 1847, the first practical application of the idea came from Frederick Collier Bakewell. He replaced Bain's pendulums with rotating cylinders, but both systems were not viable as the transmitters and receivers were not synchronised. Perfect synchronisation of two discs to the hundredth of a second was a key component of any viable system. In 1855, Giovanni Caselli developed a regulating clock as a synchronisation method and this was far superior to either, Bain or Bakewell's method. Other facsimile systems were invented by Shelford Bidwell, John Perry and William Ayrton. Each was intrinsically different from the others in the method used to transmit written text, pictures or photographs; each posed slightly different problems in finding solutions to successfully transmit the image through telegraph wires. At that time, the transmission of images or photographs was yet to be successfully accomplished by anyone and so the ability to transmit a clear image and or half-tone photograph would be a great step forward in the advancement of technology and communication.

Busy with his preparations for his trip to England in early 1890, Henry placed the Telephane system on the back burner. He concentrated on his facsimile system as he was intending to demonstrate this as well as his photographic process at the Columbian Exhibition in America in 1893.

Sometime after arriving in London and talking to a number of scientific men, Henry mentioned his proposed systems of transmitting images and it was suggested that he publish a paper on them. So on 7 November 1890, Henry's paper on "Tele-Photography" was duly published in the UK in *The Telegraphic Journal and Electrical Review,* and it immediately caused quite a stir in the scientific world. The paper was published again in the French journal *La Lumiere Electrique* in December 1890, and on 24 January 1891, it was published in *Scientific American* and caught the attention of many men of science in America. Such was the significance of Henry's paper, that two decades later on 3 September 1910, Scientific American re-published it.

In this paper, Henry describes his Telephane system in great detail. At the beginning of the paper, he stated the following about tele-photography:

"Under various names the problem of transmitting optical images by aid of the telegraph wire has at different times had attention drawn to it.

In putting forward something new in this direction, I will begin by inventing a new name, and propose calling the subject Telephany, and the electro-optic instrument the Telephane.

The art of telephony is simple compared with that of telephany. In the former we deal with a consecutive series of waves of varying rate and length, and it is the consecutive character of sound waves that lends itself so admirably to electric translation.

In telephany we are met, at the outset, with a great preliminary difficulty, having to deal with a surface or plane in which the effect appealing to the brain must be observed in all the varying character at one and the same time.

The problem stands thus: a means has to be devised whereby the varying effects on a plane surface are translated into consecutive series of currents, and by means of the consecutive series of currents to reconstruct, so to speak, a copy of the original surface; that is, we have to take an optical image, seen as a surface, translate it into a line of consecutive varying electrical currents, and by means of these produce an effect as a surface, having the characteristics of the original image.

We have here two images as surfaces having no time value, and a series of electrical currents having a time value, yet these opposing characteristics are to be presented to the brain as a momentary impression.

Before showing how this apparently impossible problem may be handled, I will explain a probable means for electrically transmitting a photograph.

If we make what photo-mechanical operators call a screen negative of a portrait, using a course screen, and from this a photo-lithographic transfer, transfer it to zinc, and transmit the result by any of the several autographic systems, we have the desired result. In fact, if we apply our knowledge of half-tone-block making to telegraphy, we are at once in possession of a means of electrically transmitting the photographic semblance of any person.

We may make a screen negative, and from that obtain a print on zinc or copper coated with sensitive albumen or bitumen, using the usual solvents, water or turpentine, as the case may be, with which to wash away the unexposed albumen or bitumen, then let the stylus of any autographic system traverse the developed image, the result at the receiving end is a facsimile portrait.

I have used the expression screen negative, as it is an understood trade name; as a matter of fact, a screen positive would be necessary. We may go further, instead of receiving the facsimile on chemically-prepared paper, as in the Caselli autographic telegraph, we can make the receiving stylus perforate thin paper (with the electric spark) by means of a constantly-working induction coil, but only put into the receiving circuit by the transmitted current. Place this paper on a lithographic stone or zinc plate, pass a roller charged with greasy ink over it, and we have a printing surface, the portrait being transmitted and reproduced for the printer, photo-electro-mechanically.

But this is not telephany; the latter must be understood as the means of transmitting images which may be in motion, as seen in a photographic camera, but not in colours.

Having spent some years in studying the problems, I designed the following system five years ago, as my Victorian scientific friends can testify. It may be of much assistance to workers in this direction. I think it offers a fair approximation to the solution of this very difficult problem; at any rate, if in its present form it is not the actual solution, I feel sure it is in the direction indicated by my method, that the successful accomplishment of telephany will result."

[Full Text, see End Notes, Chapter Thirteen, No. 1.]

There was at least one major problem with Henry's invention, however. The various chemicals used in his Telephane system were highly volatile and having a naked flame as a light source near them was very dangerous. The schematic of his Telephane System had lamp light as a light source, for at the time, electric lighting was not available to everyone. Henry could have used a light globe as his source of light instead, as it would have been safer to use than the naked flame of the lamp light.

After the publication of Henry's paper, many men of science and academics began debating his use of the word 'Telephane' to describe his system, and letters came pouring into the Electrical Review journal regarding the matter. One reader wrote, for example:

"It may appear as though we are looking ahead too much when we discuss the name of that which is at present non-existent. But, as we are told that we should first work out our ideal, and afterwards approach that ideal as nearly as hard circumstances will permit, we may as well act on the same plan in this matter of naming an undiscovered instrument. Now, as words are only symbols to express our thoughts to others by writing or speaking, it is necessary that words, when written or spoken, should express those thoughts as clearly, as shortly, and as euphoniously as possible. The series of the words Telopy, a Telopist, the Telope, I Teleope, seem to fulfil the required conditions better than Telephany, a Telephanist, the Telephane, I Telephane. If that which is expressed by these words becomes as common as telegraphy, we shall probably have some totally different word used; just as we have – a wire, and I wire. Meantime, for writing and lecturing purposes, which of the above groups of words shall we employ to express seeing at a distance? The positiveness of the assertion - 'telope,' is certainly not right, as "ope" represents nothing connected with the subject - appears crushing; but its author seems unaware that "ops" means the eye, and that "OP" is the root from which comes the words most commonly used in Greek language with regard to sight, to wit, 'omma', the eye; 'opsis', sight; the fut. aor. pass. and pft. mid. of 'horao'; and it would be not surprising if the words connected with 'skopos', which we get in telescope, and mean generally reflection based upon sight, were all derived from the same root. 'Telephane' is a sweet-sounding word, but it does not quite fulfil the required conditions. Why did so severe a maker of words write 'optikos' nom. masc. sing., and then 'phanerous' acc. Masc. Pl.? Teleopt would not do, as by itself it might easily be supposed to mean- roasted from a distance."

As confusing as the reader's letter was about what would be the correct name or term, it highlighted how even the scientific community struggled with the correct names for inventions. The word 'television' was of course still in the future, not to be used for another decade when the Russian scientist, Professor Constantin Perskyi, used it on 25 August 1900 at the First International Congress of Electricity at the World's Fair in Paris. It would still be many years before the word television became part of our everyday lives and language; not until John Logie Baird first publicly demonstrated his television system in 1926.

In January 1891, when Henry joined the London Camera Club, he was introduced to many notable men of science. This, coupled with his membership of the Society of Telegraph Engineers and of Electricians, opened many doors and he revelled in the new opportunities. Part of his aim in travelling to England was to meet and learn from the world's best men of science, many of whom were located there. It did not take long for these established scientists to recognise Henry's inventive talents, as they had been aware of his successes over the years with his battery, vacuum pump and now photographic process. Henry was duly welcomed into the throng by such men as William Crookes, Joseph Swan, William Preece and Lord Rayleigh and he was eager to learn from them all.

When Nikola Tesla arrived in London in 1892 he was greeted by William Preece (later Sir) who was the head of the British Post and Telegraph Office. Preece immediately invited him to stay at his residence, as Tesla was due to give his talk at the Institute of Electrical Engineers on 3 February and there were a number of experiments to be set up for his lecture.

The Institute's lecture hall was packed to capacity as Tesla went through each experiment. Henry and the other members of the audience sat riveted to their seats, spellbound by what they were hearing and witnessing as Tesla went through various experiments, such as demonstrating his Tesla Coil which spewed electricity bolts into the air. One statement at the beginning of Tesla's lecture easily summed up Henry's ongoing philosophy and approach to his life and work.

Extract of Tesla's Lecture on Alternating Currents:

"Each day we go to our work in the hope of discovering—in the hope that someone, no matter who, may find a solution of one of the pending great problems—and each succeeding day we return to our task with renewed ardour; and even if we are unsuccessful, our work has not been in vain, for in these strivings, in these efforts, we have found hours of untold pleasure, and we have directed our energies to the benefit of mankind."

[Expanded Extract, see End Notes, Chapter Thirteen, No. 2.]

Henry was so excited by his attendance at Tesla's lecture, that the next day he was compelled to write to the Ballarat School of Mines to share his experience of the now famous lecture. His report was published in *The Ballarat Star* on 15 March 1892.

His report, which has not been republished or seen the light of day for over 120 years, started with the final statement of the lecture Tesla gave to the American Institute of Electrical Engineers just before he came to England. Henry had read the published details of it in anticipation of his attendance at Tesla's London lecture. Henry's paper reads:

NIKOLA TESLA'S GREAT ELECTRICAL DISCOVERY

By Henry Sutton

"We are whirling through endless space with an inconceivable speed, all around us everything is spinning, everything is moving, everywhere is energy. There must be some way of availing ourselves of this energy more directly. Then with the light obtained from the medium, with the power derived from it, with every form of energy obtained without effort from the store for ever inexhaustible, humanity will advance with great strides. The mere contemplation of these magnificent possibilities expands our minds, strengthens our hopes, and fills our hearts with supreme delight."

"Such are the brilliant and prophetic words with which Nikola Tesla concluded a lecture before the American Institute of Electrical Engineers. Nikola Tesla is a young Hungarian, who for some years has resided in the United States. He has the honour of being the first man to turn out an electric motor capable of running on an alternating circuit. As the lecture referred to above was the means of attracting the whole electrical world to the wonderful experiments of Tesla, it will interest the readers of 'The Star' to hear a little of their nature, particularly as the sensational character of these experiments has been by no means diminished by his repetition of them before the Institute of Electrical Engineers, London on the 3rd of February, and before the Royal Institution on the 5th of February, he being invited to do so by both bodies.

Without going into the subject in an abstruse manner, a little should be explained in order that the non-technical readers may properly appreciate Tesla's work, and to this end it must be considered there are two systems by which electrical energy is conveyed for commercial use—first, the continuous or low pressure; and second; the high pressure and alternating. The first system is equivalent to a large water main under low pressure, and the second to a small water main under high pressure. In both cases the same amount of energy is transmitted, but the high pressure system having the smaller main is less costly in respect of mains. Now when it is considered that mains for electrical energy are solid copper, it becomes a matter of some importance to reduce the cost in this valuable metal to a minimum. Having got thus far it has followed that great attention is paid to obtaining electric energy under high pressure.

As the youngest of students is aware there are many ways of doing this, in practise however the best method is to construct dynamos to deliver alternating currents, that is, currents having a direction first one way, then the opposite. The very earliest generators were of this type so no difficulty arises in producing them. The object of having the high-pressure currents alternating, is so that by ordinary methods of induction these high-pressure currents may be transformed into low pressure at various points in the circuit; for instance, the pressure may be in the mains 10,000 volts, and in the houses only 100 volts. These high pressures, having an electro-magnetic origin, are dangerous to life if handled carelessly; and on this point Tesla's experiments are striking, for at his lecture he took with impunity a pressure of 30,000 volts, electro-statically produced. Hitherto the energy used in practise has been due to electro-magnetic induction; that is, the continuous or alternating currents have been produced by the action of magnetism.

Tesla has found quite a new field in electro-static alternation; in short, he deals with the electricity of the lightening flash. His experiments show that a rapidly alternating electro-static condition is capable of being used as a means of transforming energy from the lower to the higher form of light, and demonstrated it in a remarkable manner before the Institution of Electrical Engineers, by placing a plate of metal on the table and suspending another above his head in the air. When these plates were charged with a rapidly alternating electro-static potential or pressure, the whole space between them and some of the distance each side became an invisible source of energy. He simply took a long glass tube and brandished it in his invisible electro-static field of energy, and it became a luminous sward. It must be noted there is absolutely no connection between the globes and plates, and it needs only an extension of the experiment, and we have our ceilings covered with conducting paper and our floors carpeted with a conducting carpet; to have rooms in which energy exists, that can do no hurt to the human body, and in which, wherever we place our exhausted globe, we have light, and a veritable lamp of the gods, non-consuming. Extend the principle a little further, and we have our ceilings in the form of two sheets of glass, the air exhausted from between the glass, the conducting paper being above and conducting the carpet below. Our room would be lighted from the Heaven's mighty dome, and the source quite invisible.

Tesla made many experiments, all brilliant and novel, lamps with only a block of phosphorescent material inside, lamps with a solitary upright thread, into which the rapidly alternating charges surged back and forth, and giving out a brilliant incandescent light. He produced electro-static flames without consuming the material. These results must not be confounded with those obtained by means of vacuum tubes and the ordinary induction coil, nor must they be confounded with Crooke's experiments on radiant matter. Tesla produces the results in air and in vacua, and uses alternations or frequencies enormously more rapid than in any of Crooke's experiments, and it is this enormous frequency that makes his electro-static waves of energy have more the nature of waves of light, and which, being evolved by a falling upon a sympathetic matter, causes light to appear. In fact, he has placed the pick in virgin soil and merely turned over a few sods, but such as have been turned over reveal a wealth of scientific discovery that has not been seen or met since the immortal Faraday discovered electric-magnetic induction. It is very strange that Faraday's discovery depends on the amount of charge in a magnetic field, and that the variation or alternation of a magnetic or electro-magnetic field, evolved all the brilliant and practical results that go now to make up the weft and warp of practical electricity.

It is truly wonderful that when the principle of rapidly changing or alternating an electro-static field should be discovered by Tesla, that it should at once evolve such remarkable features, and bring the day nearer when it may be possible to have our houses illuminated without lamps, the very walls giving out light, and our homes warmed with fire. Who can foretell but that these invisible fields of energy may be controlled to such an extent, and projected into space, so as to enable communication with other planets, particularly when we consider that the all-pervading ether must exist equally on Mars and in space as it is almost a demonstrated fact that it does on Earth. The amount of grey matter on the surface of the human brain is an infinitely small fraction of the whole, and the higher the intelligence the more grey matter. When humanity has perfected itself to the stage of having the whole brain in the form of grey matter, what should be impossible?

If humanity be ten or ten hundred thousand years old, it has taken all that time to accumulate a coating of grey matter, perhaps little thicker than a sheet of paper, what possibilities are there not potentially present in the unconverted portion of the brain, and what an enormous time humanity has still to exist before this conversion is consummated. Science seems to have just stepped within the borderland of wonders, the nature of which it is hard to conceive, and Tesla is to be thanked by all engaged in research for lighting the way over this borderland."

It is clear from Henry's report, that attending Tesla's lecture was a pivotal moment in his life. To say he was excited when he wrote the paper is an understatement as it exudes the thoughts of a man whose thoughts are swinging wildly with the possibilities of what potential inventions lay before mankind. From the moment he sat in Tesla's lecture in London, his life changed forever, and his inventive genius from this point on knew no bounds. It was from that day Henry knew that nothing was impossible.

Henry Sutton and Nikola Tesla actually had a lot in common: they had the same ideals and interests and were the same age. During Tesla's stay in London, Henry was given the opportunity to personally meet him via his friends William Preece and Lord Rayleigh. It is unclear if this was before or after Tesla's lecture on 3 February or even if they had corresponded with each other prior to this visit to England. In any case, at some point during his stay in London, Henry discussed his system of transmitting images with Preece and Tesla. The latter, who would have read about it prior to his arrival in London, was very interested. With Preece's aid, arrangements were made to test the system.

To date, the only article found that mentioned the testing of Henry's system in London and the transmission of the photograph, appeared four years later when the Washington Evening Star newspaper wrote about recent inventions for sending pictures by wire. It should be noted that Henry's friend, Dr Alexander Graham Bell, was living and working in Washington during this time and may have supplied the information to the newspaper. The article, which referred to Henry as '*Professor*', noted that one of the most interesting inventions was that of Professor Henry Sutton of Victoria, Australia. It described Henry's facsimile, which could transmit a still image, and which was in essence his Telephane system.

By the time Nikola Tesla came to London in 1892, Henry had worked on his facsimile system enough to test again and obviously could not resist the opportunity to do so with Tesla, who wanted to see how well it worked. The Evening Star journalist describes Henry's invention in detail:

Pictures by Wire:

"A picture of the image on the ground glass of a camera is held before a system of lenses, which focus it upon the centre of a glass plate containing selenium. Before reaching the focus the waves of light pass through a revolving disc which separates the light rays coming from the image so they fall one after the other upon the selenium spot. By an ingenious arrangement only one of the holes can transmit the focused light at the same time. Hence upon looking through a revolving disc when it is rotating, circles of light would appear to follow one another over its

face, beginning at the centre and ending at the circumference. As the disc rotates at a very fast rate, the hundreds of light rays flashing through it singly appear to travel together in circles. The spot of selenium is connected with an electric wire, the current in which flows in vibrations continually differing according to the strength of each light ray as it comes through the disc. In this manner each light wave issuing from each shade in the picture is taken up in turn and transmitted through the wire. At the opposite end of the line the same current is received through which the light of a lamp is reflected. By a complicated arrangement of induced light rays in the receiver which are caused to flash with the same vibration as those focused in turn upon the selenium in the transmitter. Another disc in the receiver separates the rays before presenting them in turn through the eye piece, which is connected with a camera for preserving the photograph. The two discs revolve simultaneously to the exact hundredth of a second for the image to be transmitted."

The Evening Star, 1896.

The article then stated that the photograph was not transmitted very far as it was only possible to synchronize the transmitter and receiver over a short distance.

Henry was the first Australian inventor to successfully transmit a photograph in Australia, and with the aid of Preece and Tesla, to repeat that success in England. The problem that would plague inventors in the years to come, was how to perfectly synchronize the rapidly spinning discs over a long distance.

Henry's successful transmission of a photograph with Tesla in London was just a stepping stone and it went largely unnoticed except for the few observers that were there at the time. Apart from the article in the Washington newspaper in 1896, it was not mentioned again until 1908 and 1912, this time in Australian newspapers. Nevertheless, Henry's meeting with Tesla was a momentous moment in his life and he classed Nikola Tesla as a great friend with whom he kept in touch for many years.

So to settle the questions asked about Henry's work, and correct the myths – did Henry construct his "Telephane" system? The answer is undoubtedly, yes, Henry did construct a working model of his system. As mentioned above, he demonstrated his Telephane/Facsimile to Ellery in 1888 in Ballarat, and in London in 1892. However, to the question of whether Henry transmitted the Melbourne Cup race from Melbourne to Ballarat in 1885, the answer is no, Henry was never able to transmit moving pictures.

Typically, Henry was soon drawn to another more exciting scientific venture or adventure, as it would turn out to be. Tesla's lecture on

"Alternating Currents" was a turning point in science, as Tesla also talked about wireless telegraphy transmission. This was the moment in Henry's life that forever changed the direction of his scientific research. The limitations of transmitting images via telegraph lines were painfully obvious but Henry, in that moment, saw the future just as Tesla did, and immediately became convinced that it would be eventually possible to transmit images wirelessly. So, in 1893 when Henry returned to Australia, he put aside all his work on his Telephane and Facsimile systems and began experiments with wireless telegraphy in the hope that he could find a way to wirelessly transmit photographs.

From 1843 to 1885, there were many scientists who worked on the problem of transmitting pictures. Henry Sutton was one of them but unfortunately, at the time of inventing his system, the technology was not there to support it. Nonetheless, his principle of using a scanning disk, a light source controlled by a Kerr cell and his method of reassembling the image was remarkable in that it was eventually used in practical television systems nearly forty years later. Henry's transmission of an image at his home in Ballarat in 1888, and then a photograph in London in 1892, was achieved years before the German inventor Professor Arthur Korn successfully transmitted a photograph over a long distance from Munich to Nuremberg in 1907. Henry's pioneering steps were a great achievement which should not be ignored, as both his transmission of a photograph and the devising of the first feasible television system were huge steps in the quest to transmit images.

In 1906, the American inventor Lee De Forest invented the "Audion Tube" which allowed electronic amplification. This invention rendered Henry's feasible television system more practical, but it still took many years for inventors to solve the problem of synchronisation. With the advancement of technology, it was a Scottish engineer and inventor by the name of John Logie Baird who finally came up with the first working television system, first publicly demonstrating it on 26 January 1926. The image transmitted by Baird at that time was faint, and often blurred, but was clearly visible. A viewer could clearly see movement and expressions on the face of the person in the image. Baird called his television a 'Televisor'. Baird would have been well aware of Henry's Telephane system of years before because he used Henry's principles to successfully transmit the first moving pictures, as the late Dr Clive Coogan explained:

"Baird's Television used an upgraded photocell, as the photoelectric effect was superior to the photoconduction of a selenium cell. He also used Nipkow discs in his transmitter and receiver, synchronised frames and employed radio

transmission instead of a sluggish telegraph. Baird drew upon advances in the infant science of electronics and particularly by the development by Braun in Germany of the cathode ray oscilloscope – the basis of the modern TV picture tube – that modulated light much more simply and rapidly than Sutton's kerosene lamp, Kerr cell and polarisers. Baird could also use wireless transmission, where Sutton was limited by telegraph technology, with its inherently low rate of information transmission."

By 1930, Baird had mass-produced his Televisor and it became the first commercial television receiver. However, it was not a receiver as we know it today. To see the images received by the Televisor, the viewer had to look through a magnifying glass on the side of the unit, and at the same time the viewer had to constantly monitor the image and adjust the knob in the centre of the Televisor, to control the speed of the internal spinning Nipkow disc, so it synchronized with the transmission frequency. The pictures received were very small, poorly defined and flickered, but were considered entertaining and a wonder in their time.

Little was Henry to know that his work, in the years to come, would contribute to this new invention.

The nature of television improved slowly, but it was not until 1947 that the invention of the transistor revolutionised television sets around the world. This new era of electronics, begun with the invention of the transistor by John Bardeen, Walter Brattain and William Shockley, not only revolutionised television, it also set in motion great advances in science and technology.

John Logie Baird deserves all the credit bestowed upon him, but world-wide plaudits should be given to Henry Sutton who worked in obscurity and in the remoteness of country Victoria to come up with the world's first feasible television system. London and Hastings in Sussex, England have both erected plaques honouring Baird and his work. Sadly though, over the last century, Australia and the city of Ballarat have given little or no recognition to Henry Sutton and his contribution to the invention of television. This lack did not stop one of Australia's great poets from writing about Henry and his Telephane. In 1990, Les A. Murray published his poem 'The Tube' which appears in his book of poems Dog Fox Field.

In recent times there has been a groundswell of interest in Henry's achievements all around Australia.

The year 1985 was the centenary of Henry's feasible television system and just happened to coincide with the Japan Expo that was held in Tsukuba between 17 March and 16 September, and the theme of the exposition was "Dwellings and Surroundings, Science and Technology for Man at Home." The Japanese organisers thought that it was an appropriate time to look back over the history of technological progress and re-evaluate the contributions to science and technology that mankind should develop in the twenty first century. Over eleven million dollars was allocated for Australia's participation in the exposition and part of this amount covered the cost of the Pavilion fit-out - the acquisition and installation of exhibits. The Pavilion featured an overview of Australian dwellings, along with Australian advances in fields such as biotechnology, computer software, astronomy, ceramics, scientific instruments and mining technology. Other exhibits were a major audio-visual presentation and the world's largest hologram.

At the time of the Japan Exposition, the Minister of Science was Mr Barry Owen Jones who, along with Dr Clive Coogan from the CSIRO, were instrumental in highlighting Henry Sutton's contribution to the advancement of technology in Australia. A model of Henry's Telephane system was exhibited in the Australian Pavilion in company with Hargrave's box kite.

Henry's Telephane was displayed in the Origins Hall, where visitors moved through a cylindrical node space which opened up into a ramped gallery which described Australian housing, using images, models and five white light holograms. The ramp opened out into a square gallery where reproductions of nineteenth century Australian patents and various past Australian inventions including Henry's Telephane were displayed along the walls, providing an introduction to the Technology Gallery which housed exhibits that highlighted the latest developments in astronomy, medical, research, computers, solar energy and scientific instrumentation. As the visitors looked back from where they had come, they could still glimpse Hargrave's kite and see Australia's technological past, and in the front of them Australia's technological future.

In fact, 1985 was a momentous year for the invention of television. The Tsukuba Exposition highlighted not only Henry's pioneering Telephane system but also featured the first flat screen television, introduced by Japan's NHK Public Television Broadcaster's Science and Technology Research Laboratory.

Over the six months of the Exposition, two million people visited the Australian Pavilion, which was rated in the media reports as one of the two most visually inspiring of the foreign participants. The Australian Minister for Sport and Recreation and Tourism, Mr. John Brown declared that, "Australia's image as an economically and technologically developed country had been enhanced by the exhibits at the Exposition."

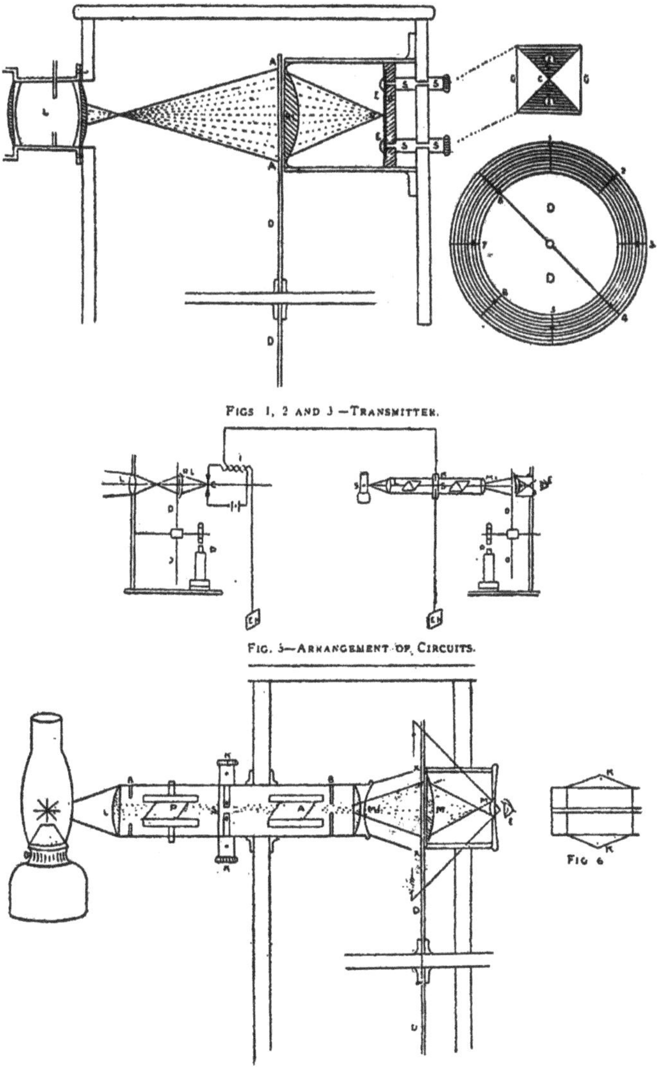

Sutton's Telephane design illustrations.

CHAPTER FOURTEEN

WIRELESS TELEGRAPHY 1892-1912

"Technology gives us the facilities that lessen the barriers of time and distance – the telegraph and cable, the telephone, radio and the rest"

Emily Greene Balch

The history of wireless telegraphy dates as far back as Michael Faraday who made the first observations in the defining properties of a semiconductor in 1833. This discovery became one of the first steps towards wireless communication. Following on from Faraday, many notable men contributed to the invention of wireless telegraphy, these men, including Henry Sutton, are known today as the pioneers of wireless communication. Henry's work in the field spanned two decades and at the time his work brought him world acclaim. What he never knew was where his work would eventually lead in the decades after his death.

Not long after his return from England in May 1893, Henry moved to Melbourne to live. He immediately resumed his experiments with wireless telegraphy, spurred on by his meeting with Nikola Tesla the previous year. He began his work with the aim of finding a way to send photographs wirelessly, and these experiments were performed long before Marconi's name was associated with wireless telegraphy and before the Australian Government, the Commonwealth Naval Forces and the Post Office began to experiment and utilise this new technology.

Even by 1897, soon after Marconi patented and demonstrated his wireless telegraphy system, there were only a few other wireless experimenters in Australia. There were no call signs or licences and very few people to contact, barring a few ships that entered Australian waters which had on board a wireless telegraphy system. It was not until 31 January 1903 that the Commonwealth Naval Forces under the direction of the Naval Commandant, Captain Creswell, first tested a wireless system for the navy's ships. This test was conducted from the naval yards at Kangaroo Point in Brisbane from the navy ship the *Gayundah* positioned in Morton Bay and the naval yard, and on this occasion a successful transmission over a distance of ten miles (approx 16 kilometres) was accomplished. By 1904, the Navy had several ships equipped with new technology, and people were required to seek permission from the Navy to be able to conduct experiments with wireless telegraphy.

In 1901, the year of Australia's Federation, the Postmaster General's Office was created but was not involved in the use or control of wireless telegraphy until the Australian Government passed the Wireless Telegraphy Act on 18 October 1905. The passing of the Act brought the use of wireless communications in Australia under the domain and control of the Australian Postal Department and the Postmaster General (PMG). The Act did allow for experimenters to continue their experiments, but they needed to apply to the Postmaster for a licence and had to pay an annual fee of £3/3s or face penalties - either having to pay a £500 fine or endure up to five years imprisonment. Due to the very high licence fees and the conditions the PMG imposed, very few experimenters at the time applied for a licence regardless of the consequences. Henry had previously sought permission from the Navy, but for reasons unknown he did not apply for a wireless licence with the Postmaster General's office and continued his experiments.

By 1905, the Italian inventor Guillermo Marconi, had patented and commercialised his wireless telegraphy system around the world. He saw Australia as a lucrative market and in November that year, Captain Louis E. Walker, a representative of the Marconi Wireless Company, arrived in Australia. Walker entered into negotiations with the Postmaster General's Office to erect wireless stations in Victoria and Tasmania to demonstrate the Marconi wireless system in the hope that the Australian Government would purchase the rights to the Marconi System. By May 1906, two wireless stations had been erected at the expense of the Marconi Company at Point Lonsdale near Queenscliff in Victoria and Devonport in Tasmania. At the invitation of Captain Walker, Australian Prime Minister Alfred Deakin, the Governor General Lord Henry Stafford Northcote, the Governor of Victoria Sir Reginald Talbot and most of the members of parliament and the ministry agreed to attend the wireless demonstration by Walker at Point Lonsdale.

Although 12 July 1906 was a sitting day for Parliament, very few members were in attendance. On that day, the government had organised special trains to transport the members of Parliament and members of the public who wanted to witness the demonstration. The transmission of wireless telegraphy messages over two hundred miles (approx 321 kilometres) from Victoria across Bass Straight to Tasmania was easily achieved by the Marconi system, and was indeed an historic event for Australia.

While the expensive exercise was a technical success, the bid to secure a contract failed, as the Australian Government refused to commit to purchasing the Marconi System. Marconi was extremely angered by the Australian Government's refusal because he had invested thousands of pounds to demonstrate his system. It was a very public humiliation for him and his anger had a marked effect on Australian wireless history.

After the Australian Government's refusal to invest in and install Marconi's wireless system, it was nevertheless decided to allocate £10,000 to erect a series of wireless stations around Australia. Henry was well aware of Australia's need for a means to send messages over long distances, both around Australia and overseas. Wireless communication was also imperative to the protection of Australian citizens and for its naval ships and he was diligently working on the problem. By 1907, he had invented and constructed a number of different apparatus for his wireless telegraphy experiments and began to develop a whole new system of communication.

During the following twelve months, he made a number of discoveries that he felt would be invaluable to the defence of Australia. His original intention, to find a way to transmit photographs, was put into the background and the development of his wireless telegraphy system became the priority.

By February 1908, Henry felt it was time to inform the Australian Government of his advances in this field. So, on 11 February, Henry wrote to Australia's Minister of Defence, Mr (later Sir) Thomas Thomson Ewing:

To The Minister of Defence, The Commonwealth of Australia.

"Sir having given some attention to wireless telegraphy, I wish to inform you, I have a system of producing ether wave storms which if arranged at distant points around our coast, will effectually cut and destroy the wireless communications of any naval power approaching the coasts with hostile intentions. In their attempts to regain the lost communication the enemy will be faced with a need to re-adjust their instruments and so completely disorganise their wireless systems

for probably several days. (A very important matter to the defender). The power to do this I think should be a carefully guarded secret and should not be known even to the wireless operators. I have constructed an ether storm apparatus and it is impossible for any wireless communication to be intelligently received within the radius of its action. The radius of action is only limited by the power at disposal."

Yours Faithfully, Henry Sutton.

Henry's letter to the Defence Minister, Mr Ewing, caught not only his attention but also that of Naval Commandant, William Rooke Creswell, who paid Henry a visit. Later, on 24 February, Captain Creswell wrote to the Minister of Defence saying that, "*I have seen Mr Sutton and will do so again. He appears to have some grounds for his claim and I plan to see him again at an early date.*"

As well as working on the idea of wireless transmission, Henry had also developed a method of jamming the enemy's wireless signals. His plan was to erect wireless stations at certain intervals around the coast of Australia that would transmit a signal from his specially designed, high powered, spark-gap transmitter simultaneously. This signal would disrupt the wireless transmissions of any unsuspecting enemy and render their communications useless until such time as they could re-calibrate their wireless signals, a process which would take hours to carry out. Being able to jam an enemy's wireless signals would be a distinct advantage in defending against any invasion of Australia. At the time, Henry's high-powered spark transmitter was the only one of its kind in the world and was highly favoured by the Australian Defence Department and the Commonwealth Naval Forces.

Some nights while investigating wireless phenomena, Henry would entertain neighbours and visitors by creating an ether wave light cloud from his high-powered spark transmitter, producing an impressive blue and violet light show for those present. Coupled with the loud noise coming from the spark gap, this display amused many a visitor to the Sutton household. Little did the neighbours and visitors know that this was in fact a wireless jamming device as well.

With the immediate interest of not only the Defence Department but also the Navy, Henry's life was to change course once again. His life was already an endless cycle of work and family obligations, and eventually his work on his wireless system combined with the daily running of the music business and commitments to the automobile club started to take its toll. His health started to slowly decline, and it was already clear to him when he wrote the letter to the Minister of Defence about his inventions that he needed to reduce his workload. He felt that he had no choice but to resign his membership and position as a member and Trustee of the Automobile Club of Victoria in order to be able to undertake the future stringent testing of his wireless system. So, early in March 1908, Henry reluctantly resigned from the very club he helped create. At the time he thought that his resignation would only be temporary, but little was he to know that a turbulent time lay ahead of him over the ensuing years.

Over the next five months, a series of tests of Henry's wireless system were undertaken by the Department of Defence at the Victoria Barracks in St Kilda Rd., Melbourne. The value of his invention was of immediate and special interest for Australia, inasmuch as tenders had been called by the Postal Department for the establishment of the instruments at several selected places in the Commonwealth. The discovery was initially kept secret while preliminary inquiries were being made as to its value and then, with the permission of Captain Creswell, Henry invited the members of the press to his office at the music store and gave an interview and a brief description of his wireless system. On 28 July 1908, the newspapers publicly announced that a further trial of his wireless system was to be undertaken at Victoria Barracks the next day.

From *The Argus,* page 5:

"Although the Commonwealth Government have invited tenders for the installation throughout the Commonwealth of wireless telegraphy it is exceedingly unlikely that any tender will be accepted. The director of the Naval Forces (Captain Creswell) and the Minister for Defence (Mr Ewing) have for some time past been watching the experiments of an Australian inventor, and are now disposed to think that he has evolved a system of wireless telegraphy far superior to anything which the Marconi, Poulsen, De Forest, or other companies have to offer. The matter has been kept a profound secret so far, but tomorrow the inventor is to be introduced to the Military Board, and on a short range apparatus established at the barracks he will demonstrate to them the wonderful results he has obtained...

The Minister has been satisfied and there is little doubt that tomorrow's demonstration will bring conviction to the Military Board. After that there is little doubt that the Commonwealth will adopt the new system and install it at the various stations recommended by the Wireless Telegraphic Conference. The peculiar properties of the system which will be of such incalculable value in time of war will still be kept secret."

From *The Sydney Morning Herald*, page 13:

"Mr Sutton, of Malvern, will tomorrow show the Military Board of the Defence Department, by practical working, a new system of wireless telegraphy, which he has claimed for it superiority over any other in existence...

The defence authorities are interviewed almost daily with war inventions, some of which are hopelessly, pathetically improbable. But in regard to this one, the view is taken that it warrants the best attention of the department. The apparatus is said to be small and portable. The inventor affirms that he has discovered a new form of atmospheric wave; entirely different from the Hertzian or the Poulsen, and with which he has obtained absolute accuracy in transmitting messages. Every portion of the wave can be utilised, and is utilised in his system. It is worked with a low power. This Australian system requires only one-sixth of the electrical power needed in the Marconi system to produce vibrations. With a spark only one-eighth of an inch long, obtained from an ordinary motor cycle coil, he transmits a clear, sharp signal, a distance of 300ft, sending complete Morse messages as fast as skilled operators can read them through a telephone receiver

This is with the small model on which the demonstration is to be made. At the inventor's laboratory in a suburb of Melbourne there is a large pole, like the mast of a ship, carrying the antennae of his wireless equipment. Down below are his instruments, and whenever a vessel of the Royal Navy or German mail steamer, or any other with wireless installation, sends messages, the bells in the office below the mast are set ringing violently, and through the telephone receiver can be heard the ticking of the wireless message sent by somebody somewhere. The vessels of the Royal Navy communicate between Sydney and Melbourne and Sydney and Adelaide. But all their messages are said to be intercepted by this suburban installation. Many a message intended to be private, it is affirmed, finds it's to the inventor's ears, although both the experts of the navy and the engineers of the German mail steamers believe that their messages cannot be tapped.

The value of the invention for defence purposes is looked for in this way in particular. There is no wireless message which cannot be caught up, and recorded by this installation. If in cipher it cannot, of course, be solved without a key. On the other hand, there is no system, however it may be tuned or guarded, which is safe against the new wireless. From its instruments messages, it is claimed, can be poured into the receivers of enemy's installations, so that the messages are simply confused medley of signals. This, however, is only one of the methods which the new system has of confounding and silencing competitors. There are said to be others more drastic and more effective, but these are being kept secret. Mr Sutton was one of the earliest experimenters with wireless. A friend of Tesla, he had the advantage of wireless expositions and demonstrations from that scientist in London long before Marconi's name was connected with it. He was the original discoverer of the system of telegraphic photography, in which, following on the announcement of his discoveries, such great strides have been made. It was while proceeding to develop telegraphic photography that he discovered the principles of the new system of wireless. He was convinced that pictures could be telegraphed without wires as well as with them, and it was in trying to solve a difficult problem which presented itself in the course of his experiments that a new vista of wireless achievement was opened up to him. He is not a man in need of money, but has made electrical research his hobby. The invention has been brought to the Commonwealth's Government unpatented, and presented to them, and Captain Creswell has already made exhaustive investigations into its merits, which, it is expected, will be satisfactorily made clear to the Military Board in the yard of the barracks to-morrow."

The next day, the experiments were undertaken at Victoria Barracks where Henry and a small contingent of representatives, some from the press, were present at the invitation of the officers of Defence. The officers that were present were the Naval Director, Captain Creswell; the Chief of Ordinance, Colonel Parnell, and other members of the military and Naval Boards. Although the press and other representatives were there, the stipulation was that the experiments should be conducted in private and only officers of the Defence Department would be present for them.

The instruments were established in the quadrangle behind the Victoria Barracks, and a series of various tests were undertaken. One test showed that the new system could be worked with less power and that messages were transmitted with absolute accuracy. As the way this is done was a secret, it was not divulged to the press or other representatives, although the Defence officers who witnessed the experiments were very impressed and reported favourably on the outcome. Henry added that messages transmitted by other systems in Australia registered on the apparatus in his laboratory and that was why the discovery became so valuable from a defence point of view.

On the day of the experiments, the portable system that Henry had designed and built in 1907, was also tested. This was the first portable wireless system in the world; it was very small, lightweight and suitable for use for field signalling. This portable wireless reached a range of four hundred yards and at a later series of tests, reached the distance of five hundred yards. In the years to come, it was further improved and was able to send a signal much further.

A report on the outcome of the trial was later forwarded to the Minister for Defence, Mr Ewing, who was out of town on the day of the trial. It would also seem that Captain Creswell, in his desire to hasten along the development of Henry's wireless system and to impress not only the public but the American Navy, did not inform Mr Ewing of the latest trial. Although Mr Ewing had been kept informed of the ongoing testing of Henry's wireless system, when questioned by the press he stated, "Definite steps have been taken during my absence and I will see Captain Creswell tomorrow and go fully into the whole question. At present I am unable to say anything further on the matter." Captain Creswell bathed in the personal glory and media coverage of the announcement of Henry's wireless system and this was, in many ways, a slap in the face to the Postmaster General who was waiting on funds from the government for the installation of their wireless telegraphy stations and systems. Nonetheless, a report from Captain Creswell was also sent to The Postmaster General, Mr Samuel Mauger (pronounced Major) who later stated that he would arrange for further trials to take place next month on the return of the Chief Electrical Engineer, Mr John Hesketh, from Broken Hill.

On 4 August, Henry was asked to give a further demonstration of his wireless system, to the Prime Minister, Alfred Deakin, at Victoria Barracks. At this time, Henry pointed out to the Prime Minister that in time of war, an enemy's wireless instruments could easily be put out of action. He then showed how it could be done, mentioning that because it was so easy to do, the German Government had bought the British Admiralty's stock of homing pigeons, which the Admiralty had disposed of when they installed wireless telegraphy on their ships. Henry informed the Prime Minister that, in his opinion, his wireless telegraphy discoveries were already known to the Germans as they had done a great deal of research work in that direction. Consequently, it was obvious to him that any wireless system still needed to be backed up by a stock of reliable pigeons to carry communication in time of war between ships at sea and the land defences. Minister of Defence,

Ewing, accordingly asked the Commonwealth representative in London (Captain Collins) to obtain information about pigeons.

Not long after the demonstration, the Prime Minister received information from Captain Collins that according to the German estimates for 1908-09, a sum of £6,000 had been budgeted for the maintenance of a pigeon service. There were depots at all headquarters and fortresses, while two hundred private homing societies had placed their birds at the disposal of the Government. France, too, had a complete pigeon service and a system of co-operation with private owners. In all other European countries except Switzerland and Norway, pigeon lofts were actively maintained to keep the birds in operation as part of a communication system.

Henry's information had proved correct and due to the confimation received from Captain Collins, Minister of Defence Ewing immediately agreed with the need to have homing pigeons as part of Australia's defences. He went on to make the necessary arrangements with private bird fanciers to place their pigeons at the disposal of the Government when needed. On finding that there were about 20,000 homing pigeons in Australia, Ewing immediately budgeted a sum of money to provide trophies for homing competitions, so as to encourage the breeding of reliable birds in Australia. The breeders who competed for the trophies would have to place their lofts at the disposal of the Defence department should the occasion arise.

When World War I began in 1914, Australia utilised homing pigeons and they proved their worth many times over. During World War II they were still utilised, as even modern technology equipment could fail. In 1942, with the war on Australia's doorstep, pigeon breeders once again came to the fore and set up a network that could carry messages, and later that year the Australian Corps of Signals Pigeon Service was established. A testament to their invaluable service was their support to the operations on the Kokoda Track, as they could fly over the mountains and jungle where transmitting messages by radio was very difficult due to the inhospitable terrain.

The announcement to the press of Henry's wireless system was a strategic one for the Director of the Naval Forces, Captain Creswell. Prior to July, the Australian Government had invited the American Great White Fleet to visit Australia and they were currently on their way, nearing the Pacific Islands off the coast of Samoa. The Fleet, consisted of 14,000 crew on sixteen battle ships which had been fitted with Lee De Forest's wireless system before leaving the United States.

The Australian public was still unaware that the tests of Henry's wireless system by the Navy were actually to determine the range of his system and its ability to transmit a wireless message to ships at sea. As the Fleet sailed closer, Henry once again tested his wireless system; this time in the evening at his own wireless installation at his home in Malvern. To the amazement of all present, Henry's transmission was picked up by the flagship of the Fleet, *USS Connecticut*, then steaming off Pago Pago, Samoa. You could imagine the surprise of the wireless officers, receiving Henry's wireless transmission from Australia when they were some 3,200 miles (approx 5,150 kilometres) out into the Pacific. That this was a directly transmitted message which had not been relayed through any other station and which had travelled over a greater distance than had ever been previously achieved, made it a truly historic occasion. The longest distance, until then, for transmitting a direct wireless message was 2,900 miles (approx. 4,667 kilometres), established in July 1908 when messages were sent between the *USS Connecticut* and Point Loma in California after the Fleet had left the United States.

Henry's long range transmission had caught the attention of the American Fleet's wireless officers, so when the Great White Fleet arrived in Sydney, Henry and Admiral Sperry of the American Fleet sent wireless messages to each other.

These massive ships sailed through the heads into Port Phillip Bay on 29 August 1908, providing a unique spectacle as they steamed towards Hobson's Bay.

While the Fleet was visiting Melbourne, the American Navy officers met with Henry in person, and they spent many hours together. When they parted, the officers were suitably impressed with Henry's discoveries and inventions.

One piece of wireless equipment was of particular interest to them - his loud and sensitive wireless receiver which today is known as Henry's portable wireless. They were intrigued that this small piece of wireless technology could do away with the need for the familiar high pole and large, bulky outdoor equipment. Henry was able to demonstrate to the naval wireless officers that he could successfully receive and transmit messages with the small portable wireless. He made such an impression on them that many of them kept in contact for a number of years after the Fleet's visit.

When the American Fleet returned to the United States they utilised some of Henry's wireless equipment and scrapped the De Forest System, which had not performed as well as they had hoped.

After the American Fleet's visit to Australia, the world soon heard of the advances Henry had made in wireless telegraphy. The navies and armed forces of America, Japan, France and Britain all came to utilise some of Henry's wireless inventions in one form or another.

However, the secret of how Henry's wireless system worked was never divulged, nor was it used in its entirety by other nations. Despite the fact that the wireless system could have been a source of great wealth for him, Henry insisted it was only for the benefit and protection of the Australian Government and the Australian people and he sought no reward or personal accolade for his invention.

In September 1908, it was announced that there were to be no provisions made in the postal estimates for the installation of wireless telegraphy and the following report was printed in the newspapers in *The Argus* on the 21st September 1908:

No Federal Stations.

"It is understood that there will be no provision made in the Postal Estimates this year for the installation of wireless telegraphy. It was originally intended that wireless stations should be established at Port Moresby, Good Island, Thursday Island and Fremantle. These were to be semi-military stations. Commercial stations were to be established on the south of Victoria, on King Island and on the north coast of Tasmania-probably at Low Head. Tenders were called for the six first named stations and tenders from five of the leading companies were received. Not one of these tenders however provided for the erection and installation of the wireless system. All that the companies cared to do was to sell to the Government the necessary equipment leaving to the Postal department the responsibility of installation. In these circumstances none of the tenders have been accepted. There have been also invited tenders' for the commercial installations proposed. These do not close for some weeks hence, but the Government policy in regard to wireless has undergone a change and it is improbable that any tender will be accepted. It has been found that the money available for urgent necessary postal works is insufficient to supply the needs of the Commonwealth and the view taken by the Commonwealth is that other services should be properly equipped and the public demand met before such an undertaking as wireless telegraphy is embarked upon.

There is another reason which weighs with the Federal Cabinet. A new system of wireless has been developed by Mr Henry Sutton of Melbourne. This has already been described in "The Argus" and a series of experiments have been undertaken by the naval and military authorities. These have been satisfactory and though the chief electrical officer of the Postal department (Mr Hesketh) has not yet been called in to advise on the subject, it is known that the minister for Defence (Mr Ewing) is most favourably impressed. He has communicated his views to the Cabinet and other Ministers agree with him that it would be a foolish policy to accept the tender of any company until Mr. Sutton's system (which he proposes to present to the Commonwealth) has been thoroughly examined and tested. It is probable that extensive experiments will be undertaken by the Telegraph department before the end of this year with the object of testing the Sutton system. Mr. Sutton will be given full opportunity to demonstrate the utility of his system and should his claims be justified steps will be taken to install it wherever needed."

Unfortunately for Henry, on Sunday 27 September, Mr Henry Walter Jenvey from the electrical branch of the Postal Office was puzzled when he experienced a lot of interference while conducting wireless experiments. A few days later, when Henry made a request at the Post Office, Mr Jenvey became aware that the interference was because Henry had been conducting some experiments of his own from his Malvern home.

Up until 30 September 1908, the Post Office and the Postmaster General had not bothered to investigate or try out Henry's wireless system as they had stated they would in late July. Instead, they chose to ignore him and his ongoing experiments. Henry, who was oblivious to the fact that he had interfered with Jenvey's experiments, thought it would be best for his demonstration to the press on 30 September, to have another person with knowledge of wireless telegraphy present. He went to the Melbourne Post Office to ask if one of the Post Office's electrical engineers could attend the demonstration. His timing could not have been worse and even he was not prepared for what came next. His message was passed on, and a little later the Postmaster General rang Henry regarding his proposed experiments. Then the Postmaster's secretary Mr (later Sir) Robert Townley Scott rang Henry at the music store and unceremoniously read him the riot act in relation to breaching The Wireless Telegraphy Act of 1905 by not having applied to the department for a wireless licence, threatening to put Henry in jail if he continued his experiments.

Henry was shocked and shaken by the angry and threatening phone call, so at lunchtime he went into the Melbourne Argus newspaper office to inform them what had happened and advise them that he could not conduct the experiments as organised due to the fact he did not have a licence.

On that same morning at the Federal Postal Commission, the Chairman, Mr Hume Cook, asked the Postal Departments chief engineer, Mr John Hesketh, his opinion on the commercial viability of wireless telegraphy for the Postal Department. Hume Cook asked, "*Are there any new systems which look like being more commercially successful such as the Sutton System?*" Mr Hesketh replied "*I have read of the Sutton System in the public press but I know nothing more about it than what I have read.*"

News spread quickly about Mr Scott's threats to Henry and the following article appeared in the Melbourne Age newspaper under the following heading on the 1 October 1908.

How inventors are treated; Ridiculous Departmental Obstruction:

...Of course it is impossible for an inventor to test anything by which the ethereal forces are to be made subservient unless he has a certain freedom in the use of the atmosphere. Being unwilling to run the risk of fine or imprisonment, Mr Sutton, at an early stage in his operations, made enquiries, and was told by a high officer of the Defence department to "go ahead," as it would be "all right," whilst the Postmaster-General also informed him, as he understood, that no official notice would be taken of his proceedings. Since the publication of some wonderful successes achieved by the inventor, however, such as the tapping of wireless messages which were sent from New Zealand to the American fleet in Sydney harbour, and the designing of an installation that will send wireless messages between Victoria and Tasmania at a cost much below the departmental estimate, Mr Mauger has said that he cannot now remember giving even an implied permission for the carrying on of wireless tests at Mr Sutton's "station" at Malvern, and has warned him not to break the law. The only remote justification for this seems to have been, not the interception of the American fleet messages, but the hushing up, accidentally as it happens, of some department experiments...

Application was made to the Postal department for the loan of a telegraph operator for the purpose. There was a prompt refusal, Mr Sutton was told that he had been illegally experimenting, that he had confused departmental messages with his Malvern apparatus, and the department would not grant him any privileges or facilities for yesterday's intended exhibition. On the other hand, he was not forbidden to demonstrate; but was specifically warned not to break the law. The result was that, Mr Sutton claims the right to experiment and demonstrate freely within the walls of his house or business premises, like any university professor, not knowing precisely the law and being intimidated by the menacing observations of the secretary of the central administration and the Postmaster-General, he decided to postpone the exhibition.

What he was going to show, and that he had invented it is undoubted, is a receiver capable not only of reproducing wireless messages audibly, but of conveying sufficient surplus power through a galvanometer to turn on an electric light, or a bell signal when a message is announced, or drive a small motor. No other such instrument, he says, has yet been produced which can do all these operations at once, and this fact is attested by electrical text books. From Mr Sutton's latest experiences it is evident that he is greatly confined in the exercise of his genius as an inventor by the absurd attitude of the Postal department, which, having first "cornered" the Government rights to deal with electricity as a messenger, now apparently wants to prevent all progress. This, of course, is not merely a private, but a public wrong."

[Full Letter, see End Notes, Chapter Fourteen, No. 1.]

The Postmaster General's Department inevitably became angered and frustrated by the publicity surrounding Henry's wireless system and the Government's decision not to fund the postal Department's wireless system. This coupled with the Defence Department's continual praise about Henry's wireless system, became a huge bone of contention within the Postal Department.

With rumours flying around Melbourne about the Postmaster's treatment of Henry, the Postmaster General, Mr Mauger was asked by the journalist from the *Argus* newspaper, the post office's position on the matter, and he stated:

"There must have been some mistake on Mr Sutton's part. Mr Sutton said the Postmaster-General had rung up during the morning and had said that he had intended to give the demonstration but understood he could not do so without a licence. Mr Mauger admitted that this was correct but referred Mr Sutton to Mr Scott, secretary to the Postmaster-General. When however he learned that Mr Sutton proposed to give the demonstration in a room he said that it might be given without any licence at all; that he had thought Mr Sutton intended to give it outdoors."

He then stated that; *"In any case there would be no difficulty in obtaining a licence and that far from blocking investigation his department would do all it could to encourage it."*

The last sentence of Mr Mauger's statement would later prove otherwise, but with this apparent public about-face, and in an attempt to clear up what was going on, the Postmaster General and his Chief Engineer, Mr Hesketh, visited Henry at the family's music store on 6 October to witness a demonstration of his wireless system and discuss his application for a wireless telegraphy licence. The Postmaster agreed that if the Government were to purchase the rights to Henry's wireless system, that his department would reimburse Henry for the cost of the equipment. Henry was happy to give his wireless system free to the Australian Government; the reimbursement of the cost of the equipment that he supplied was all that he asked. After coming to this agreement with the Postmaster General about his wireless system, Henry was issued a "*Provisional Wireless Licence*" by the Post Office on 7 October.

Henry could not conduct any experiments in wireless telegraphy unless he agreed to the required conditions of the licence. However, there were some conditions to which he could not agree. His invention was only useful if it was kept secret and the Postal Office insisted on inspections and a clear breakdown of how his system worked. As a result, the Postmaster General refused to sign off on Henry's wireless licence agreement, and it remained in limbo - under 'Seal' - and the Postmaster proceeded to write to the Defence Department regarding Henry's experiments.

The Post Office and Postmaster had been well aware of Henry's ongoing experiments since July, 1908, and said and did nothing in relation to them, not even to inform Henry that he was required to obtain a licence to continue his experiments. After the Government's notification of no funds for their wireless system, it was clear to everyone including the public that Henry had now become a target of the Post Office's frustration and ire.

In the following days, Henry tried to appeal to the Defence Department to intervene to get the Postmaster General to permit him to continue his experiments without interference, but their hands were tied as the Post office had jurisdiction over wireless telegraphy. A few days later, the Acting Secretary of Defence received the Postmaster General's letter, which was in effect a 'please explain' letter. The letter from the Post Office Secretary Mr Robert Scott dated 15 October, 1908, read;

Subject: Henry Sutton Wireless Telegraphy experiments.

"I have the honour to inform you it is understood that Mr H. Sutton, of Sutton's Proprietary Limited, Bourke Street Melbourne, is carrying out certain experiments and giving demonstrations in wireless telegraphy in this city with the cognisance of your Department, and as it does not appear that a licence such as required by the Wireless Telegraphy Act 1905 has been issued him, to ask if you will be so good as to state whether you are aware of any circumstances in connection with the case to justify the conditions of the Act referred to not being observed."

The receipt of this letter angered the Defence Department, and so the gloves were off and the department went into bat for Henry and his wireless system. There is nothing more telling than the series of letters that went back and forth between the Department of Defence, the Prime Minister's Office, the Postal Office and Henry over the next four years. The war of words, obstruction and red tape was a testament to spite and sheer short-sightedness on the part of the Post Office hierarchy , who wanted to prove their department was the more powerful.

The response to Mr Scott's letter from Captain Creswell on 21 October read:

"Mr Sutton represented to me that he had made discoveries in Wireless Telegraphy of Naval value, and that he was in hopes of still more valuable and important developments which he would place at the disposal of the Government.

- *It seemed to me that Mr Sutton had grounds for his representations, and it was my wish that he should be encouraged in the continuance of his experiments. I was under the impression that the Postal authorities be equally desirous of any success he might obtain and would further his efforts.*
- *The requirement to experiment under license is I understand one from which the Admiralty is exempt. Vessels of war, British and Foreign and of the German Mercantile Marine exercise Wireless Telegraphy without licence, their waves covering a good portion of Australia.*
- *The terms of the licence require that the premises of the investigator or experimenter may be opened to inspection at any time and his investigations would be open to the view of any expert able to understand them. This at once cancels its defence value. Any new process or method or principle is liable to discovery. I am sorry to say that Mr. Sutton has informed me that he has ceased all further experiment and investigation, - the only terms under which he can work under the licence being of a kind he would not agree to. So far as I am able to judge, his discoveries were of value and their further development would have*

been of great service. His discontinuance may quite well be of serious public loss and I am not able to see that there was any sufficient occasion for putting the Act in force. Any Postal Wireless operations are not yet in existence. There is no public Wireless service to be interfered with. Vessels in our ports, as already stated, work without license and their effect would not be less, probably far greater than any caused by Mr. Sutton's apparatus.

- *The pressing need of enforcement of the Act is primarily to protect a service which as yet does not even exist. Its effect is merely to check investigations possible to be of national value.*

In my opinion, Mr Sutton should be given carte blanche to continue his investigations without interference."

The situation was at a stalemate. In November the Government changed hands and Mr Andrew Fisher became Australia's new Prime Minister. On 13 November, Mr Josiah Thomas became the new Postmaster General and Senator (later Sir) George Pearce became the Minister of Defence.

During the eight long months that it would take the Post Office to respond to Captain Creswell's letter, Henry was caught between a rock and a hard place. To try and expedite things along during this time, he wrote again to the Department of Defence and Captain Creswell. Henry's letter was a gift to Australia and the Australian people and the Department of Defence knew it. Dated 12 December 1908, to Captain Creswell, read:

"Sir, I have the honour to definitely offer my system to the Government for use, primarily for Defence purposes, and if the Defence department deem it desirable, for Post Office use. As you are aware, having discovered certain new principles rendering Wireless more simple and efficient, I considered it would be more valuable for Defence work, that these principles be retained by the Defence Department as secrets rather than as patents, as in latter case publicity would lessen the advantage and value for Defence work. I therefore have decided not to patent or form a syndicate, but to offer the system to the Defence Department, conditionally that the Government reimburse to me what they considered should be expended on the purchase of exclusive rights to a Wireless system. These conditions were proposed and verbally accepted at the interview between the Postmaster General and myself, in my office, on October 6th 1908. As the Government now have quotations from other sources, it is better able than myself to ascertain the financial value to the Commonwealth of an exclusive Wireless system. Having laid down practical principles of the system after a great deal of

labour and expense, I think the time is ripe for the Government to step in and apply same to use. As there are parts I deem it wise the Defence department retain, I would suggest that department take in hand the matter of extending the system."

Captain Creswell then wrote to The Naval Board regarding Henry's very generous offer.

Subject: Mr. Sutton's Wireless Telegraphy.
"With reference to Mr Sutton's offer, the apparatus he has submitted was examined by Sub-Lieutenant Cresswell, who was of the opinion that it gave evidence of justifying Mr Sutton's claim to a simpler and less expensive system of Wireless, and of great possibilities. I would suggest that Mr Sutton be thanked for his offer, and asked if without in any way committing the Commonwealth Government to the purchase, he would have any objection to assisting in the making of certain experiments to test his apparatus in the Bay. The torpedo boats could be used for this purpose."

At the Naval Board's meeting held on 16th January, 1909, the matter of Henry's wireless system was held over until the new Invention Committee was appointed. With the delay on further experiments, Henry wrote to Captain Creswell on 4 February 1909 to request the return of his wireless apparatus, and it was clear Henry was still planning ahead to undertake further experiments:

"Dear Captain Creswell, -- Will you kindly let me have my wireless apparatus. I have one power set finished suited to my 10 inch coil. It does not follow that this set is suited to the coils you may have available. So that before going further it would be necessary to know the length of spark and amperage of the Government coils. Some time ago I put myself in communication with the best sources of practical information in New York and London, re: Wireless Telegraphy and will shortly have all that is available of any value. In addition I will receive from Germany in about six weeks the necessary apparatus enabling one to carry out "Wireless Telegraphy" experiments according to my method. Australia is not going to be left behind in this matter: we are very much further ahead than the public imagine."

Seeing the value of Henry's wireless system, the Defence Department in the end did not wait until the implementation of the Inventions Board and it would be only one month later on the 1 March that there was some movement within the Defence Department regarding Henry's wireless system. Lt F. G. Cresswell from the Naval Militia had a meeting with Captain Creswell about the Defence Department trialling Henry's wireless system.

Subject: Henry Sutton Wireless Telegraphy.

"Sir,-- With reference to the interview I had with you today in connection with Mr Sutton's new system of Wireless Telegraphy which he has submitted with an offer to the Department of Defence, and which he proposes should be given a trial, I would suggest that Mr Sutton be communicated with in the following terms. That he be requested to submit an estimate of the probable cost of fitting up (2) complete tuned installations to work up to a range of 25 miles. The estimate should include the cost of erecting the necessary antennae and omit the cost of suppling two (2) 10 inch spark coils, Polarised Relays, Telephones and Recorders as I think these instruments could be obtained on loan from the Postal Department. Mr Sutton should also be requested to state whether his patent rights cover his system to the extent, that in the event of the Department adopting and using his system as a means of Electrical communication, whether the Department would be held liable for the payment of Royalty to any existing Wireless or Ether Wave Telegraph Company."

As Henry could not undertake his own experiments, the Defence Department took it in their hands to undergo some trials of Henry's system on their own with the view of adopting the new system. During this time, Henry at least had a reprieve in his work load but none the less he still would have had an advisory role in connection to the trials. At least the Defence Department could undergo experiments without the interference of the Postal Department.

Meanwhile, Australia's political upheaval continued, and on 2 June 1909, Alfred Deakin was reinstalled as Australia's Prime Minister. Deakin then appointed Sir John Quick as the new Postmaster General and Mr (later Sir) Joseph Cook was appointed the Minister of Defence. The political events of the past several months and the unprecedented eight months it took for the Post Office to respond to Captain Creswell's letter was not helpful to Henry's cause. The Post Office's long delayed response served as a lesson in thwarting the enemy, but Mr Scott finally wrote to the Acting Secretary of Defence on 15 June. To add another complication and delay, Captain Creswell unfortunately did not get to see Mr Scott's response letter until nearly a whole year later on 2 June 1910 as he was absent from Australia at the time the letter was received.

If Henry was hoping he might make some progress with the new Postmaster he was very much mistaken. Sir John Quick gave an interview about wireless telegraphy inventors in August 1909, in which he specifically targeted Henry and the Defence Department under the leadership of the Defence Minister, Mr Joseph Cook.

However, there was another underlying reason why postal department put up barriers to Henry's wireless system and had previously ignored his experiments, and that was that the Post Office's electrical engineer, Mr Henry Walter Jenvey, was in the process of developing his own wireless system in the hope that the post office would utilise it in the future. Henry became a thorn in Jenvey's side and the engineer was behind a lot of the blocking and negative rhetoric within the Postal Department that was aimed at Henry's experiments and wireless system – then and over the next few years.

On the subject of inventors and wireless telegraphy the new Postmaster General stated that:

"The Postmaster General, Sir John Quick, recently defined the position of the Commonwealth Government with respect to inventors and others who may wish to avail themselves of the advantages of wireless telegraphy or even to develop improvements on the systems already before the public. As stated by Sir John Quick the position is as follows; The Commonwealth has by statute complete control of 'wireless' in any shape or form and no person can engage in practical or experimental use of it without a licence from the Federal Government. A recent application for permission to engage in unrestricted experiment was referred to the Defence Department. That department consulted the Admiral in charge of the Australian Naval Station and he reported that it would be most undesirable to allow any 'wireless' station to be erected not under the control of the Commonwealth. This view is accented to by the Commonwealth authorities and while there is no objection to grant a licence on any reasonable application, the government intends to retain absolute control of this means of communication."

The year dragged on. The government was still in a quandary in relation to the installation of wireless telegraphy. Henry's anguish and frustration was diverted for a time when his wife, Annie May, once again became pregnant. The impending birth of another child was eagerly anticipated by both Henry and Annie May. Later that year, on 30 November 1909, Annie May gave birth to another son whom they named John and whose birth brought a lot of joy into the Sutton family home. Regardless of what was happening, Henry still went to work at the music store every day. His love of the family business never lessened, even when he was distracted by the ongoing events around his inventions. Apart from the few years in England, Henry never ignored his music store responsibilities. Clearly Richard and Mary's legacy was safe with all their children.

Henry's brothers and his family shared his frustration about the ongoing wireless situation and their support and encouragement continued throughout the long ordeal. Even though he could not conduct wireless telegraphy experiments at his home or at the family music store, he still worked on other aspects of wireless telegraphy and continued to invent new wireless equipment. Some ongoing experiments were still possible, and the arrival of a very expensive and important piece of equipment from Germany in March, which allowed him to continue his research.

In October, Henry's wireless licence was due for renewal. At first it still seemed a pointless exercise, but after reaching an agreement with the Post Office, Henry paid his £3/3s and on 7 October was issued a provisional wireless licence by the Postmaster General, Sir John Quick, pending the adoption of new clauses to the Wireless Licence agreement. It was issued on the understanding that should it be necessary, it would be withdrawn and any necessary amendments substituted and the licence would be in force for twelve months. The Postmaster General signed off on Henry's wireless licence on 17 October. Henry's provisional licence enabled him to establish and maintain a wireless station at his home which worked by dynamos and accumulators (batteries) having a maximum range of 250 miles (approx. 402 kilometres). The description of Henry's wireless apparatus was called a Potential Detector Syntonic and the power and voltage of Henry's wireless station was 1Kw for telegraphy and 2Kw for radio/telephony.

It was stipulated in the licence agreement that the licensee's sole object was to conduct demonstrations or experiments in wireless telegraphy. All public rights were safe guarded and special precautions were undertaken to prevent interference with naval or military signalling. Provisions as to secrecy were stringent: the licensee was not allowed to divulge to any person, or make any use what so ever of, any message or knowledge transmitted by naval or military signalling or any other Government system of wireless. Transmission or delivery of messages was forbidden except with the express approval of the Postmaster General or his duly authorised officer.

Just as Henry was issued his wireless licence, the Postal Department was granted £20,000 for the installation of two wireless stations, one to be built in Sydney, New South Wales and the other in Fremantle, in Western Australia. The Postmaster then advertised for tenders for a wireless system, which would be installed in the new stations when they were built. Henry at this point must have been shaking his head in disbelief, as he had continually offered his system free to the Australian Government. As Henry was never intending to commercialise his system, he did not contemplate putting in a tender for the wireless stations. Tenders for the erection of the two wireless stations with a range of 1,250 miles (2011.68 kilometres) closed on the 1st March, 1910, at which point five companies had submitted their tenders:

Australasian Wireless Limited:	£4,150, per station.
British Radio-Telegraph and Telephone Company:	£9,317.10, per station.
De Bingham Telegraph and Telephone Company:	£7,000, per station.
Lepel Wireless Syndicate:	£7,088, per station.
Marconi Company:	£19,020, per station.

Following a long standing tradition, the Australian Government accepted the cheapest tender - that of Australasian Wireless Limited, which in actual fact turned out to be the Telefunken Company. It was questioned at the time how the two stations could be erected at such a cheap cost, but the Telefunken Company had behind it two powerful electrical undertakings: Siemens and Halske and Allgemeine Elektricititats-Gesellschaft, with its world-wide organisation. The competition for wireless telegraphy dominance was already in full swing around the world and it was at this point that the Marconi Company expressed their intention of suing Australasian Wireless for infringement of patents.

Mr Guglielmo Marconi of the Marconi Company had been keeping abreast of what was happening in Australia since 1906, when he was thwarted in his attempt to secure the contract in Australia for his wireless system. His representative, Mr Walker, continually sent Marconi updates on what the Australian Government was doing in relation to wireless telegraphy. Marconi was no doubt intensely angered by the constant announcements in all the Australian newspapers about Henry's system of wireless being considered for use by the Australian Government. He was ferocious in protecting his wireless invention and considered anyone else's system to be in breach of his patent rights. His litigation regime was incessant all around the world and thousands of pounds were spent on protecting his patents. Anyone with a new system was a target and more than likely Henry would have been in his sights if the Australian Government was to go ahead and use his system. Instead, Australasian Wireless became his target.

As part of Henry's plan of action, he decided to lodge his first wireless patent to protect his invention. He was granted a provisional patent titled *"Improvements relating to the production of hertzian waves."* on 20 December 1909. Henry later went ahead and lodged another wireless patent this time it was for *"Improved means for preventing 'arcing' of the gap in the production of high potential electric oscillations"* This provisional patent was granted on 15 February 1910. Then on 19 March 1910, Henry was granted another provisional wireless patent which was titled *"Improved means for detecting acoustic, electric, or like waves."*

This latter patent would later prove to be an important discovery. Within months, it would be known to the world and even Henry could not have imagined what his wireless experiments and patents would eventually lead to in the decades that followed.

[Full Texts, see Patents, Chapter Fourteen, Nos. 1, 2 & 3]

This was a busy year for Henry and although he was caught up in red tape and political battles between the Postal Department, the Defence Force and now the threat of Marconi, not to mention the constant changing face of the Australian Government, his wireless work continued. Since the arrival of the American Great White Fleet in 1908, Henry had been approached by many other navies around the world in relation to his wireless inventions. His reputation in the field of wireless had gained a lot of attention and in April 1910, the Japanese Navy also paid a visit to Australia.

On 7 April, the Japanese cruisers the *Aso* and *Soya* arrived in Hobson's Bay and moored near Williamstown and Port Melbourne in Victoria as part of their six month navy cadet training course. Two days later, Captain Creswell met with Rear Admiral Ijichi on the *Aso* and during their conversation the Japanese Admiral requested if his ships wireless officers could pay Henry a visit.

During the week that followed, Henry demonstrated many of his wireless inventions to the Japanese officers who showed great interest in them, and consequently, the Japanese Navy sought to utilise some of Henry's wireless inventions themselves. Clearly, Australia lagged behind in regards to wireless telegraphy and even though they were faced with Henry's great improvements, the Australian Government moved slowly in this area of what was becoming a global necessity for every ship and country.

In May, 1910, Halley's Comet passed by Earth and so Henry set about the task of performing a series of wireless experiments related to his work which he thought would be interesting. On 20 May, 1910, *The Age* newspaper published Henry's letter stating the results of his experiments:

"Sir, you will be interested to learn that a number of experiments were made today, to know if the cometary matter we are supposed to be in touch with is associated with radiant form of energy in the way of electric oscillations, that is, etheric oscillations between the limits of 200,000 and 3,000.000 per second (the limits of the experiment). The experiments were made by aid of oscillograph, tube, Neon tube, carbon dioxide tube, cymometer and potential detector. In no case was a trace of electric or etheric oscillation found, though the instruments used for their detection are marvellously sensitive to etheric disturbance. If there be no radiation having a definite etheric oscillation constant, from comets or other celestial bodies, there can be no danger that electric disturbance can be set up on the earth by resonance. As far as they go, the experiments I made yesterday show no etheric disturbance whatever. If the earth's natural period of electric vibration (or oscillation constant) should at any time coincide with a celestial body actually radiating electric waves having a similar period (and the radiation reach the earth)., the whole of the earth's electricity would be set in action by resonance, and – well, that would be the end."

The general public may not have understood the contents of Henry's letter but at least in regards to Halley's Comet passing the Earth, they were all safe. However, Henry's not so subtle warning at the end of his letter might have been a source of much conversation among readers and may even have alarmed a few of them.

Up until May 1910, Henry's wireless work had been continuing with the Defence Department and his need to progress his experiments without the constant interruptions from the Post Office led him to consider not re-applying for a wireless licence when it was time to renew it in October. Every time Henry wanted to change something in his system, he was required to get the approval of the Post Office and submit his equipment to inspection by them. It was becoming increasingly impossible for Henry to effectively continue his work and maintain the secrecy of how his wireless system worked, which was the preference of the Defence Department as well.

In May, Henry expressed the view to Captain Creswell, that under the circumstances he was considering not re-applying for his wireless licence and stopping all work on his system. News of Henry's position and the suggestion that he would cease his wireless work was immediately brought to the attention of the Department of Defence. Under the instructions of The Secretary of Defence, Brigadier General Samuel Augustus Pethebridge (later Sir), Major Cecil Henry Foott from the Naval Inventions Board paid a visit to Henry on 27 May. He lodged the following report to the Secretary of Defence on the 1 June:

"In his workshop Mr Sutton showed me certain pieces of his apparatus and made certain experiments and explanations. Although of course, it was not possible in the limited time at my disposal, and in the confinement of a building, to carry out any exhaustive tests. I consider that Mr Sutton's principles of wireless telegraphy are well worth further and very careful consideration. The position in which Mr. Sutton stands is, apparently this:--He is a private inventor not desirous of financial gain from his experiments, but carrying them out as a hobby apart from his business. He does not desire to place his inventions with a syndicate for exploitation but is willing (as his letter of the 12 December 1908) to hand over his inventions to the Dept. of Defence practically at our own price. Mr Sutton however does not wish to go to the trouble of taking out a licence as required by the provisions of the Wireless Telegraphy Act of 1905. He points out, and I think with some reason that the publicity enhanced thereby, due to inspection of his apparatus by that Dept. would be prejudicial to its secrecy and on that score would deprive it of a practical part of its value for Defence purposes. I am of course not in a position to state whether it would be advisable to override or mitigate the provisions of the Act referred to, even if it were possible, but I suggest for careful consideration the following factions which appear to me to influence the situation.

- *No power can compel Mr Sutton to take out a licence- he has the option of merely abandoning or disposing of his invention.*
- *If he merely abandons it, his research work is lost to us, and if he disposes of it we are in a worse position.*
- *The research work in wireless and the striving after greater efficiency therein is not confined to the British people.*

A consideration of these factors leads me to submit the following suggestions for dealing with this case, although in so doing I am fully aware that they may be impossible or undesirable:-

- *The question of license under the W.T. Act 1905 be worded in favour of this Dept. in regard to W.T. Experiments of a purely Naval and Military nature, provided perhaps that notice of intention to experiment be given to the Postal Dept.*
- *That in the case of such Naval and Military experiments, the inspection of plant etc by officers of the Post and Telegraph Dept be dispensed with, and such inspections to be made only by authorised Officers of the Naval or Military Forces.*
- *That the Dept. of Defence, if necessary take out a licence for wireless experiments and that Ministerial approval be obtained for the provisions of para. 2 (non inspection by P & T Officers) to be applied in its case.*

As a concrete result of my interview with Mr. Sutton he will in the course of a few days submit his requirements, for a series of experiments, in regard to stores and appliances, most if not all of which I gather from my interview will be easily supplied from engineers equipment at present on hand. But under the conditions that appear to arise from the Wireless Telegraphy Act, it will be impossible to carry out any such experiments under purely Departmental Control, and I question whether Mr. Sutton will permit investigation by other than Defence Dept. Officials. In conclusion I strongly recommend that the Naval Dept. take part in all these experiments, if possible conjunctly with the Military Forces. The initial work done by Lt. Cresswell must be of a nature too valuable to be lost."

It was also at this time, on the 2 June, that Captain Creswell was finally given the letter from Mr Scott the Postmaster's Secretary, which Scott had sent the year before on 15 June 1909. It read:

Subject: Henry Sutton Wireless Telegraphy.

"In returning the accompanying papers, which you forwarded to your office on the 23rd October last, relative to certain experiments and demonstrations in wireless telegraphy which it was understood were being conducted in Melbourne by Mr H. Sutton, of Sutton's Proprietary Limited, Bourke Street, Melbourne, I have the honour to inform you that this Department is most desirous of furthering any efforts made by Mr Sutton or other inventor, but, being responsible for the administration of the Wireless Telegraphy Act 1905, it is at the same time compelled to see that any assistance given to an individual inventor shall be given subject to conditions imposed by that Act and the Regulations there under,

and also that the assistance shall be such as can be given to anyone else within Australia making similar claims to important discoveries.

- *With regard to paragraph 3 of the comments of the Commonwealth Naval Director, it may be stated that the Wireless Telegraphy Act 1905 does not apply to ships belonging to the King's navy, but it does apply to the use of wireless telegraphy by any other ships when in Australia (vide section 6 (2).*
- *If Mr Sutton does not feel disposed to agree to the terms of the form of licence which it has already been proposed shall be issued to him, to enable him to pursue his investigations in a regular and legalised manner, then it would appear that it is reasonable to expect that he himself shall notify the Postmaster-General of the directions in which he desires modifications to be made in that form of license.*
- *It is observed that the Commonwealth Naval Director considers carte blanche should be given to Mr Sutton to continue his investigations without interference, but his opinion in that respect is in direct conflict with that of His Excellency the Vice-Admiral on the Australian Station, who on the 23 October last, stated "it would be most undesirable to allow any wireless station to be erected which is not entirely under Government control." It is that Government control which the licence proposed to be issued to Mr Sutton is designed to secure.*
- *I may add that to single out Mr Sutton as the one individual to whom carte blanche is to be given is a step which it would be found difficult to justify, and unless the Wireless Telegraphy Act 1905 is to become a dead letter and everyone be allowed to do just as he pleases, there appears to be no reason for any alteration in the attitude already adopted by this Department."*

Captain Creswell was very understandably angered and frustrated with the Post Office's attitude and the Defence Department promptly wrote to the Postmaster General requesting a copy of Henry's wireless licence agreement. On 20 June, a copy was sent to the Secretary of Defence with the following accompanying letter, the Postal Department's way of exerting their authority and it read:

"I forward herewith copy of a licence to use wireless telegraphy for experimental purposes, which has been issued to Mr Henry Sutton, of 290-292 Bourke Street Melbourne. 2. The licence in question has been issued provisionally pending the adoption of a Wireless Telegraphy Licence embodying certain new clauses, on the understanding that should it be necessary to do so it will be withdrawn and any necessary amendment substituted, and is to remain in force for twelve months as from the 17th October 1909, subject to the powers of revocation shown therein.

Mr Sutton has been informed that if at the end of that period he should desire to have the licence extended it will be necessary for him to make application accordingly."

Increasingly angered and frustrated by the situation and the letter, Captain Creswell wrote a very terse response letter to the Postal Department on 21 June:

Subject: Henry Sutton Wireless Telegraphy.

"Secretary:- To be licensed by the Post Office to carry out one of the most important defence services – rapid transmission of intelligence – was not proposed by the Post Office. No licence is necessary. If Mr Sutton's ideas and investigations are of value, and such appears to be the general opinion, he could carry out his Wireless Telegraphy investigations, free of the inquisitional inspection to which he takes exception, by being attached to this Department with an honorary Commission, his laboratory and apparatus being placed under "Defence" for experimental purposes and for communication with the Naval and Military Establishments at Williamstown, Queenscliff, etc."

The question of the Defence Department possibly granting an honorary commission in the Australian Navy was put to Henry by the Defence Minister, Senator George Pearce. Henry stated that if the Defence Department deemed it necessary, he would happily accept an honorary commission in the Australian Navy. Under the Wireless Telegraphy Act, with amendments to permit Henry to carry out wireless telegraphy experiments without the interference of the Postal Department, it would mean that he could alert the naval wireless telegraphy officer of his experiments, and have his equipment inspected by an authorised officer of the naval or military forces instead of the Postal Department's engineer.

In the interim, Captain Creswell suggested that a permanent wireless officer should be appointed in the Navy as the advance in the use of wireless had been considerable. The new Destroyers using wireless telegraphy were due to arrive in Australia in a few months, and Captain Creswell thought that the navy should therefore be in a position to receive wireless signals.

Henry Sutton with his
portable receiver.
Photograph courtesy of the
Wright/Sutton Family archives.

Sir John Quick
Postmaster General
2 June 1909 - 29 April 1910
Image courtesy of the National
Library of Australia

Prime Minister Alfred Deakin
Image courtesy of the National
Library of Australia

Naval Captain William Rooke Creswell
Image courtesy of the Australian War
Museum

The Great White Fleet
USS Conneticut in the lead.
Photograph courtesy of the United States Naval Historical Foundation

CHAPTER FIFTEEN

THE RECOGNITION OF DR BELL

"A man, as a general rule, owes very little to what he is born with – a man is what he makes of himself."

Alexander Graham Bell

As the Defence Department and the Post Office were finding ways to win their fight, Henry was granted a wonderful reprieve from the stress of the situation. His long-time friend, Dr Alexander Graham Bell, was on a world trip and scheduled to visit Australia. When he was in Melbourne, they made arrangements to meet.

Dr Bell arrived in Australia on 13 June 1910 on the mail steamer the *R.M.S. Makura* accompanied by his wife and the well-known Canadian aviator Mr F.W. Baldwin and his wife. For the occasion of his arrival, Henry furnished the Melbourne Age newspaper with a very vivid account of Dr Bell's telephone invention and described its first public demonstration back in 1876.

"His was the first electric vocal telephone, and it was born in 1876. Just then the great Centennial Exhibition was being held, and a small out of the way corner was allotted to what was considered a crank idea. Many an eminent man passed it by with a smile; even the judges took so little interest they left it to the last moment of the day appointed for a trial. When they turned up, the nervous inventor's description was received with incredulity as first one and then another picked up the home-made affair and put it down in doubt. Among the judges were the famous Professor Henry and Sir William Thomson (Lord Kelvin of later date). In the middle of these doubts Dom Pedro, Emperor of Brazil, in passing, recognised Dr Bell, whom he had previously met in connection with a deaf mute institute. The Emperor shook hands. Dr Bell then asked him to hold the receiver to his ear while he spoke in the transmitter. Dom Pedro startled exclaimed, 'God

it speaks!' Instantly Professor Henry rushed for it, then Sir William Thomson. The excitement was extraordinary. The apparatus was removed to the judges' annexe, and the most despised little exhibit became the great centre of attraction. Sir William Thomson cabled the news to England and the world knew that a persevering man had succeeded in transmitting speech electrically by wire."

Henry knew the story intimately. Even thirty five years later, it was as if Dr Bell had just told him. He was still full of respect and admiration for the man and his discovery.

Dr Alexander Graham Bell

Dr Bell was not only scheduled to meet Henry in Melbourne; he was also catching up with some relations and a long-time friend and past girlfriend, Mrs Marie McBurney (*née* Accleston) whose husband had recently died. Samuel McBurney had met Marie in England while on a visit from Australia. In a twist of fate that showed what a small world it can be, he was a music teacher whose method was based on the Tonic Sol-fa system. This used a Braille system, which Samuel taught at the Blind Institute on St Kilda Road in Melbourne, and it paralleled Dr. Bell's work with the deaf. As well as that, Mr and Mrs McBurney were close friends of Henry's sister Elizabeth and her husband Henry Finch Rix, and they were acquaintances of the families of Henry's brothers Alfred and Walter.

Bell's meeting with Henry was a private affair, which suited them both as it left them free to enjoy each other's' company. They met on 15 August, and they both travelled to Ballarat on the 7.40am train from Spencer Street Station (now Southern Cross Station). During the four hour journey, both men finally had time to become more personally acquainted. After arriving at the Sutton family's music store, Henry showed how he had networked his telephone system around the building, demonstrating to Bell a more sophisticated telephone system than even he had seen before. He was, of course, hugely impressed. The two men spent time catching up on the ideas that each of them had developed over the years. Bell's current research was aviation, a subject dear to Henry's heart and that, too, would have been one of the main topics of conversation on the trip to and from Ballarat. Another topic of conversation was about Henry's current research on wireless telegraphy and his recent world acclaim. After refreshments and lunch with Henry's brother, Frederick, and his family, Henry and Bell boarded the 2 pm train back to Melbourne. It is quite clear that the two men had a long conversation about Henry's work with wireless telegraphy and his ongoing battle with a Post Office Department that was clearly stifling the creative and innovative talents of not only Henry but other Australian inventors as well. Such blatant interference in scientific research was unheard of in the United States and this clearly hit a nerve with Bell.

Dr Bell had been invited by the Postal Royal Commission to give information on his view of the Australian telephone service and exchanges. As the inventor of the telephone, he felt obliged to assist. The Royal Commission had begun in 1908 and was looking into the practices, management, finance and organisation of the Postmaster General's Office which had been in disarray for many years. The Commission was nearing the end of hearing evidence into the practices of the Postal Department at the time of Bell's arrival in Australia, and his input was much sought after.

On Tuesday 16 August, at Parliament House in Melbourne, Bell gave evidence to the Royal Commission regarding the telephone exchanges he had visited while in Australia and telephone systems in America. What he told the commission revealed how he viewed the treatment of inventors and inventions by the Australian Government and his views of Australia's patent system. His testimony was influenced by his personal observations and by information about the Australian Patent System learnt from Lawrence Hargrave, whom he had visited in Sydney. Hargrave was so opposed to the Australian patent system, he'd written and published a "*Memorandum on Patents*" in which he expressed his very vocal views on the state of the Australian patent laws and their effect on inventors.

Bell's subsequent conversation with Henry in relation to his wireless system and his ongoing treatment by the Postmaster General's Office was also echoed in his testimony. Although cloaked around the Postal Department's telephone system, his answers spoke volumes in relation to the Australian Government's obstructive treatment of inventors and highlighted the detrimental effects of the current patent system on Australian innovators and experimenters.

An extract of Dr Alexander Graham Bell's Testimony at the Royal Commission on the Postmaster General's Department 16th August 1910:

Question: *"Did you see a selector invented by Mr Godfrey, of the Railway Department, here?*

Dr Bell: *"Yes, I was very much interested in it; it is a very ingenious instrument, and works very well on a set of twelve stations. But it would require very much modification to adapt it to a telephone system. This is the case of an Australian inventor who should be encouraged in his endeavours."*

Question: *"Supposing that inventor devised an automatic system, how could it be tried?"*

Dr Bell: *"In this country the Government has a monopoly, and the invention cannot be tried unless the Government gives the inventor an opportunity. All these new ideas require a great deal of adaption before they can become commercially successful. We have a number of systems, such as that of Mr Godfrey, in America on the step by step principle. I do not know that this one has had advantage over those in the United States of America except that it is an Australian production, which should give it an advantage in your eyes. You want a proper system of encouraging investors and a department to test the value of inventions."*

Question: *"Do you consider that wireless telephony is likely to develop greatly in the future?"*

Dr Bell: *"Yes, it is in use now in America experimentally, and speech has been transmitted 12 or 14 miles. When an invention begins in that way one cannot tell how far it will go. Wireless telephony is in the same stage as wireless telegraphy was some time ago, but it has not yet entered the field of practicability. In any case, it is unlikely to enter into competition with the present system of telephones on account of the difficulty of securing privacy."*

Question: *"Did you form the opinion that Australia has not yet passed out of the infant stage?"*

Dr Bell: *"I am of opinion that the laws in Australia are not favourable to inventors. You seem to base your patent law on the English model, which is not favourable to inventors. In America no patent is granted to any other man than* *the first inventor, and then only after an examination has been made by the officials of the patent office to find out whether the supposed inventor was really the first inventor of the matter claimed. A man never knows whether he is the first to invent any instrument. He may know that the invention was original with himself, but does not know what others have done. The same thing may have been invented before he was born; therefore, the patent office makes an investigation to find that out; and only if he is the first inventor will he be given a patent. When I investigated the patent system in England some thirty years ago, I found that a man, whether the inventor or not, could obtain a patent for almost anything he chose, if he paid the fees, and after getting the patent it was not worth much in the way of protection, for the onus of proof that you were the first inventor was thrown upon you. You would have to prove to the satisfaction of the court that you were the first inventor before your patent was of much avail. No patent was of value until it had passed through a lawsuit - an expensive process. In America, the patent granted by the patent office is held by a Court of law as prima facie evidence that you are the first inventor, and the burden of proof that you are not is thrown on the other side.*

The result of the American patent law can be seen in the stimulus which it gives to inventors. It would be to the interests of Australia to stimulate invention, and protect inventors in the ownership of their creations. One of the great dangers in a country like Australia is that Government ownership is going to interfere with the improvements of inventors by stopping financial rewards that are before them."

[For a longer Extract, see End Notes, Chapter Fifteen, No. 2.]

Dr Bell was not subtle in his view regarding inventors in Australia, and many people at the time were very well aware to whom he was alluding to in his testimony. It is safe to assume his references regarding inventors was to some degree an embarrassment and a slight slap in the face to the Postmaster General, Mr Thomas, given the public's sentiment around the department's treatment of Henry over the past two years. The very next day, in fact, Thomas was mocked in the press regarding Bell's testimony as he publicly tried to focus on the good points of Bell's statements. However, the opinion of the public and the newspapers was an overwhelming condemnation of the Postmaster General and his running of the Postal Department. Henry would have been humbled by Bell's testimony, as he was well aware of its true meaning. Over the next few months, public pressure began to mount against Thomas in relation to the Post Offices' position regarding wireless telegraphy licences and wireless inventors.

At the time of Bell's visit, Henry was a member of the Victorian Institute of Electrical Engineers and after his visit to Ballarat, Bell accepted an invitation from the President of that Institution, Mr Alfred Henrich Jackson, to attend their first official Association dinner that was to be held on 19 August at the Café Francais in Swanston Street, Melbourne. That night the Lord Mayor of Melbourne, Lieut-Colonel Burston, was in attendance along with Josiah Thomas, the Postmaster General, whom Henry probably avoided, and many other notable members. Dr Alexander Graham Bell, as the guest of honour, presided over the evening with the President of the Institute, Mr Jackson. The menu for the annual dinner showed some humour, as only fitting on such an historic occasion:

"In strict accordance with the rules of the Fire Underwriters' Association"
"Complaints- See Murray"

Hors d'Oeuvres Varies- Complete with starter.
Consomme Printanier- Specific Gravity, 3.14159.
Fried Fillet of Whiting- 2,600 Megohm Grade.
Calf's Sweetbread and Mushrooms- Tested to 10,000 volts A.C.
Lamb Cutlets en Aspic- Crumb Plated.
Poulet Roti- Best Annealed.
Asperge en Branche- Earthed.
Pommes Rissoles (Double Cotton Covered).
Cauliflower au Gratin- (Over Compounded).
Pouding a la Francais- No Eddy Curra(e)nts.
Glace Vannille- Safe for Bare Conductors.
Café Dessert Varies.
Rectifiers and Transmitters (Bell's)
Corporation Fuses supplied to stand fifty per cent. Overload.

This was the first official dinner of the Victorian Institute for Electrical Engineers, which had only formed in 1905, and what an honour it must have been to have Dr Bell as their first guest. He was received with great applause as he spoke of his experiences in Australia and recalled stories about the history and development of his telephone.

"It seems like a dream to me that I had anything to do with it at all. I am not a commercial man, my interest in it evaporated when it went into business hands. So it seems to me that you are talking to someone else when you speak to me about the inventor of the telephone." As he spoke, he recounted almost word for word what Henry had previously stated in Age newspaper just before his arrival in Australia, and he finished his speech by talking about the advances made since 1876 regarding the telephone. Many other speeches were made during the evening and festivities were finished off by the Postmaster General, Mr Thomas, who proposed a toast to the electrical profession. Thomas then thanked Dr Bell, perhaps with tongue in cheek, for the great help he had given him during the past week.

On Saturday 20 August, at the invitation of the Melbourne Lord Mayor, Lieut-Colonel Burston, a number of well-known Melbourne residents, including Henry, met at the Lord Mayor's Office to wish bon voyage to Dr Bell who left Melbourne later that day on the steamer *Rotomahana,* bound for Hobart. This personal visit by Bell was a highlight of Henry's life, and it is quite clear that he held the man in high regard and that Bell was a contributing influence and inspiration for many of his own inventions.

On 11 October 1910, Henry wrote to the Postmaster Generals Office requesting a renewal of his wireless licence. Nearly three weeks later, on 31 October, Henry's letter and licence agreement was sent to him to sign. At that time, Mr Hesketh from the post office sent the Crown Law Department, a copy of both. Hesketh's question to the department was about the wording of the licence agreement, specifically whether any amendments should be made and if Henry's licence should be a renewal of the existing licence or an extension of the existing licence, or whether a complete new licence should be issued. This question was then referred on to the Crown Solicitor Mr Charles Powers, who decided to refer the question on to the Attorney General's Office.

On the 7 November, the Attorney General's Office responded to Powers, who passed the information onto Hesketh.

The Attorney General saw the situation as being that as Henry's first licence issued on the 7 October 1908 was still under seal and that the extension or renewal of his original licence application in 1909 was, in effect, not in place as neither party had signed the licence agreement. It was his opinion that a new licence should be executed. He also suggested that amendments should be made to clauses 2, 9 and 13 so as to make the licence a continuing one subject to determination at any time by the Postmaster General by twenty-four hours' notice. This would obviate the necessity of having fresh agreements or renewals made should the licence be continued after the twelve months now proposed to be granted.

There are no details of Henry's reaction to the news that he needed to apply for a completely new licence and that his previous licence was, in effect, null and void at the time of issue. It is not hard to gauge the Defence Department's reaction; the single mindedness of the Postmaster General was no doubt infuriating. After being informed of the proposed changes to his wireless licence, Henry refused to sign the agreement. As the news spread about the situation, one man stood out and took a stand on behalf of Henry and all wireless experimenters and the Post Office's treatment of them.

Lieutenant George Augustine Taylor, who was a member of the Aerial League and a member of the New South Wales Intelligence Corps, along with Mr Walter Henry Hannam and others, had formed the Wireless Institute of Australia in Sydney on 11 March 1910. On behalf of the Institute's members, Taylor and Hannam had been lobbying the Postal Department in relation to wireless telegraphy experimenters and for the post office to reduce the fees for wireless licences. Lieutenant Taylor was, at the time, also in the process of applying for his wireless licence from the Postal Department and was currently having the same battles as Henry in relation to his wireless telegraphy experiments. He was also undertaking discussions with the Defence Department regarding the use of wireless telegraphy in aeroplanes and for military purposes and was currently in Melbourne, as part of his travels around the country lecturing and campaigning on "*Wireless and National Defence*". He heard about Henry's treatment by the Postal Department and was outraged. Lieutenant Taylor's anger finally boiled over and he vented his anger in his letter to the Editor which was published in the *Argus* newspaper on 18 November 1910:

Letter to Editor: George A. Taylor.

"Sir- Wireless telegraphy is to Australia as a gift from heaven. Its extensive use in this great continent of magnificent distances is more necessary than perhaps in any other country; hence it is surprising that an investigator in this vital national necessity should be threatened with a five hundred pound fine and confiscation of apparatus, unless after considerable trouble he obtains a license that will cost him three pounds and three shillings per year and is only granted under restrictions that are not only absurd but contravene common justice and gravely restrict individual freedom. For three pound and three shillings per year he is permitted to spend, say, one hundred pounds on apparatus "for the sole object of conducting demonstrations for experiments." "Sketches of all appliances must be sent in and working explained," and when one considers that an experimenter is always

seeking to improve his apparatus, one can understand what progress he can make when he has to give notice of any proposed alteration of his instruments. That is to say, if he proposes to alter his apparatus, he must send in notice, and wait for reply. In the course of time a reply comes along, and he makes the proposed alteration: but it is possible that he may find he can get the result another way; hence, under the regulations the same rigmarole must again be gone through. No doubt the postal authorities, as sensible men, are liberal enough not to insist on the regulations being obeyed to the letter, but there they stand; hence it is no wonder that the Wireless Telegraphy Act of 1905, with its drastic regulations, had almost smothered Australian invention, that when our continent four years later was shocked to the heart at the loss of the Waratah with 200 lives vainly calling for that wireless that would have saved them. Australia woke up, asked for this great national necessity, and had to get it made in Germany.

Meanwhile Australia lags in the background; ships are arriving from overseas fitted with wireless; yet there are only a few wireless land stations that have survived the regulations and can take messages. It is time a change was made; the present grip one official in the post-office has upon the throat of Australian invention should be relaxed. The absurd license fee should be abolished; no doubt it was to prevent indiscriminate interference by ignorant amateurs, but a qualifying examination prior to issue of a license should suffice without any license fee. There is no fee required in Great Britain.

The present regulations are also absurdly conflicting; there are three authorities to whom the experimenter is responsible. These regulations should be immediately amended and freest scope given to Australian investigators in wireless transmission.

Australia holds a national reputation for the originality and fecundity of invention, even in wireless. There are two Australians whom the world gives the homage that is denied them in their own country. The valuable improvements in wireless apparatus by Henry Sutton of Melbourne are now world famed, whilst the work of E. H. Kirkby formerly of Melbourne now of Sydney is also deserving of more appreciation by Australia. I may mention that the whole of my wireless apparatus was made for me by the latter gentleman, every ounce of it being Australian made. The national importance of wireless telegraphy calls for its control by a commission armed with scientific, military and business knowledge, instead of by a departmental officer, who, though an estimable gentleman, apparently cannot give his best attention to a great national necessity that now seems to be but a side line in his multifarious public duties. A commission as suggested should draft practical regulations, consider all tenders for installations and give Australia the best opportunities for developing the wonder science of modern days."

Lieutenant George Augustine Taylor
Image Courtesy of National Library Australia

All Henry had wanted to do was gift to Australia and its citizens his wireless system, but instead he received years of frustrating and at times disheartening grief. Nikola Tesla's statement that the inventor "lives and labours and hopes" rings so true, and even Henry struggled to understand the behaviour and attitude of all involved.

In an effort to ensure Henry and the Defence Department were able to continue the experiments with Henry's wireless system, a number of avenues were pursued by the Defence Department. It was during this time that the Australian Government had invited Admiral Sir Reginald Henderson from the British Navy to Australia with the view of seeking his expertise and advice on the naval defence of Australia. Captain Creswell had a vision for Australia and the Australian Navy, and seeking the advice of Admiral Henderson was a step towards that goal. Henderson was to report on such things as naval depots, dockyards, naval bases and schools as well as the use of wireless telegraphy and sites for wireless telegraphy stations around Australia. He brought with him the British Navy's wireless expert Lieutenant John Ambrose Slee, whose expertise would prove invaluable in determining Australia's wireless telegraphy needs. Henderson, who had arrived in Fremantle, Western Australia, on 6 September, had been undertaking an extensive tour of Australia, evaluating the country's defences including Australia's urgent need for wireless stations to be erected.

In February 1911, at the request of the Defence Department and Captain Creswell, Admiral Henderson was asked to review and report on Henry's wireless experiments and wireless system. He'd no doubt heard a lot about it, as by this stage Henry's wireless work had received world acclaim and had been the talk of many of the world's navies. Henderson sent his wireless expert, Slee, to visit Henry at the family music store in Bourke Street on 15 February 1911.

The next day Slee sent the following report on his findings:

"Sir, I have the honour to inform you that, in pursuance of your verbal orders, I visited Mr H. Sutton on Feb. 15, and he very kindly showed me his wireless Telegraph apparatus. He informed me that his delicate health prevented him from carrying out experiments in the open, and that his results were all obtained in the laboratory. He has apparently made many experiments with great care, and has made several deductions from them, which if properly followed up by extended practical trials might prove of value to the Commonwealth. At the same time, I must state that several of the points explained to me by Mr Sutton are re-discoveries on his part, and when his apparatus are re-constructed for the larger powers which will, in my opinion, be found necessary for practical working, I anticipate that some, if not many, of the theoretical considerations on which his designs are based will be disproved."

Although Lieutenant Slee considered that a number of Henry's designs and theoretical considerations might disprove the validity of his experiments, he felt that Henry's demonstration of his wireless system had some merit and so his recommendation was to take action and give practical tests of the system, in the open on board the Navy ships. On 25 February, the Director of Engineers, Major Cecil Foott, recommended to the Minister that a party of one officer and three Petty Officers or N.C.Os be selected from Naval and Military Forces in Victoria, jointly for the purpose of carrying out experiments with the following plan:

"That Mr Sutton be asked to lend sufficient wireless plant to the Department to carry out the trials. That the trials take place near Melbourne, and as the powers of the plant are demonstrated and the operators gain knowledge, the radius be enlarged, say from Williamstown to Port Phillip Heads."

It was at this time that the British Admiralty became very interested and after Lieutenant Slee's report and some subsequent tests by the Australian Navy, the British Navy experts became acutely interested in Henry's equipment and wireless system.

While the Defence Department arranged for more wireless tests on the boats, they also explored the possibility of Henry's honorary commission in the Navy and had contacted the Governor General, the Right Hon. William Humble Ward, about it. On 11 March, in order to ascertain the circumstances and necessity of awarding Henry an honorary commission, the Governor General wrote a letter to the Vice Admiral, George King Hall, who was in charge of the Australian Station and, at the time, in Port Adelaide on the navy ship *HMS. Powerful.*

The usual procedure was that the Postmaster General was responsible for giving the details of any new wireless telegraphy licences that had been issued to the Governor General, but the details would only be of who had been issued the licence and when. In order to answer the Defence Department's inquiry, the Vice Admiral needed to send the Governor General a copy of Henry's wireless licence agreement, and he did so on 24 March.

On 7 February 1911, Mr Henry Walter Jenvey, the Postmaster General's electrical engineer, who'd had high hopes of putting forward his own wireless system, retired from the Post Office.

To a large degree, Jenvey has been responsible for thwarting Henry's experiments and his attempts to put forward his wireless system to the Australian Government. Then with the growing pressure from the Defence Department, along with George A. Taylor's letter which was published on the 18 November 1910, and coupled with months of lobbying by the members of the Wireless Institute, the Postmaster General finally made some concessions. Finally there was a great victory for wireless telegraphy experimenters and operators in relation to their licences.

In the end, there was no need for the Governor General to grant Henry an honorary commission. On 10 March 1911, the Postmaster General announced that he would, in future, give every encouragement to Australian wireless experimenters.

The Postmaster General, Mr Thomas, promised that the present regulations in regards to private stations would be revised and a substantial reduction in fees would be made. He proposed to lower the licence fees for wireless telegraphy licences to £1/1s for the first year, with a nominal fee not more than 7 shillings and 6 pence for each subsequent year. He also stated that he would be happy to receive suggestions from the Wireless Institute regarding the best way to encourage Australian wireless research, and he would consider these ideas when drafting the new regulations. He then went on to state that, "I fully recognise the national importance of the subject and it will not suffer in my hands!" You can imagine the incredulity of everyone involved to the Postmaster General's about face, as well as the sighs of relief.

Buoyed by the Postmaster General's announcement, which he saw as an easing of the licence conditions, Henry finally signed his wireless licence agreement, which had been issued on the 23 February 1911, and back-dated to 7 October, 1910.

In mid-1910 the Postmaster General's Department issued two letter call signs to wireless experimenters, all of which had the prefix X with no distinction between the various states and whether they were experimenters or commercial wireless operators. Henry Sutton's wireless licence was No. 2 and his call sign was XLM. This call sign, which had previously been unknown to wireless historians, can now be added to the official register of Australian call sign history.

The lowering of the licence fees was a great outcome, not only for Henry, but for all other experimenters and it resulted in many other wireless telegraphy experimenters applying for licences. In February 1911, only eleven licences had been issued. By August, thirty-one experimental licences had been issued. In a win for all wireless experimenters, the new Postmaster General, Mr Charles Edward Frazer, abolished wireless licence fees altogether on 26 October, 1911.

On 3 March 1911, Henry Sutton's long-time friend and his greatest advocate, James Oddie, died in Ballarat. News of his death was undoubtedly a sad blow to Henry and, as he took the long train ride up to Ballarat for his funeral, he would have reflected on his past memories of the great man who had achieved and given so much to him and the city of Ballarat.

The funeral was attended by many well-known Ballarat residents, but for Henry the loss of his lifelong friend was akin to losing a father. The mutual respect he and Oddie had for each other was a testament to how deep and rich their relationship was. After the death of Henry's father in 1876, Oddie took up the mantle of surrogate father to Henry and he supported him and his scientific endeavours over the ensuing years. Oddie not only had unwavering support for Henry, but also for the people of Ballarat. Today he is commonly referred to as "The Father of Ballarat".

Henry resumed his wireless experiments in March 1911 and over the next several months felt free enough to express his views. He brought to the attention of the public many aspects of wireless telegraphy that evolved over the rest of the year. One of his views was in relation to the suitability of

Melbourne for wireless telegraphy. Henry considered the city to be "particularly well situated for such a purpose - much better than either Sydney or Fremantle. Melbourne, geographically, has the run of the main shipping route - Fremantle to Sydney, the South African route to Sydney, the Launceston route and the whole of Bass Strait; in fact all the really big traffic routes."

Later, Henry spoke about wireless stations and the "all red line" telegraph system that linked much of the British Empire.

"The information this morning that the British Postmaster-General has stated to the conference that an all red wireless series of stations across the Pacific is impossible owing to the British Government being tied to the Marconi system and the over sea Dominions having adopted another system will not I think prevent the linking up of the Pacific by wireless. Wireless has nothing to do with all-red routes and the most ultra-vermilion speech cannot affect the fact that wireless refuses to be chained to any particular route, country or creed."

When the issue of wireless rights came to public attention Henry stated:

"The question has several times been asked in the House of Representatives---What rights will be left to the wireless companies at present operating land stations in the Commonwealth after the Government stations are taken over? Outside territorial waters the Government has no rights between ship and ship and as there are already somewhere nearly 100 ships interstate and overseas fitted with wireless, the revenue must eventually become considerable. The owners of the systems appear to have the best of it in their arrangements with the ship owners. It is stated the operators and apparatus are hired to the ship owners, the latter bearing all the expenses and the wireless companies taking all the revenue".

[Full Texts, see End Notes, Chapter Fifteen, No. 3, 4, & 5.]

On 28 June, the Postmaster General, Josiah Thomas, announced that the office of Director of Wireless Telegraphy for the Commonwealth had been filled. Admiral Henderson had given a list of six candidates to the Prime Minister, Mr Andrew Fisher, and out of the six candidates the wireless expert Mr John Graeme Balsillie was chosen for the position. It was a controversial appointment as the British Radio-Telegraph and Telephone Company, where Balsillie had previously worked, had used Balsillie's wireless system and been sued by The Marconi Company. In February 1911, Justice Parker in the British court had ruled that the Balsillie wireless system, had infringed on Marconi's wireless patents.

The 'Parker Judgement' affected the wireless systems used by many wireless companies in countries around the world. At the time of the judgement, there were twelve different wireless systems in commercial use internationally, and Australia was on the verge of investing in its own wireless system, which could be subject to future litigation.

Mr Balsillie, who arrived in Australia from England on 17 September, was hired by the Australian Government to advise and oversee the erection of Australia's wireless stations and advise on the various systems that were presented and on the best system to use. Currently, the Australian Government was locked into the first two wireless stations in Sydney and Fremantle which were using the Amalgamated Wireless/Telefunken System, but under the Parker Judgement this system was now liable to litigation by the Marconi Company. The ongoing world-wide ramifications of the Parker Judgement was not lost on all involved, including Henry. His system, if used, could also be subject to future litigation by Marconi. Since the judgement, the Marconi Company had been busy lodging writs against many wireless companies around the world. On 21 September, Henry felt compelled to express his views on the Parker Judgement and the current state of affairs:

"The Marconi Company having won an action in which Mr Justice Parker gives the rights to the transformer to the Marconi Company. In addition he defines legally what constitutes a transformer. The Commonwealth stations are to have the quenched spark system. The quenched spark method of producing waves was discovered by Wien in 1906 and he generously gave it free to the world, hence the use of that system is free. The Telefunken engineers devised a particular way of quenching the spark and it is that particular way of destroying the spark that constitutes their system. It is possible the Marconi Company may ask the Commonwealth Government whether they do or do not use a transformer between the spark and the aerial. In that case Mr Justice Parker's definition of a transformer becomes a very important matter. The reason the Lodge 1897 prior patent may affect the Marconi 1900 patent is that the Parker judgment gives Marconi the right to a principle (not a particular apparatus). This right to a principle constitutes the Marconi patent a master patent and it is just here that I see the effect of the Lodge patent; for the principle involved exists very decidedly in the Lodge apparatus. It is possible the disclosure of this principle (tuning the resonance) in the Lodge patent enabled Marconi and others to perfect the principle of resonance in the apparatus constituting his 1900 patent. Marconi is entitled to a patent on his 1900 apparatus as an invention but the principle involved in that invention belongs to Sir Oliver Lodge, and therefore it is quite

possible the Lodge patent now extended seven years may eventually be declared a master patent."

Mr Marconi, who kept abreast of what was happening in Australia, was sent news clippings and regular letters and telegrams about what the Australian Government was doing in relation to wireless telegraphy. Still seething about their rejection of his wireless system in 1906, he once again approached the Postal Department, this time proposing to erect two permanent wireless stations in place of the smaller stations his company had earlier erected at Queenscliff in Victoria and Devonport in Tasmania. He was turned down by the new Postmaster General, Mr Frazer, because the proposed wireless stations would have an insufficient working range. The Postmaster's dismissal of his new proposal, on top of the humiliation of the 1906 refusal, enflamed the already tense situation regarding potential litigation. No doubt, Marconi was trying to circumvent the vetting process and get in the back door with his wireless system before a decision was made by the Australian Government about which system they would use.

Marconi was ferocious in protecting his wireless patents and considered anyone's system to be in breach of his patent rights. His litigation regime was incessant all around the world and tens of thousands of pounds were spent yearly on protecting his patents. Anyone with a new system was a target and likely Henry would be in his sights if the Australian government was to go ahead and use his system. The Parker judgement had not only cemented Marconi's position in England but opened the doors globally, and the Australian Government would soon learn firsthand the legal ramifications of not using the Marconi's system.

While the Australian Government agonised over the ongoing state of the wireless telegraphy situation, Henry wrote a letter to Andrew Fisher who was again the Prime Minister. Having completed more experiments and testing, Henry decided to go straight to the top instead of going through the channels of red tape of the different departments. On 17 August, Henry informed Mr Fisher of his new discovery regarding the qualities of the rectifying power of a variety of Australian minerals Henry's letter to the Prime Minister Fisher, read:

"Dear Sir,

I have much pleasure in being able to inform you that after long research into the behaviour of various minerals as wireless detectors. I find a peculiar variety of a common Australian mineral stands easily first. When I first discovered it had the desired properties, I got several friends I have who are working in the mines to search for this special variety. In one case these good helpers have sent me half a sack full to test. I also got my metallurgical friends to assist in the search. The result is we have found a source in a distant Queensland mining town. The practical Wireless results are, we in Melbourne can hear the 'Powerful' when in Brisbane harbour, the 'Luvuka' 1400 miles, the 'Suevic' 1,750 miles, the 'Karoola' 1,200 miles. Operators off the big Liners say we receive twice as loud as they do. Of course, we have my own special circuits. Still it is a valuable fact to know, this extremely sensitive variety of mineral exists in Australia (and no doubt all over the world). The mineral is superior to all the known kinds, in that in addition to its greater sensitiveness, it does not lose its power, as other extremely sensitive minerals do."

The special mineral Henry was referring to was found at the Mount Morgan Mine in Queensland. The Prime Minister's Office sent Henry a letter on the 23 August, acknowledging receipt of his letter and forwarded Henry's letter to the Postmaster General. After a number of inquiries and discussions with relative parties, the Prime Minister, wrote back to Henry on 11 October that,

"With further reference to your letter of the 17th of August last, relating to a variety of Australian Mineral which is said to have special properties as a wireless detector, I now have the honour to invite you to be so good as to submit to the Postmaster General a sample of the mineral or minerals or combination of minerals which you find are most suitable for radio-telegraphy, with a view to a test thereof being made at the Melbourne or Flinders station, as the case may be, and the commercial value of the minerals in question investigated."

Further letters were exchanged between the Prime Minister's office and Henry over the next few months and during this time Henry sought to protect his new discovery and detector. On 29 September, Henry lodged another patent, this one for "*An improved detector of electric oscillations for wireless telegraphy and like purposes.*" The particularly sensitive detector was that which enabled him to interpret signals from points as far away as Suva (2,431 miles, approx 3,912 kilometres) and the Bluff in New Zealand (1,334 miles, approx 2,146.8 kilometres).

[Full Text, see Patents, Chapter Fifteen, No. 1.]

On 13 October, Henry wrote again to Prime Minister Fisher, who then forwarded Henry's letter to the Postmaster General. Henry's letter stated;

"Sir,

I have the honour to acknowledge receipt of your letter dated the 11 Oct, re: my latest Wireless detector. The Matter being both of scientific and commercial importance, I decided to publish the facts in Europe (as I have done in previous instances). In order to secure a date and protect myself, I last month applied for a Commonwealth patent. Re: submitting a detector for investigation. When the Melbourne or Flinders Street station is ready, I shall be pleased to do this. It must however, be some time before a station can be erected at either of the above sites. I would, therefore, suggest meantime, that the detector be tested at my private station, and feel sure the Postmaster General will be pleased with the fine results we are getting. With the Postmaster General's permission, I will be glad to show Mr Balsillie what is being done, and feel sure, my station can give valuable data for the erection of the Post Office station."

A week later, Henry forwarded to the Prime Minister, six pages of some of the test findings about distances of transmission of his wireless detector which were taken from between July and October, 1911, the details were then forwarded to the Postmaster General and in part read;

"Re: The wireless receiving tests made last year at my station; These tests were made for me by Mr A. E. Sutton in order to determine the value of a new detector I had devised. It is not generally known in Australia that for several years my station has been engaged in important research work, some of the results of which have gone into use in several navies, and other of the results published in technical works. The range of my investigation has covered both transmission and reception, several new methods being devised for producing electric waves, and several methods of receiving electric waves, including novel detectors.

An automatic system was devised to enable the operator to listen at the same time as sent. This works very effectively, also methods of protecting the detectors. During the last year a brilliant discovery was made, the technical details of which are reserved for publication in Europe. I have discovered the cause of certain phenomena that hitherto has defeated scientific research. On this discovery, I have based a wireless detector which is proving itself easily the best and most satisfactory. It has three times the sensitivity of the next best. Between the months of July and October, tests were made every night and the results tabulated."

What was to follow was a series of stalling letters from the Postmaster General's Department and on the 27 October, the Postmaster General's

Secretary, Justinian Oxenham wrote to Henry thanking him for his kind offer and when the new station is built, Mr Balsillie would be glad to come and look at his new detector. To say this was a stalling tactic is an understatement the letter was merely a polite fob off by the department. During this time Henry's wireless licence was due for renewal and as per usual the Postmaster General started making it difficult again for Henry and demanded certain conditions before renewing his licence. Henry complained about the conditions although he eventually complied and was granted a renewal of his licence on 3 November, 1911.

On 4 December 1911, Marconi announced that he had withdrawn his legal action against the Telefunken Company in England. Henry no doubt saw this as a good sign and hoped that the Australian Government would finally take up his offer of his wireless system. On that same day, he wrote another letter to the Prime Minister, and once again offered his wireless system to the Commonwealth. His letter read:

"Sir, I have noted with disappointed interest the difficulties continually surrounding the introduction of Wireless telegraphic communication into the Commonwealth. Therefore, in the event of your ministry taking up an independent position, I shall be pleased to place at the disposal of the Commonwealth my private system. This system has very little, if anything, in common with what are known as the Commercial systems. My system being original from the Morse key to receiver, that is, the whole transmitting and receiving system is novel, and involves some principles that do not so far as I am aware exist in other systems, certainly not in either the Marconi or Telefunken.

Personally I consider the Parker judgement must be stretched to cover my transmitter. This of course, is a matter of opinion. I make this offer with the feeling that in the event of your Radio Department adopting anything from the results of my several years' investigations, your Ministry will recognise the value of the work."

A copy of Henry's letter was sent to the Postmaster General's office on 5 December, which was then passed on to the wireless expert Mr Balsillie. The timing of Henry's letter and when the Prime Minister passed it on is highly significant as it was only days later on 8 December the Postmaster General, Mr Frazer, announced in Parliament that they would be using Mr Balsillie's wireless system for the Commonwealth wireless stations. The chosen system (not the same system used by The British Radio-Telegraph and Telephone Company) was yet to be proven or patented, but regardless of this fact the Postal Department was fully intending to use it.

Mr John Graeme Balsillie
Postmaster General Wireless Telegraphy Engineer

Henry was mortified when he read about the Postmaster General's announcement in the newspapers on 9 December. Upset and confused by the situation, Henry sought to clarify the matter and so on 15 December Henry once again wrote to the Prime Minister, Mr Andrew Fisher:

"Sir, On December 4th, I wrote you a letter making an offer of a wireless system to the Commonwealth. At the time of that offer, I was not aware that an arrangement had been made with the Engineer for Radio Telegraphy (as per press reports December 9th and later). I have no desire to embarrass the departments, but I have thought it strange that except for acknowledgement, December 5th, from your office, that the offer had been received and submitted to the Postmaster General. I have heard nothing further of the matter. In view of the Postmaster General's announcement in Parliament on December 8th, it appears to me that either Parliament or the Press should have been informed of my offer of December 4th, which offer was sent in good faith. There is no doubt my most recent work is in advance of my work as reported by Sir Reginald Henderson's expert. At the instance of the Minister of Defence early this year, it was then acknowledged to me that in many ways I was in advance of the Admiralty, the advice being that it should be developed in the open by the Defence Department. I erected an outside station myself and the later developments have taken place in this station. Under the circumstances of the correspondence during the past three months between myself, your own and the Postal Department, I think it advisable I make optional on my part the various offers made by myself during this latter period."

Due to the Postmaster General's announcement, and while awaiting a response from the Prime Minister's Office, Henry sought to protect his new inventions and lodged his fifth wireless patent on 14 December, which was titled "*Improved means of producing electric oscillations for wireless telegraphy and other purposes*".

[Full Text, see Patents, Chapter Fifteen, No. 2.]

The Prime Minister's Office once again forwarded a copy of Henry's letter on to the Postmaster General's Secretary Justinian Oxenham and to Mr Balsillie, noting that the letter was in relation to Henry's letter and offer on 4 December. Balsillie wrote in a memo to the Postmaster General's secretary:

"I do not quite understand the communication from Mr Sutton more particularly the last paragraph in which he states "I think it advisable I make optional on my part the various offers made by myself during this latter period", but submit that he be thanked for his communication of the 15th December and it be pointed out that the policy of the Postmaster General in respect to radiotelegraphy has been decided in respect to system to be employed, but should Mr Sutton care to bring before the notice of the Postmaster General any novel apparatus comprising his system the P.M.G. will be pleased to consider same and if necessary have same tested."

Even though both men had indeed been informed of Henry's wireless work and previous offers, they both gave the impression that they knew nothing about either. Two weeks later, on 29 December, 1911, the Postmaster General's Secretary, Oxenham, wrote a letter to both Henry and the Prime Minister. Henry's letter stated:

"Sir, I am directed to thank you for your communication of the 15th, addressed to the Right Hon. The Prime Minister, which has been forwarded to this office, and to inform you that the policy of the Postmaster General in respect to radiotelegraphy has been decided as to system employed, but that should you care to bring before the notice of the Minister any novel apparatus comprising your system, he will be pleased to consider it and if necessary have it tested."

A copy of the above communication was sent to the Prime Minister with a letter stating that they had responded to Henry's letter. A certain amount of outrage and confusion followed the announcement by the Postmaster General that the Commonwealth would be using the Balsillie system for its wireless stations. This was not only in the political field but also amongst the public and in the press. Daily reports in the newspapers questioned the use of Balsillie's mysterious wireless system which no details had been furnished even to members of Parliament.

The confusion and mystery was not lessened when, mere days after the announcement, the Postmaster General Frazer stated that it was only known that the system was devised by the Federal expert and he was not prepared to say what system it was. This statement came as a surprise as, for many months, the speculation was centred on either the Marconi or Telefunken system being used, although these had been ruled out by the Postmaster General's Office.

The timing of the Postmaster General's announcement is significant. Frazer, who had come to the position of Postmaster General on 14 October, was not aware of the background on Henry's wireless system. He was influenced, and most likely pressured, by Balsillie to use his as yet unproven wireless system. The announcement came only days after Henry's offer was submitted to the Postmaster General and Balsillie. Whether or not the announcement was a strategic move by the latter to block Henry's offer of his wireless system to the Commonwealth is up for debate, but the timing is very hard to ignore. Clearly, though, Balsillie was an opportunist and he could have also seen that redemption from the Parker Judgement could be found if he produced a new wireless system that was used successfully by the Commonwealth.

Within days of the Postmaster General's announcement in Parliament, questions were asked in the House of Representatives and the Senate about Balsillie's mysterious wireless system, as his previous system had been judged to have breached Marconi's patent rights. MPs were angered by the possible legal threat to the Government if it were to use an unproven and unpatented wireless system in case it, too, breached the patent rights of either the Marconi or Telefunken system. The Postmaster General's position worsened when he refused to give any details on the proposed Balsillie system. By Tuesday 12 December, newspaper reports stated that Mr Frazer did not know if Balsillie had had a system of "wireless" in operation in England or elsewhere, or if he had patent rights to the selected wireless system before he came to Australia. In fact, Balsillie had not.

Mr Frazer went on to state that the Australian Government would acquire, free of charge, all the rights for the Commonwealth to use the Balsillie system. Any necessary legal defence of the system against actions for alleged infringements of patent rights, would be a matter for the Commonwealth. Balsillie, personally, would be free of any liability of the use of the system by the Commonwealth.

Frazer's obvious lack of knowledge about his wireless expert showed how little he knew about wireless telegraphy matters and the department he was put in charge. This did not sit well with the members of Parliament, who urged him to explain the situation and the precarious position he had put the Australian Government in.

On 19 December, during the sitting of Parliament which ran very late and well into the early hours of the next morning, Frazer was grilled once again on his wireless decision. Mr Irvine, the Member for Flinders, continued his criticism of Frazer by bringing to the notice of the House Members that if the government failed in the defence of the Balsillie system the government would not only have to pay compensation but they would be prevented from going on with the erection of the wireless stations:

Mr Frazer responded: *"Well, I make this statement that we are prepared to take that risk."*

Mr Irvine: *"But it was a risk the country was taking."*

Mr Frazer: *"The government has taken all reasonable precautions to protect the people."*

Mr Irvine: *"Has this invention been submitted to the ordinary test of the examiners in the patent office?"*

Mr Frazer: *"Does the member think we are going to lay the cards of the Government on the table?"*

Mr Irvine: *"What I ask is, has this been submitted to an examiner for report."*

Mr Frazer: *"Do you think we are a lot of dunces?"*

Mr Irvine: *"I do not desire to express an opinion on that!"*

With that the Members of Parliament erupted in laughter, but the question still remained unanswered and Mr (later Sir) Joseph Cook then stated that Irvine had done a great service to the country in raising this question. He went on to state that the government had no right to take any risk in infringing patent rights as it was an immoral act and the government was the very authority that should guard patent rights. Mr James Fenton then asked Frazer to make an effort to tell the truth, where upon Mr Edward Anstey stated, "No one should be asked to make that effort at this time of the morning!" Regardless of the early hour of the morning, truth in

Parliament was fundamental in pursuance of good governance but no truth was forthcoming from Frazer, nor would it in the months and years that followed.

Henry was right; either the Parliament and/or the press should have been informed of his recent advances in wireless telegraphy and his most recent offer to gift free of charge to Australia a working patented system of wireless telegraphy. The furore on the subject continued into the new-year with many questions remaining unanswered. At least, under the pressure from the Prime Minister, the Postal Department was compelled to look at and test Henry's wireless system which was now patented and had been developed to work in combination with other systems. On 3 January, 1912, Henry with somewhat lifted spirits wrote a response to the Post Office's letter:

"Sir, In acknowledging with thanks your communication dated the 29th in reply to mine of December 4th addressed to the Right Honourable, the Prime Minister. I would say the whole of the apparatus and combination of the same, comprising my wireless system is original and has been systematised to work in combination. I would further add that both transmitter and detector are protected and the latter, after passing the investigations of the examiners as to novelty, has also passed the stage of acceptance and now only awaits the final stage of issue. I have to thank you for the information that the Hon. Postmaster General will be pleased to consider and if necessary test my apparatus. For your information, I would add that with my resonance transmitter circuit and using 80 watts primary current, my station has communicated with the S.S. "Karoola" off Wilson's Promontory (140 miles approx 225 kilometres)"

Three days later, Balsillie acknowledged receipt of Henry's letter. The Postmaster General's secretary, Justinian Oxenham wrote back to Henry on 12 January; yet another dismissive letter thanking him for his communication and information on his wireless system. But Henry was not done yet and plans were made to bring more public attention his wireless system.

It had been a long and tumultuous journey for Australia to grasp the national importance of wireless telegraphy but finally Australia was installing wireless stations nationally and the Melbourne wireless station was nearing completion. The Government first considered building the Melbourne station near the Exhibition Building in Spring Street but that was quickly ruled out. The site chosen was in the Government Domain opposite the

Melbourne Church Of England Grammar School on Domain Road. By the end of January, the wireless station building had been completed and was being fitted out and the 182 feet high (approx 55.47 metre) wireless mast, made of Oregon pine, was being erected.

On 9 February, at 3.30pm Australia's first Wireless Station was officially opened in The Domain, Melbourne, by Governor General Lord Thomas Denman. The many dignitaries present including Lady Denman, Premier of Victoria John (Jack) Murray, State Ministers, Federal Members and Officers, as well as the Lord Mayor of Melbourne, Thomas James Davey and the Lady Mayoress, were all greeted by Postmaster General Frazer. A number of press representatives and other invited guests also attended the opening ceremony.

Balsillie, the Commonwealth wireless expert and inventor of the wireless system, assisted with the sending of the first messages. Lord Denman opened the Melbourne wireless station by handing the wireless operator the first communication, which was addressed to Vice Admiral King-Hall, the Commander-in-Chief of the Australian squadron, who was on the navy ship the Drake which was anchored in the Derwent River in Hobart:

"I am glad to send you the first message transmitted from Commonwealth wireless station Melbourne. I trust that this new means of communication may prove of material advantage to the naval forces of the Commonwealth, and emphasise the cordial co-operation between the Commonwealth Government and the Admiralty."

The next communication was to the Governor of Tasmania, Major General Sir Harry Barron:

"I send you cordial greetings and good wishes, through the medium of the Commonwealth wireless station at Melbourne. On the completion of the Hobart station, I look forward to increased facilities for communication with Tasmania and a substantial advantage to the trade and commerce of your State."

While the guests were awaiting responses from Hobart, Mr Frazer entertained them during afternoon tea in a marquee which had been erected on the lawn. The Prime Minister, Mr Fisher, proposed a toast to the Governor General's health and the Governor General, in turn, proposed a toast to wireless telegraphy, the Government and the officers connected to the enterprise. Unfortunately for Frazer and Balsillie, things did not go according to plan and by the time the festivities had come to an end there had still been no response to any of the wireless messages sent.

The wireless officers kept sending the messages for over six hours and it was not until 10pm that night that it was reported that the *Drake* had picked up parts of the Melbourne wireless stations message and a response had finally came in from Vice Admiral King-Hall. The night wore on and there was still no response from the Hobart wireless station. Although Frazer announced that the opening of the station was a success, this did not prove the case. The press and public must have thought that it would have been quicker to use homing pigeons to send the messages to Tasmania.

By morning it was discovered that the *H.M.S. Drake* had sent word that they had not received the Melbourne Station's message at all on 9 February; in fact, it was only during the tests on the previous day that the ship picked up parts of any transmission sent. Frazer was faced with a barrage of questions and ridicule, both on the length of time it took to try and get a message through and also on the failure of being able to transmit a message to the Hobart station or the *Drake*. In desperation, Frazer cited many different reasons to explain the delay in sending the message and still tried to insist that the wireless station was a great success. However there was more humiliation in store for him and Balsillie.

Up until this point, the Postmaster General's Department had not tested Henry's wireless equipment as the Prime Minister had directed them to do. Now that the Balsillie system had failed in its purpose to communicate, it was time for Henry to act. In this he was secretly aided by the Editor of The Argus newspaper and the plan had been made since the 18th of January.

On the morning of 10 February 1912, after witnessing the long delay and failure of the Melbourne station, the Editor of The Argus newspaper published the well-planned article about Henry Sutton's very successful wireless system, and noting his successful work over the years. This article was republished in newspapers all over Australia and New Zealand in the following days.

Wireless Experiments: A Melbourne Inventor.

"A number of improvements in methods of transmitting and receiving have from time to time been invented by Mr Sutton, some of which have gone into use in various navies, while others have been described in technical works. In the 1911 edition of Dr James Erskine-Murray's standard handbook of Wireless

Telegraphy for example it is mentioned that a new detector constructed of two pieces of galena, was discovered independently by both Mr Sutton and the well-known English expert Dr W. H. Eccles. After Dr Eccles had announced his discovery he learned that Mr Sutton had previously patented the same device in Australia, having invented it a year or two before March 1910 - the date of the patent. With the disinterested courtesy of the true man of science Dr Eccles at once gave all the credit to Mr Sutton and warmly congratulated him on his success. Mr Sutton's inventions are concerned with the production as well as with the reception of the electric waves and in his own operating room may be seen at work an ingenious automatic system which enables the operator to listen at the same time as he sends messages.

Mr Sutton has written a full account of the discovery but is reserving it for publication in a European technical journal. Upon the discovery however he has based a new detector which was recently patented in the Commonwealth and is showing remarkably good results. It is a great advance on the patent of 1910 and the discovery makes its multiplication easy. From the records of its work Mr Sutton has reason to claim that it is somewhere about three times as sensitive as the next best in use."

[Extended Text, see End Notes, Chapter Fifteen, No. 6.]

After the official opening of the Melbourne Station, officers of the Central Postal Administration had doubts about the capacity of the station and did not hide their disappointment at the failure of the Melbourne wireless station to communicate with Tasmania and publicly declared so.

Henry, having heard on his home wireless system of the constant difficulties the Melbourne station was having communicating with Tasmania, decided he needed to help. On the 12 February, he took his wireless equipment to the Melbourne Wireless Station to offer his assistance. Balsillie, who was still trying desperately to communicate with Hobart, had little choice but to accept Henry's assistance - if only not to lose more face and credibility in front of the wireless operators. So on that day, under great secrecy, Henry's wireless receiver and new detector was installed in the Melbourne Wireless Station. On 13 February 1912, wireless communication was finally established between Melbourne and Hobart. This information was never made public and until recently had not been known in the history of Australian wireless. At the time though, a few people were aware of this information and on the 7 March, Henry wrote to Editor of *The Argus* and informed him of what had occurred. His letter was a private one and was never published or made public. It read:

"Dear Sir, So that you may be kept advised as to the progress of my last wireless detector, I take the present opportunity of informing you (not for publication) that on the 12 February, noting the receiving difficulties of the Domain Station, I loaned that station one of my new receivers, and this has been so successful that from the 13th February to the present, all the work of the station has been carried on with it. The initial difficulties of the opening days of the station being instantly removed: With this detector, the Domain Station was the first land station to receive Macquarie Island, which happened at 9PM, February 14th. My private station was closed that night till 11pm, and received Macquarie at 11.15pm. So the Domain station "hoist me with mine own petard" Please treat this as confidential."

When communications were finally established, the officer in charge of the wireless station reported to the Postmaster General that the station was now working with complete success. No doubt it must have been a proud moment for Henry after all the years of grief, even though the public and the world never knew.

While Balsillie received praise in the press for the eventual success of the Melbourne wireless station, Henry's role in that success was never divulged. For Henry, though, it was not about the glory, or the money; it was all about helping his beloved country Australia.

On the back of his success at the wireless station and after receiving the Commonwealth acceptance of his patent, Henry sought legal advice. Under patent regulations at the time, you were required to demonstrate or prove that you were going to commercially work your invention. Henry felt the time was right to once again approach the Postmaster General about his wireless system. The terms which his legal team advised were not exorbitant and were quite reasonable given the market place. Although Henry had offered, many times, to give his system for only the reimbursement of costs, he had learned his lesson and he felt that some reward should come his way and in setting a price for his invention, it fulfilled his patent obligations. For the Australian Government, it was still a bargain offer and Henry set out the terms on the 18th April when he wrote to the Postmaster General:

"Sir,

The Commonwealth patent for my new Wireless detector being now sealed and in my hands, and a full technical description of the investigation leading up to the invention having reached London for publication. I have the honour to submit to the Postmaster General the first offer of the right to use the patent in the Post Office Radio department, on the terms as specified below:-

As the Domain station has had on loan one of these detectors, in regular use, since the 13th February, the officers of the department will be aware of the detector's capabilities, in the way of increasing the range and serviceability of the Wireless Station. The communication sent to London, will, I have no doubt, excite considerable scientific and commercial interest, as the investigation has been a lengthened one and discloses many new facts concerning the subject. The communication, will, I think, still further show, that successful Wireless research can be carried out in Australia. My private station has now received up to 3,000 miles with this detector. The terms on which the patent may be used in the Commonwealth Land Stations are – A Royalty of £100 (one hundred pounds) per annum, per station, and a fixed sum of £6,000 (six thousand pounds) for the right to use the patent."

The amount per year quoted by Henry was very low and the fixed sum of six thousand pounds was to recoup the cost of his research and equipment. To say Henry was shocked by the return letter is an understatement as it simply beggared belief. Henry, who was now very ill, was clearly shaken by the letter from Mr Charles Bright, the Acting Secretary of the Post Office, the letter stated:

"Sir,

With reference to your letter of the 18th ultimo, offering to this Department, on certain terms, the right to use the patent wireless detector invented by you, I am directed to inform you that satisfactory results have all along been obtained with your patent detector, but no appreciable increase in the range of signalling has been obtained through its use. When compared with the ordinary crystal detector, the chief feature of your patent is reliability, but the very small gain occasioned through its use could not warrant the Acting Postmaster General even considering an offer as contained in your letter under notice.

2. The detector no longer interests the Acting Postmaster General, and if you will be so kind as to call at the Wireless Station, Domain, Melbourne, the detector which you were good enough to lend the Department will be returned to you."

Up until that time, there had been no questions about Henry's detector and clearly it had served the wireless station well until then. It should be remembered at this stage, that the public and the courts did not know about Henry's wireless equipment being used. If the Postmaster General was to pay for Henry's wireless equipment, the public would soon find out and many questions would be asked. It is most likely that the Acting Postmaster General, Mr Charles Bright, had sought legal advice and was told that if he continued to use Henry's detector the Department could be liable to pay the amounts Henry had stipulated. Henry went to the Wireless Station immediately upon receipt of the letter and picked up his detector, but he did not go down without a fight and on the 7th May, he wrote a very curt letter back to the Acting Postmaster General.

"Sir, I have read with interest, your letter of 6th May, but cannot agree with same in re the qualities of Wireless detectors. Prior to my detector being used in the Melbourne Domain Station, the detector there used was Mr Pickard's patent 'Perikon' detector, coupled to a tuned transformer circuit of the pattern used by Ducretet, Telefunken Company, and others. I venture to say, that when my detector is coupled to such a circuit, a considerable practical of range and reliability is inevitable. This, has already been demonstrated outside of Australia. You must, therefore, pardon me taking up the defence of the technical ability of the detector. At your request, I have removed the detector from the Domain Station, and note the Acting Postmaster General's decision, that he takes no interest in the matter."

The response to Henry's letter came on the May 20, once again dismissing Henry's patent. The letter which contained information from Mr Balsillie, also stated that it was not Mr Pickard's patent '*Perikon*' detector coupled to a tuned transformer circuit of the pattern used by Ducretet, Telefunken and others which had been previously used at the Melbourne Wireless Station. Henry who helped set up his detector at the Melbourne Station at the time, clearly knew what type of detector was being used as Mr Pickard's detector had been patented and in commercial use since 1906, regardless of what had been written in the letter. Due to the secrecy behind what was being used at the station no one in the public or government knew anything about Mr Balsillie's system which had yet to be patented. Certainly the Australian Government did not want any details leaked to the public either due to the Marconi litigation.

Over the next week or so there were many telephone conversations between Henry and the Postmaster General. Henry took it upon himself to present his case, in writing to this time, to the Postmaster General. Henry's letter dated 29 May, 1912, read;

"Sir, I have the pleasure in handing you a copy of the 'Electrician' of April 19th, 1912. Page 66 contains an account of the research carried out by myself, and which resulted in the production of the new detector. The detector accompanying the communication to the Editor of the 'Electrician' was handed to Dr Eccles, whose authority on these subjects is unquestioned. Dr Eccles has kindly written and congratulated me on the sensitivity and powers of the detector, and states it is more sensitive than the most sensitive he has to compare it with, and that he is about to test it for endurance. I would further add, that the financial part of my letter of April 18th, 1912, was written under advice, on the same day as I received the sealed patent, in order to cover the patent act provisions re; working patent, and leave me free to handle the detector in any way I pleased myself."

In response to Henry's letter, on the 4th June, Balsillie rang Henry and stated that he was prepared to advise the Postmaster General to offer him £10 for his patent. Balsillie's gall was astounding and for a gentleman, his offer was totally insulting and he knew it.

Henry's last and final letter was addressed to Mr Balsillie and it stated:
"Should the Government suitably reward the time, experience, and expense of these investigations, I would be prepared to allow free use of the discovery and associated invention to the Commonwealth departments and the Commercial Shipping by freeing the patent".

The lack of interest by Basillie had another unexpected outcome. The following article appeared later in the year.

An Amateur Scientist Beats Commonwealth Wireless
Sydney Sun, 5 August, 1912.
"The death of the late Mr Sutton, music seller and amateur scientist, of Melbourne, recalls a curious little incident in connection with the wireless telegraph station in the Melbourne Domain. The late Mr Sutton, who was a fine type of the pushing but yet thoughtful young Australian, was an amateur electrician of no mean order, and his wireless installation at his residence in the eastern suburbs could with ease pick up messages that passed the big, expensive

station unheeded. As a matter of fact, it is admitted that only at night can any good results be got from the domain wireless station, even Hobart messages, from less than 300 miles away, being dumb to the government station during the day. Among other matters in connection with wireless, Mr Sutton had discovered that the use of certain pyrites in the manufacture of receiving instruments had a wonderfully stimulating effect. Another man who is regularly in the trade, discovered where Mr Sutton got his pyrites, and entered into trade in a small way by purchasing the pyrites at 1 shilling per pound and doling it out at 2 shillings and 6 pence for a small quantity, the purchasers being the wireless station. Suddenly the seller, a well-known metallurgist in Richmond, declined to sell except to Mr Sutton, and now that gentleman has gone the department is in a fix, more especially as the supply is very limited, and as yet the exact source of supply cannot be traced. The whole incident well illustrates the ineptitude of the wireless people, an ineptitude that, according to all accounts, is of surprising magnitude, and makes the station practically useless."

The difficulties with the Balsillie system were eventually sorted out but it was not until October 1912, that it finally passed the patent process. The Australian Government and the Postmaster General, Mr Frazer, always maintained that all of the apparatus in use at the Melbourne station were the complete invention of Mr John Graeme Balsillie. Given the political climate and the impending litigation by the Marconi Wireless Company, you can almost understand the Government not making public Henry's role in establishing Australia's first wireless station. The Postmaster General did not want to add fuel to the fire as to what was actually in use by the government. Henry at least, had proved a point with the Postmaster General and Mr Balsillie, who can be imagined to be embarrassed by the poor performance of the Balsillie system, although they never publicly said so. The ramifications of the use of Balsillie's wireless system is a matter of wireless folklore as it spawned the start of what has become known in Australia as the 'Wireless Wars'.

The Australian Government did however, pay the price for utilising the Balsillie wireless system. On 14 February 1912, immediately after the Melbourne wireless station opened, a writ citing breaches of patent rights was issued against the Australian Government by the Marconi Wireless Company, which had by this stage merged with the Telefunken Company. The Marconi Wireless Company were desperate to know if any part of Balsillie's wireless system utilised any of their patents, but for the next two and a half years, the Australian Government refused to disclose anything about the Balsillie system.

Instead, the Australian Government continually informed the Marconi Company and the courts that every part of the wireless system in use in the Melbourne and other wireless stations was the complete invention of Mr Balsillie. This misnomer was repeated almost daily - in the Australian newspapers, to the courts, to the Australian people and especially to the Marconi Company. The history of the Wireless Wars has been retold by many people over the years and can be read elsewhere. In short, the long protracted legal case went on until 1914, by which time it became apparent that the litigation was doomed due to the threat of the approaching war. The Australian Government, whilst never publicly announcing that it had or had not breached any patent rights, settled with the Marconi Wireless Company on 18 July, 1914. Only ten days later on 28 July, World War I began. The Commonwealth Government agreed to pay £5,000 for the use of the Marconi and Telefunken patents and the two companies agreed not to attack the validity of the Balsillie system.

The Marconi Wireless Company remained blissfully unaware of Henry's detector and receiver being used in Australia's wireless stations, regardless of the fact that on two occasions in 1912 information was published in the *Australian Mining Journal* by the Australian geologist and biologist, Mr Thomas Sergeant Hall. Henry had previously sought the expertise of Mr Hall in understanding and identifying the various qualities and rectifying powers of Australian minerals such as iron pyrites and the crystal galena. It is hard to fathom how the Marconi legal team and Marconi's representative, Mr Ernest Fisk, missed these published articles considering the lengths they went to, to try and uncover the details of the system used. Nonetheless, even the Australian public and the courts did not pick up on the published information about Henry's wireless equipment being used. It could be argued, then, that the Australian Government, the Postmaster General and Mr Balsillie were all guilty of perjury in saying that every part of the wireless system in use at the Melbourne Station at the time was invented by Mr Balsillie. How different Australian wireless history would have been if the public had known about the successful use of Henry's detector at the opening of Australia's first wireless station, as Henry's contribution would now be Australian wireless folklore instead of lost to Australian wireless history.

Regardless of the fact that the Postmaster General and Australian Government did not in the end take up Henry Sutton's wireless inventions, it did not mean that no one was interested in them. On the 1 July 1912, Henry received a message from the Defence Department stating that Major Reginald Harvey Henderson Boys from the British War Office wanted one of Henry's new detectors for trial. Conveniently, he now had a spare detector – the one that was no longer being used at the Melbourne Wireless Station. Various navies, including the British Imperial Navy, had utilised many of Henry's wireless inventions over the years and now the British Army were interested. The timing of Major Boys' request is extremely significant as it was right at this time that the British Royal Engineers had begun to take wireless telegraphy seriously.

In May 1912, the British Army formed an Army Signal Service as part of the Royal Engineers and had set up a wireless telegraphy section at the Aldershot Military Base in England and had begun experimenting. In the process, the Royal Engineers set up two new corps within the army - one was the Army Signal Corps and the other became the Royal Flying Corps. Both corps needed wireless telegraphy equipment as, up until this point, little emphasis had been placed on its usefulness. It was clear to Britain that war was looming and a few months later in August 1912, the British politician Sir Henry Norman set up a Wireless Committee to look into the merits of various wireless inventions in order to determine what would be suitable for the British Army. The committee members were, Major Reginald H .H. Boys DSO. RE, Mr E Russell Clarke, Barrister-at –law, Mr H. A. Madge, the wireless expert from the British Naval ship the "Vernon", Colonel R. D. Whigham, DSO of the General Staff, and Commander Arthur Edward Silvertop. The committee became known as the Norman Committee and over the next year the committee members looked at the value of the various wireless inventions available. Henry's new wireless detector was of great interest to the Royal Engineers after the publication of the details in the British Journal the *Electrician* in April 1912. Unfortunately it is not known what actually transpired between Henry, Major Boys and the British Engineers, although future research may one day uncover more details about Henry's step into British Military wireless history.

The first details published about Henry's wireless telegraphy equipment and inventions appeared as a large article accompanied by a full page of photographs in a May 1910 edition of the Melbourne Leader newspaper. Published two days after the Japanese Navy left Australia, it's possible the article might have been another attempt by Henry to highlight his wireless telegraphy work which was still being hampered by the Postal Department. At the time of the Japanese Navy's visit, the navies of many other nations were very interested, or were already using, some of Henry's inventions. These included both the Australian and American Navies. The article and photographs highlighted the extent of Henry's years of work as well as his wireless inventions. It gave the public some insight into his success, because up until this point his wireless system had only been vaguely reported in newspapers. However, the details of how it worked were not divulged in the article, which merely described each piece of wireless equipment in the photographs and gave some insight into the equipment's function:

Research in Wireless Telegraphy:

"We publish among our illustrations this week pictures showing some of the instruments designed by Mr Henry Sutton, a most indefatigable worker in this branch of scientific research. He has kindly supplied explanatory notes which will enlighten the reader as to the nature of what is being done."

[Full Text, see End Notes, Chapter Fifteen, No. 7.]

[See numbered pictures on next page]

Picture 1: A. An Arc Lamp and Blast Magnet. **B**. A Variable Condenser and Inductance Box. **C**. Inductance Coil. **D.** Transmitting Microphone. **E.** Receiving Wireless Telephone.

Picture 2: A. Universal Receiver. **B**. Extra Sensitive Receiver. **C.** Portable Transmitter.

Picture 3: Henry Sutton listening with his Portable Wireless Detector.

Picture 4: High Powered Transmitter Producing Fundamental Tuned Waves. **A**. Ten Inch Spark Coil. **B.** Tuning Apparatus.

Picture 5: 300 Meter (approx. 328 yards) Wave Military Set. **A**. Coil. **B**. Key. **C.** Transmitter. **D.** Receiver. **E.** Field Receiver.

Picture 6: Wireless Research Instruments. **A.** Professor Ruhmer's Oscillograph. **B.** Professor Flemings's F.R.S. Cymometer. **C**. Professor's Ramsay and Dewar's Rare Gas Neon Tube. **D**. Thermo Galvanometer.
E. Marconi Milliammeter. **F.** Marconi Special Rheostat.

After Vice Admiral Henderson's report in February 1911, and the change in attitude of the Postmaster General in relation to wireless experimenters, Henry needed a larger wireless tower to undertake the more extensive experiments he was now able to do. As it happened, Henry's home in Erskine Street, Malvern, was ideally suited to erect a larger tower. The house fronted onto Erskine Street and backed onto Murray Street which at the time was clear of any buildings. To the rear of his property, there were still

1
A
B
C
D
D
E
2
A
B
C
3
4
A
B
5
A
B
C
D
E
6
A
B
C
D
E
F

large blocks of land that were yet to be developed and so there was enough space for a larger tower.

Up until 1911, Henry had not installed a powerful transmitting plant at home because he felt that wireless telegraphy belonged to the Government and he preferred to spend his money on his research instead. When Henry was granted permission by the Defence Department to develop his wireless system in the open, he changed his mind and set about erecting a larger outdoor wireless station and aerial. Apart from the tests carried out by the Navy on board their ships and at Victoria Barracks, Henry's wireless experiments up until then where all conducted on a small scale indoors. Now it was time for him to expand his experiments to prove the worth and capabilities of his wireless telegraphy discoveries. So early in 1911, at great expense to himself, Henry erected a much larger outdoor wireless station at his home in Malvern. His wireless aerial was initially only a tall mast, 50 feet (15.24 metres) high, but in order to undertake more research he extended it to 102 feet (31.09 metres) in height and converted it into an umbrella aerial. With the land elevation at his home, the new aerial was 250 feet (76.2 metres) above sea level and had 2,000 feet (approx. 609.6 metres) of steel wire cable. The aerial mast was 18 inches (45.72 centimetres) at the base and tapered to 5 inches (12.7 centimetres) at the top.

"I have designed my aerial with a special object," Henry commented, "and that is the ribs have such an indination (indentation) as to cut the wave front. Instead of the aerial being flat, the ribs have an angle of 45 degrees to the two central vertical air wires. There are 16 ribs in groups of 4, the ribs being 100ft long and the four wires of each group arranged as a rectangular cage with 3ft sides. The aerial appears very efficient both in sending and receiving messages. My transmitter is a resonance transmitter circuit using 80 watts of primary current."

Henry's laboratory with his wireless equipment was positioned in an upstairs room at the back of the family home. The walls were lined with shelves full of his research logs, as well as many books, scientific journals and electrical treatises. The transmitting and receiving apparatus was connected to the umbrella aerial outside, which could be viewed from the window.

In wireless telegraphy, many minerals were utilised over the years in the construction of wireless detectors and these detectors became known as cat's-whisker crystal detectors. The first use and patent of a crystal detector

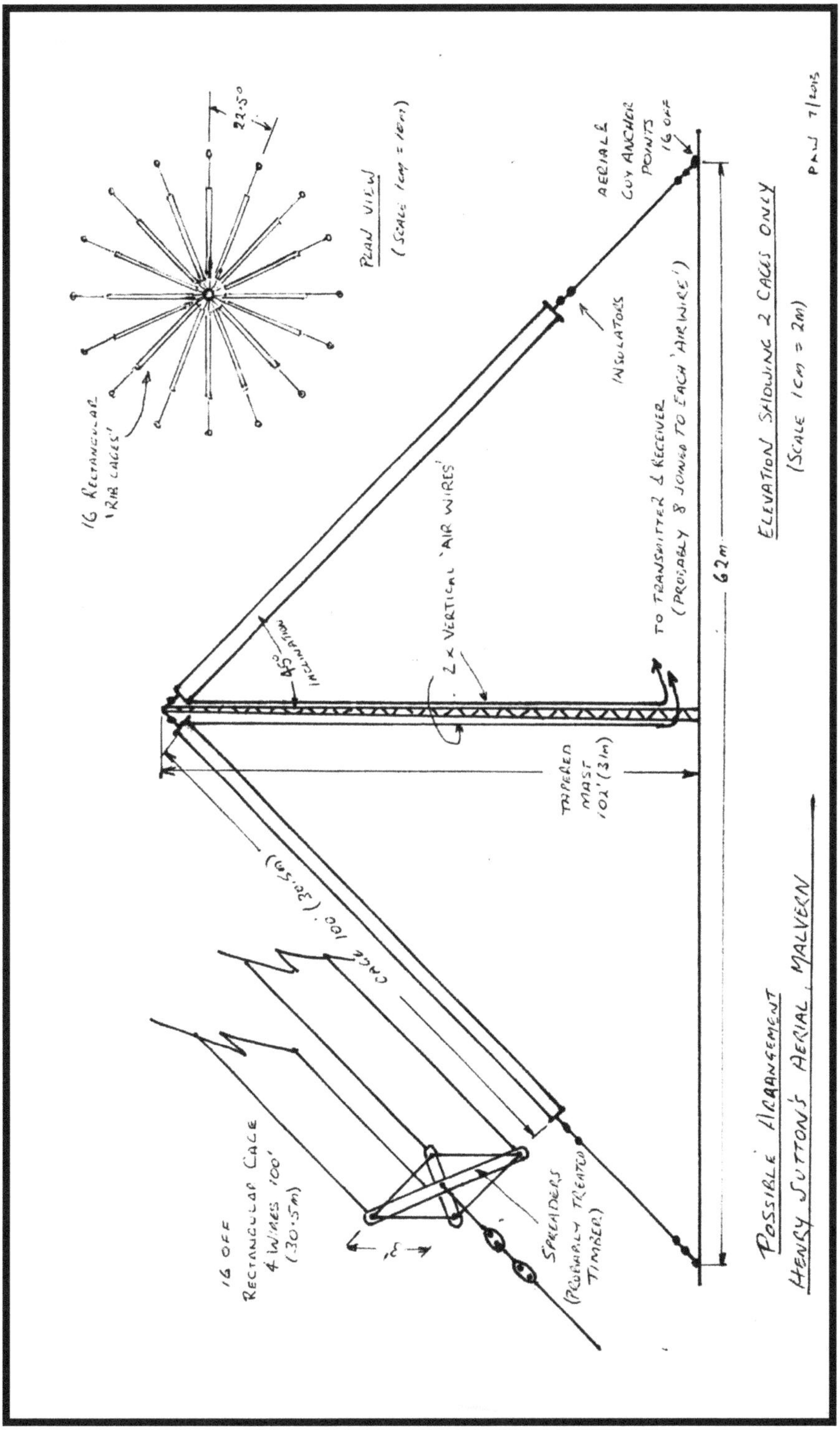

Henry Sutton's Wireless Tower. Drawing Courtesy of Peter Wolfenden

was in 1901 by the Indian professor of physics Jagadis Chandra Bose, who demonstrated the use of the mineral galena (lead sulphide) crystals and a metal point contact. In 1906, a number of other scientists also lodged detector patents.

The German physicist Karl Ferdinand Braun had discovered and demonstrated the rectification effect of certain minerals in 1874. While he had found no useful application for his discovery until the advent of wireless telegraphy around 1901, he patented a detector which was based on psilomelane, a hydrated oxide of manganese.

In the United States, Louis Winslow Austin filed a patent for a tellurium-silicon detector, General Henry Harrison Chase Dunwoody patented a carborundum detector, and Greenleaf Wittier Pickard filed a patent for a silicon metal rectifier.

To rectify is to convert AC (alternating current) into DC (direct current), the by-product is not only DC but sound waves as well. Minerals are often semi-conductors or used as rectifiers, in early wireless sets. The wire or 'cat's whisker' as it became known consisted of a thin wire that lightly touches a crystal of semiconducting mineral, such as galena, to make a point-contact rectifier. The crystal surface functions as rectifying junctions and is only useful at certain points of the crystal. What a rectifier does in relation to wireless is convert the radio frequency into an audio signal that can be heard in headphones or speaker. The crystal detector is very sensitive to the contact between wire and crystal and would need to be adjusted for it to rectify. In the evolution of wireless telegraphy, the mineral galena became known as the first semiconductor. During his years of working with wireless telegraphy, Henry experimented with many minerals but he had the most success with iron pyrite (iron sulphide) which has the chemical symbol FeS2 and galena (lead sulphide) which is the natural form of lead sulphide and has chemical symbol PbS. Even though galena detectors were already in use and had been for quite a while, he used his discovery to amplify the radio waves and make his detector more sensitive. This in turn led him to lodge his five wireless patents and it would be the one he lodged on 19 March, 1910, that provided him with the distinction of being the first in the world to discover that crystals could be made to oscillate wireless waves.

In September, 1910, Henry received his August issue of the journal *The Electrician* from England and was surprised to read that the well-known English wireless expert, Dr William Henry Eccles, had published a paper titled "*An Oscillation Detector Actuated Solely on Resistance Temperature Changes.*" Unfortunately for Eccles, Henry had already observed the phenomenon and, based on that discovery, had patented his galena to galena detector. Henry, feeling he needed to set the record straight, wrote to *The Electrician* on 29 September explaining that, in actual fact, he had made the same discovery before Eccles.

Upon reading Henry's letter, which included evidence of his claims, the Editor of The Electrician contacted Eccles for a response. Given the evidence, Eccles had no choice but to lodge a formal public apology to Henry for claiming the discovery as his own. Unfortunately, the letter of apology only went so far. In order to cover his personal and professional embarrassment and the humiliation of having to apologise, Eccles then attempted to claim the high ground and save face by pointing out faults in Henry's technicalities. None the less, on 16 December, *The Electrician* published Henry's letter along with Dr Eccles's not so humble apology.

To the Editor of The Electrician:

Sir: *In your issue of August 26th is published an abstract of Dr Eccles' communication to the Physical Society on "An Oscillation Detector Actuated Solely by Resistance Temperature Changes"; the detector being galena to galena contact. It is a coincidence that Dr Eccles and myself appear to have been conducting researches about the same time with the one object. I was not satisfied that crystal detectors are actuated by thermo-electric effects; therefore, I searched for detectors with loose contacts of the one mineral and found galena (sulphide of lead) most effective and suitable for wireless purposes, both in actuating a relay and in telephonic circuit.*

On March 19, 1910, I applied for a provisional patent in the Commonwealth, which was accepted. Dr Eccles' Physical Society Paper is dated May 27, 1910. My results were known to officers of two Government departments here in 1908, and actual details of my galena to galena contacts were known to officers of the Defence Department in 1909.

I found the resistance of galena contacts reduced under the action of electric waves and, further, found, this reduction was due to pressure caused by increase of temperature expanding the galena at the points of contact. The experiments proved to me that an electric current will heat up and produce an increase of mechanical pressure between a pair of galena contacts. It was further shown that to get the best results in the telephone receiver using galena to galena contacts as a detector of oscillations, there is a relationship between the initial mechanical pressure and the E.M.F. of the steady current, the weaker the steady (or applied) current, the more delicate must be the initial pressure, and the E.M.F., the loudness of the sound increased. The E.M.F. could go up to 4.5 volts if the pressure of contacts were adjusted. The detected wireless signal could then be heard with the receiving telephone one foot from the ear.

The experiment with which I was able to demonstrate that galena contacts expand under action of temperature due to an electric current heating up the junction is simple.

In Fig. 1, A B are two insulated No. 28 copper wires. These are twisted for about 1 in., the ends bared and conductively attached to two cubes if galena, g. The other ends of these wires pass through a thick piece of cork, C, glued to an upright. The cork is to deaden mechanical vibration and the twisting of the wires is to prevent the transmission of any mechanical vibration that may pass the cork. The galena contacts are then in a mechanically damped condition.

The circuit is arranged as in Fig. 2 in which g is the galena contact, B the battery, P the potentiometer, and T the telephone. By adjusting the mechanical pressure at g and the E.M.F. by the potentiometer P, a position is found at which, on closing the circuit, the telephone gives out a musical note (whistle), showing that an alternate increase and decrease of pressure occurs at the junction.

This could only come about by the temperature at the contact expanding the minute points of galena and kicking the contacts apart by the sudden pressure; the contacts return, become again heated and again kicked off with rapidity sufficient to produce the high musical note. The action being precisely the same as the Trevelyan rocker experiment in which a hot fluted copper bar is placed on the edge of a block of lead and heats the contact, the sudden expansion of the lead throwing the bar up, the hot bar falls and touching the cooled lead, again heats it, the alternate heating and cooling of the lad keeping the bar rocking. The bar

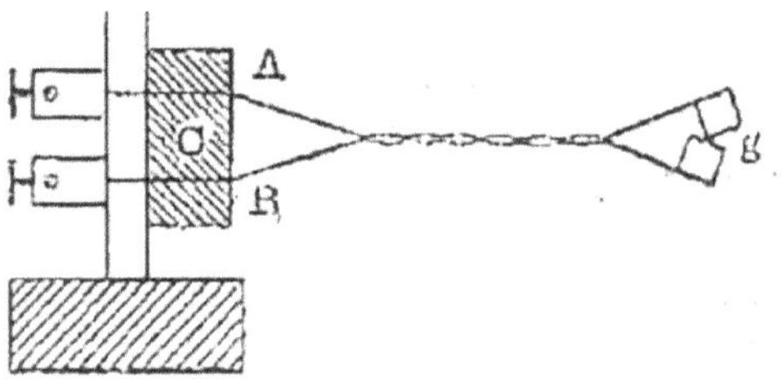

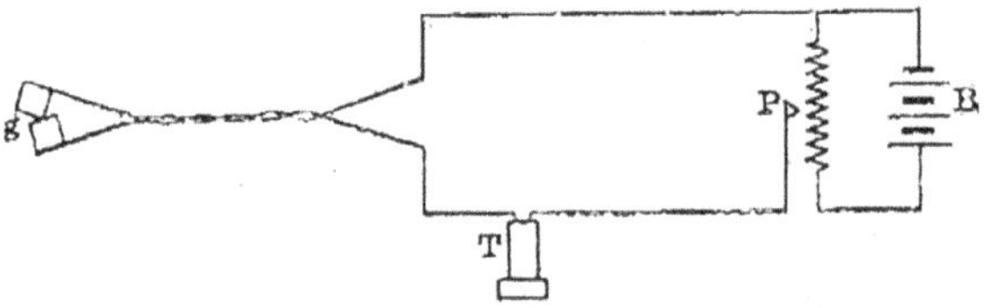

Fig. 2.

Galena Detector Figure 1(top) and Figure 2 (below)

then gives out a musical note owing to the high rate of vibration. I found that a pair of galena contacts arranged in delicate flexible contact, as in Fig. 1 acts the same as the hot copper-lead rocker.

In the case of galena contact used for detection of electric oscillations, I find the initial mechanical pressure must be greater than that at which the rocker effect occurs. It seems to me, that, in the case of the galena to galena detector, the reduction of resistance is due to pressure of contacts produced by expansion of the minute point of mineral at the contact, the latter being heated up by the oscillatory current. When receiving with this detector, the first group wave produces a considerable coherence, weakening the effect of the following group waves. This initial coherence is sufficient to keep a relay closed, and although the first group wave has reduced the resistance sufficiently to hold the relay closed, there is still sufficient de-coherence and coherence with each subsequent group wave to affect the telephone clearly, so that with the galena to galena detector we get the important dual effect of the first group wave ringing a bell or lighting a lamp, and the subsequent group waves giving the message in the telephone in Morse code. This experiment I made in the presence of the Postmaster-General and the Chief Electrical Engineer to the Post Office, October 6, 1908, and prior to that the apparatus was in the hands of the Defence Department for private trial. I enclose an authorised copy of my provisional patent, of March 19, 1910.

Henry Sutton, Melbourne, Victoria, 29th Sept. 1910.

"Our correspondent also enclosed a photograph showing wireless telegraphic apparatus, including galena receivers. We forwarded a copy of this letter to Dr. Eccles from whom we have received the following communication:

To the Editor of the Electrician: Dr Eccles

"Sir; I thank you for giving me the opportunity of being the first to congratulate Mr Sutton on the fact that his patent for a galena detector ante-dates my Paper. I may add that it would have been a matter of regret to me if the Paper had, by prior publication, injured a meritorious inventor. Your correspondent's object was technical, mine was scientific; but while achieving a great technical success he offers an explanation of the mode of action of the galena detector. He suggests that the fall of resistance at the contact is due to the increase of mechanical pressure there through thermal expansion produced by the current. This possible effect, and other conceivable ones, I include in the "negative temperature coefficient of resistance of a contact" used in my paper, but inasmuch as I have failed to find a useful oscillation detector constituted of a contact between

substances possessing positive coefficients in the bulk, I believe we always have to deal, in good detectors, with bulk negative coefficients of resistance."

[Full Text of Dr Eccles' letter, see End Notes, Chapter Fifteen, No. 8.]

It was clear that, at the time, both Henry and Dr Eccles had differing points of view. Their experiments took different approaches but, none the less, it was Henry Sutton who was the first to discover that crystals could be made to oscillate and who invented the galena to galena detector. As history would have it, over a century later, it was the publishing of Henry's letter and Dr Eccles apology letter on 16 December that can now shed light on Henry Sutton's position in wireless and electronics history.

Dr William Henry Eccles,

After setting the record straight, Henry continued his experiments with galena and iron pyrite. During the following months, he further advanced his discovery and his wireless system and by 1911, he came up with an even more sensitive wireless detector and receiver which had the capacity to transmit and receive messages over even longer distances.

Given the fact that the Postmaster General announced, late in 1911, that they were going to use the Balsillie System of Wireless Telegraphy and not his own, Henry lodged two more patents to protect his latest discovery and new detector.

[Full Text, see Patents, Chapter Fifteen, Nos. 1 & 2]

On 6 March 1912, just after the Melbourne Wireless Station opened and when his receiver and detector were working well, Henry decided it was time to publicly announce his findings. He wrote another letter to the Editor of *The Electrician* in which he noted his ongoing wireless experiments. The letter was published on 19 April under the title, "*Production of rectifiers by physical means.*"

Although he had discovered that crystals could be made to oscillate wireless waves, neither he nor experts such as Dr Eccles could explain why. He did put his latest theories forward in his letter to the journal.

Production of Rectifiers by Physical Means:

"Sir: So far as I have been able to trace, no explanation has hitherto been found for the rectifying action that certain metallic crystals have on electric oscillations,

and although the effect is largely utilised in detectors used in wireless telegraphy, the phenomenon is always treated as mysterious and of a somewhat elusive nature. Given an understanding why this valuable effect takes place, the field is opened for great improvements in wireless telegraph detectors and receivers.

Whilst investigating this peculiar property in various minerals, I discovered that the rectifying power could be mechanically induced in a mineral which in its native state had not the property. I found the rectifying property could be given in a highly exalted degree by altering the physical form of the crystalline surface."

[Full Text, see End Notes, Chapter Fifteen, No. 9.]

Henry's letter created quite a stir around the world. One well- known American wireless expert, Mr Greenleaf Wittier Pickard, in a letter to the Editor of *The Electrician*, challenged Henry's findings. Over a series of months, due to the delay in letters arriving in England from both Australia and America, the two men engaged in an informative debate on the history and rectifying powers of minerals. Henry, who always seemed to have the last say in any debate, wrote his last response to Mr Pickard on 20 June, 1912. If Mr Pickard responded, it was not published, but their letters clearly indicate the ongoing struggle by many an experimenter and inventor to understand the rectifying nature of crystals.

[Full Texts, see End Notes, Chapter Fifteen, Nos. 10 & 11.]

For many years after Henry's discovery, the theories and debate about how and why crystals could oscillate wireless waves continued amongst the world's leading wireless telegraphy experts.

Henry's wireless discoveries and inventions were applauded around the world, but apart from the occasional article in Australian newspapers, his work was totally ignored and went unrecorded by Australia's scientific community.

Thomas Sergeant Hall wrote a couple of very small articles that appeared in the Australian Mining Journal, but nothing else about Henry's wireless system or discoveries was published in Australian journals.

Hall was a friend of Henry's and was a member of the Royal Society of Victoria and a lecturer in Biology at Melbourne University. He also wrote the column *Scientific Notes* under the pseudonym "Physicus" for *The Australasian* newspaper. Well aware of Henry's ongoing experiments and his important discovery of the rectifying properties of crystals in the field of wireless telegraphy, he wrote:

"Some light, however, is thrown on the subject by a discovery made a few

months ago by Mr Henry Sutton, of Melbourne, who found that, in one case at least, the rectifying power depended on the mechanical treatment of the crystal. A native crystal of disulphide of iron is, according to Mr Sutton's researches, devoid of rectifying power. If, however, a portion of the native surface be knocked off, and the inside material exposed, places can readily be found which, in conjunction with a gold point, give a rectifier of much greater sensitiveness than silicon and gold. The right position for the gold point is indicated by the presence, in the fracture, of ridges which are easily seen under the microscope. With such a rectifier and a telephone, Mr Sutton claims to have received signals from a ship nearly 2,400 miles (approx. 3,862 kilometres) away. If this claim be substantiated, he has hit upon the most sensitive detector yet invented, and has incidentally discovered a fact of no small scientific interest."

[Extract from full Text, see End Notes, Chapter Fifteen, No. 12.]

Henry Sutton experimented with wireless telegraphy for two decades and, although his aim in the beginning was to find a way to transmit images wirelessly, his discoveries and inventions waylaid his efforts. In the course of the two decades since his first introduction to wireless telegraphy by Nikola Tesla in 1892, Henry Sutton made many important advances in wireless telegraphy.

Just a few of Henry Sutton's wireless inventions:

- Invented a method of jamming wireless signals.
- The Sutton Portable Field Signalling Wireless Set (first portable radio)
- A complete portable wireless receiving station.
- A special microphone transmitter capable of taking the current of a dynamo.
- The Sutton Universal Wireless Receiver.
- A sensitive receiver for long distances.
- A portable testing transmitter.
- A portable tripod receiver
- A portable detector.
- The Sutton Oscillator.
- The Sutton Wireless Valve

Not all of Henry Sutton's wireless equipment, inventions and discoveries have been listed, as many were kept secret and information on them is limited or unavailable due to the secrecy needed at the time. As for his initial idea of transmitting images wirelessly, he was to state in 1908 that he was at least ten years away from being able to wirelessly transmit an image. This statement clearly implies that he intended to spend the next ten years, or more, working on wireless telegraphy. It demonstrates his commitment to an idea that he originally thought possible back in 1892. Henry's vision did finally come to fruition in 1920 when the French photographer and inventor Edouard Belin wirelessly transmitted a written note. Within a year, photographs could be transmitted wirelessly.

The legacy of Henry's wireless telegraphy experiments and inventions did not end with his death in July 1912. In September of that year, at the annual meeting of the British Association for the Advancement to Science held in Dundee, Scotland, Professor John Ambrose Fleming opened a discussion on the present state of wireless telegraphy. His paper was titled *'The Scientific Theory and Outstanding Problems of Wireless Telegraphy'.*

Fleming, who had invented and patented the Fleming vacuum tube in November 1905, formed a committee of a number of well-known radio-telegraphists, meteorologists, physicists and mathematicians to discuss and further advance the unsolved problems of wireless telegraphy. The committee focused on radio waves and rectifiers, including Fleming's oscillation valve (vacuum tube) as well as Henry's gold point galena to galena detector, which was especially commended at the time as being one of the best rectifiers. The committee worked for the next two years on trying to solve the various problems of static and radio waves and set about trying to understand how they were transmitted, but due to the outbreak of World War 1 the committee was disbanded.

In the decades that were to follow, the true worth of Henry's work eventually came to light. In 1994, the late great British wireless historian, author and editor, John Patrick Hawker wrote an article about the history of the invention that changed the world: *The Transistor.*

From the beginning, Mr Hawker's article is intriguing and it will be a surprise to many Australians to learn that Henry Sutton's discovery and invention, which he patented in 1910, had something to do with the invention of the transistor. To understand Henry's role in the evolution of the transistor, it is necessary to review the pre-history surrounding the inventive steps that would eventually lead to the creation of the modern transistor on 23 December, 1947.

Henry Sutton's discovery that crystals could be made to oscillate, was an important step forward in the evolution of wireless technology, although the value and potential of his discovery were not fully understood at the time. History has recorded that Dr William Henry Eccles was the original discoverer and inventor of the crystal diode, but as he himself noted in *The*

Electrician on 16 December, 1910, the honour of the discovery that crystals could oscillate which gave birth to the invention of the crystal diode actually belongs to Henry Sutton.

Originally called a rectifier, because of its ability to convert alternating current (AC) to direct current (DC), the point contact rectifier was improved upon and eventually renamed a '*diode*' in 1919 by Eccles who coined the term from the Greek root *di*, meaning 'two', and *ode*, a shortened form of 'electrode'. The word electrode, derived from the Greek words *electron* (meaning amber, from which the word electricity is derived) and *hodos* (meaning way, path, road, or connection), was first used by the eminent scientist Michael Faraday. Coincidently, or not so coincidently, when Eccles coined the term *diode* in 1919 this was the same year that Henry Sutton's name and work disappeared from the wireless history books, including from Eccles own references and books on wireless telegraphy.

The crystal diode (semiconductor), and later the transistor, were the inventions that changed the world and began a new era in the technology of the 20th Century. They were the origin of modern day electronics. The work done by Henry Sutton and later by Eccles, laid down the foundations for future experimenters, but it would take many years for their work to be examined more closely. This was largely due to the invention of the vacuum and audion tubes which became widely used around the world because they were cheap to manufacture, were much more stable as a rectifier and more reliable. The vacuum and audion tubes were electronic amplifying tubes that worked so well that further, in depth experiments on oscillating crystals were set aside, not only by Eccles, but by other experimenters as well.

There was a resurgence of interest in oscillating crystals after 1923, when a Russian engineer by the name of Oleg Vladimirovich Lossev reported his work on oscillating, amplifying zincite crystal detectors. Piqued with curiosity by Lossev's work, Captain Henry Joseph Round and Noel Myer Rust decided that there might be some merit in looking back and repeating some experiments of some of the earlier experimenters. They decided to start with Henry's and Dr Eccles oscillating galena experiments and in doing so found that many other minerals were found to have similar properties. Round and Rust finally realised that there was no practical value in continuing their experiments, as an enormous improvement to the ability to broadcast would have to be made first.

As time wore on, the exploration of the mysteries of semiconductor physics occasionally raised its head as the quest for practical applications were constantly sought by scientists, but in the meantime the reign of the vacuum tube continued. Over the next few decades, people began to look back as wireless technology was not advancing to any degree and it began to make sense to see if technology could take another direction.

When the Japanese bombed Pearl Harbour in December 1941, the United States entered WWII. America's involvement in the war led them to launch an immediate programme and investigation into the properties of semiconductor materials for the use in radar as the traditional tube-based receivers did not work well at the higher frequencies needed.

By 1943, over thirty laboratories in the United States were funded by the American Government to develop semi-conductor devices. When the war ended in 1945, Bell Telephone Laboratories were given authorization to continue their solid state research as it was felt that there were still gains to be made in that field. A solid state physics group under the supervision of Dr William Shockley and Dr Stanley O. Morgan was then set up with a sub-group which included Dr Walter Brattain and Professor Gerald L. Pearson. Professor John Bardeen joined the sub-group a few months later and the decision was made to limit all the semiconductor research work to the minerals silicon and germanium.

Bardeen, Brattain and Shockley at the Murray Hill, New Jersey Bell Telephone Laboratory. 1948
Photograph courtesy of Nokia Bell Laboratories Archive

After many failed directions in research, Professor Bardeen proposed a new hypothesis which turned out to be of crucial importance. He put forward the theory of "*surface states*". The theory was that a semiconductor surface must be in equilibrium even before any electrical contact is made to it so that any electrons that were trapped at the surface must be neutralized by a space charge region equal and opposite to the electrons' charge at the surface.

Dr Brattain and Professor Bardeen resumed experiments under this new theory with some ideas from Dr. Shockley who continued to pursue field effect concepts. From November 17 to the 23 December 1947, Professor Bardeen and Dr Brattain observed that when two gold point electrical contacts were applied to a crystal of germanium, the output power was larger than the input, the very thing Henry had discovered in 1907. This was the basis for his gold point galena to galena detector which bought him world acclaim. William Shockley saw the potential in this and worked over the next few months greatly expanding the knowledge of semiconductors.

On the 23 December 1947, the group at Bell Laboratories placed two electrodes next to each other on a germanium ingot. A current was injected into the base by a first emitting point and collected by a neighbouring oppositely-polarized point. The current could be controlled by means of a small voltage applied between the emitter and the base. Next day, Walter Brattain recorded the findings in his notebook. The demonstration had shown, that with this arrangement, an audio-frequency power gain of 18 or more was possible when measured from a single tone of 1000 hertz. The shared excitement of the people present must have been palpable as they witnessed the first successful demonstration of what was to become known as the invention that changed the world - the *Point Contact Transistor.*

The transistor did not become an immediate success. It would not be until the mid-1950s that there were sufficient improvements and potential uses that its true value was cemented in technology history. In 1956 Professor John Bardeen, Dr William Houser Brattain and Dr William Shockley received the Nobel Prize in Physics *"For their researches on semiconductors and their discovery of the transistor effect."*

The invention of the transistor by Bardeen, Brattain, Shockley and other Bell Laboratory researchers does need to be applauded but at no time did Bell Laboratories or the trio of scientists acknowledge the pioneering work of others in the field of semiconductors. It is very clear that the three men did not achieve their success alone; history has constantly proved that no one principle or invention is invented in isolation and each advance in science and technology has its own personal history. It is the view of some of the world's best wireless historians such as John Patrick Hawker (UK), Dr Harry E. Stockman (US) and others that Bardeen, Brattain and Shockley did not invent the transistor alone - they re-invented it by drawing on the previous work done by Henry Sutton and others. History may have not given recognition to Henry Sutton for the pioneering work that sparked the invention of the transistor but nonetheless it does not diminish the extraordinary achievement of Australia's greatest inventor.

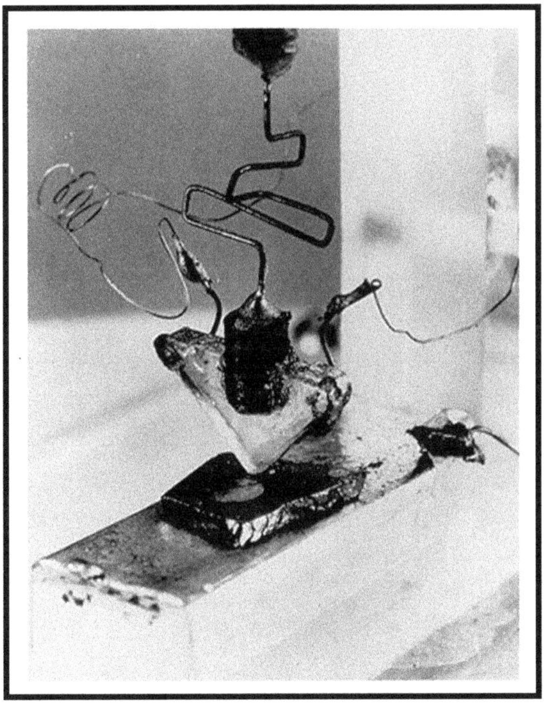

First Transistor by Bell Laboratory in 1947.
Photograph courtesy of Nokia Bell Laboratories Archive

CHAPTER SIXTEEN

A LIFE OF INVENTIONS

"To raise new questions, new possibilities, to regard old problems from a new angle, requires creative imagination and marks real advance in science"

Albert Einstein

Over the last 140 years, many people have written about Henry Sutton and his patents while expressing different points of view regarding his supposed refusal to patent most of his inventions. Various reasons were put forward, one being that Henry thought the fruits of science belonged to everyone and everyone should freely benefit from them. History has often recorded that Henry only ever applied for two patents and many people over the years chastised him for not taking out more and profiting from them. There is an element of truth in all these views, given the information that was at hand at the time. In fact, Henry Sutton took out over thirty-five patents during his lifetime, and although one could argue that some of these patents were for the same invention, submitted in other countries, the total remains.

Early in Henry's career, as a child scientist and inventor, he was not in a financial position to afford to patent any of his inventions such as his continuous current dynamo, or his telephones. The cost of submitting and maintaining a patent was very expensive and at £50 a submission it was far out of the reach of a budding young scientist; it was really only worth doing if you intended to commercialise and promote your invention. As a child, science and inventing was merely a hobby for him and, yes, as a young man he felt the fruits of science belonged to everyone as did some other inventors of that era. Henry's experience with sixteen of his telephones being patented by others, though, must have struck a chord and over time he came to realise the benefits of patents.

When Henry started inventing, Australia had a very small population and in many respects the cost of patenting an invention and then commercialising it was not financially viable. The question then became whether to pay all that money to patent an invention if there was no financial gain from doing so. This remained true for most Australian inventors and inventions for most of the second half of the 1800s and is reflected in the relatively small number of patents lodged by Australians during this period. By 1890, patent laws in Australia began to change, becoming more stringent, and so after Australia's Federation in 1901 the number of patents lodged by Australian inventors began to rise. With the establishment of the Commonwealth Patent Office in 1904, the need to patent an invention in every state of Australia was essentially eliminated; an inventor now needed to lodge only a single application for a patent that would apply throughout the entire country. With this improved patent application, approval and registration process, there was a sharp rise in Australian inventions and patents during the following years, often instigated and financed by Australia's rapidly growing manufacturing industry.

Finding most of Henry Sutton's patents was difficult because in 1947, the Australian Patent Office decided to throw out all but about one hundred patent files from the period between the late 1890s and early 1900s. The Patent Office considered that they were just 'tidying up' and only needed to keep what they thought were important patents. There was no guarantee that any of Henry's wireless patents, for example, had survived the 'Great Patent Purge' at all. Fortunately in Henry's case, there were references to his patents in other places, and so, after a prolonged search through many volumes of patent books, amounting to thousands of pages, his wireless patent records were located.

Henry Sutton's first patent was for his secondary cell battery, which he lodged in Britain on 21 July, 1882. He lodged it only in England, probably because he exhibited his battery there at the 1882 Crystal Palace Electrical Exhibition in London. In doing so, Henry was not seeking to gain financially and he still made it available for everyone to experiment with and use commercially.

The worldwide success of his battery, and the commercial use of it by many around the world, led Henry to change his position slightly on patenting and he started to see the value in protecting his inventions. By 1886, Henry was in a better financial position and it was at this point that he began to lodge patents for his inventions. The next patent he lodged was for improvements to telephonic circuits and, although he did not put this patent into commercial use, it was a small step towards inventing a commercial product that was financially worth patenting.

Henry's first commercially viable patent was his halftone photographic process which was lodged 20 October 1887, and 27 April 1888. This was his first worldwide patent and after his photographic process proved to be a commercial success in England in the 1890s, he was now fully aware of the importance and value of patents. The lessons of the past and of others benefiting from his inventions had been learned; when Henry invented a number of bicycle design improvements during the period 1894 to 1899, instead of patenting them, he either formed a syndicate or sold them directly to overseas manufacturers.

By 1898, Henry had taken out patents around the world for his new combustion engine and, although research to date has only uncovered the details of a few of these, it is known that he lodged patents in South Africa and Europe and possibly other countries. In the United States, Henry's combustion engine patent was split into two patents which covered his combustion engine inventions, ensuring their commercial success.

With his wireless patents, Henry went back to his roots and wanted to gift to Australia the fruit of his twenty years of hard work and not commercially benefit from any of his wireless inventions. However, as circumstances unfolded, he was forced into a position to protect his inventions and between 20 December 1909, and 14 December 1911, he lodged five different patents relating to wireless telegraphy, which covered all of his wireless inventions.

Although Henry has been ridiculed by historians over the years for his refusal to patent his inventions, they may well now reconsider their views, for he clearly did patent many more than just two inventions over his lifetime. Henry's wireless patents were the last patents he lodged and they remain testament to his life as a man of science and inventor.

On 28 July 1912, at the young age of 56, Henry Sutton died suddenly of chronic nephritis and heart failure. His health had been in decline in the past year and he knew he had not long to live. Henry saw his wireless system as his legacy to his beloved country and, in the decades that followed his death, his wireless work did indeed prove to be of great value, not only to Australia, but to the world as well.

One of Henry's last letters was written 12 July 1912, just sixteen days before he passed away. It was sent to the Editor of the Melbourne *Argus* newspaper and was published the following day. The subject, of course, was about Australia's wireless stations and was a short letter that read:

"Sir, as a matter of scientific interest, I am able to state that last night my private station succeeded in taking a measurement of the actual strength of the incoming wave from the Pennant Hills wireless station. After passing over 446 miles (approx 717.76 kilometres) of land (Sydney to Melbourne), the current registered at my station was 20 microamperes, which means that Pennant Hills should be heard over 4,000 miles (approx. 6,437 kilometres) and gives an idea of the capabilities of the station when on full power."

Although the Australian Government did not take up the offer of the Sutton System of Australian Wireless, he was justifiably proud of his small contribution to Australia's first wireless telegraphy system.

For the Sutton family, Henry's death was a great shock. As the eldest son of Richard and Mary Sutton, he had been the head of the Sutton family after their deaths. Loved and respected by all of his brothers and sisters, his death was a cruel and tragic loss to the family. For his young wife, Annie May, it was a devastating blow and for Henry's children, Arthur, Edward, Albert and John, it meant the loss of their much loved and adored father.

Henry Sutton died amid no great fanfare or acknowledgment. He died as he lived: a simple and humble man of science, father and husband.

Where his worldwide friends such as Bell, Tesla, Preece and Lord Rayleigh mourned the loss of their great friend, here in Australia no accolades were bestowed on Henry's life and genius, nor was his great contribution to Australian science and innovation acknowledged in the annals of Australian history. Instead, Australia's first world renowned scientist and inventor slipped from the pages of history and up until now little had been known or written about this great man and his extraordinary life. For the relatively few people who were interested in the histories of the various fields of science, engineering and technology with which Henry was involved, he was always quietly revered and some certainly recognised that he was Australia's greatest scientist and inventor. Few people, in Australia or in other countries, have so successfully worked in as many fields of science and engineering. Even a century after Henry Sutton's death, anyone who reads about his body of work is inspired and left in awe of this great man who unselfishly gave so much to science.

The most glowing of the eulogies which were published just after Henry Sutton's death was from the Ballarat School of Mines. This was published in the school's student magazine and represented a great respect for Henry's career as a scientist, inventor and lecturer there:

Obituary:

"It is our painful duty to record the death of Henry Sutton, who was a native of Ballarat, and for some time Instructor in Electricity and Magnetism at this School…

Henry Sutton had always a very active and fertile brain, and his greatest joy and almost all his leisure lay in the direction of science, and especially in its practical applications. He touched many subjects; he had many interests, and excelled in them all, whether it was drawing or photography, or pianoforte construction, or practical electricity. He read heavily and omnivorously in science, and as a very young man he used to set himself tasks to be finished in given times. He kept the flywheel of his brain spinning, and thought out his work systematically and persistently; until he had developed his subject as far as his quiet, steadily-working inventive powers would carry him. His methods of work seemed to have much in common with those of Edison."

The Ballarat Courier newspaper also gave Henry a glowing tribute. It spoke of his triumphs such as the carbonising of manila fibre, whereby the resistance is set up in light globes creating electric light, of his being the first person in Australia to set up a telephonic communication in Australia and with the aid of Tesla transmitting a photograph by telegraph.

[Full Texts, see End Notes, Chapter Sixteen, Nos. 1, 2, 3, 4, 5 & 6.]

There were only a few, very small tributes bestowed in newspapers and journals outside of Ballarat. Only one Australian journal reported the death of Australia's first notable scientist and inventor and this was only because the article in the *Mining and Engineering Review* was written by Mr Thomas Sergeant Hall, a friend of Henry's. The English journal *The Electrician* published the sad news of Henry Sutton's death on the 13 September 1912.

In the various obituaries, there were statements about his achievements and character as an inventor and man of science.

Henry Sutton's funeral was held on 29 July 1912, and a long procession of vehicles travelled from his home in Malvern to Brighton Cemetery where Henry was laid to rest. On the day he was buried he was not a scientist or inventor; he was a dear husband, a much loved father, an adored brother and a great friend. Everyone who knew Henry personally felt the loss of this extraordinary man. The Syme family of *The Age,* lodged a small tribute to Henry in their newspaper, commenting that his death would be widely regretted.

With Henry's passing, life continued for his family. A few years after his death, his wife Annie May remarried. She had two children with her second husband, but later divorced and eventually moved to Adelaide to be near her son John. Annie May died in Adelaide in 1974.

The question is often asked "*Did any of Henry's children inherit the inventive talents of their father?*"

The many talents of Henry, and of the Sutton family, have been passed down through the generations in the form of art and design, music, technology and engineering and inventing. All four of Henry's sons were interested in wireless/radio, and that interest has been handed down through the family. The stories of Henry and his inventions have likewise passed down through the generations and these family stories will continue to be retold to further generations of the Sutton family.

While debate continues as to whether Henry Sutton's *magnum opus* might have been his telephane/television or his wireless system, one could argue that it is really the entire body of his work that is Henry Sutton's true gift to the future.

Henry Sutton was a quiet man whose contributions to science, engineering and technology spanned many decades and left an indelible mark in the many fields to which he turned his inventive mind. The depth of his scientific endeavours is chronicled in the thousands of pages of journals around the world; his letters and published papers are testament to his drive and desire to show what the inventions of science can offer the world. His philosophy was that "*Eight hours in a day will not lift a man in this world*" and so he worked tirelessly, freely giving to the world many of his inventions. He was renowned as one of the greatest teachers at the Ballarat School of Mines and today his legacy as a scientist, engineer, inventor and teacher is instilled in each person who takes the time to read and learn of this extraordinary man and his contribution to science and technology.

It is a travesty that Henry Sutton slipped largely unrecognised between the pages of Australian scientific history for over a century. In the last few years, though, there has been a groundswell of renewed interest in Australia's first notable man of science that will, hopefully, continue to grow so that Henry

Sutton's achievements will be officially recorded and he can finally take his rightful place in Australian and world history.

Henry Sutton's contribution to science will one day be taught in schools across the nation, enabling his legacy to Australia, and the world, to live on.

The last known photograph of Henry Sutton,
with his sons Albert and John, at his home in Malvern 1912

PATENTS

CHAPTER TWO:

No. 1: Telephonic Patent:

Patent Title and Record: Improvements in electric circuits for telephonic purposes. *"Patents and Patentees, by Sands and McDougall"* Volume 21, 1886. Victorian Patent No. 4784, 26 October, 1886. p.127. (No Drawings)

Patent:

"Designed to enable the usual battery attached to the transmitter to be dispensed with, and in place thereof the necessary electric current is derived from a circuit composed of a gas main or service pipe and a water main or service pipe, such circuit being connected with a telephonic transmitter known as a battery transmitter, or the circuit may be formed of either gas or water main when connected with an earth plate capable of conducting electricity."

CHAPTER FOUR:

No. 1: Battery Patent:

Provisional Patent Title and Record: Improvements in Secondary Batteries or in Apparatus for Storing Electricity. *"British Commissioner of Patents Journal, Class 53, Abridgement Class, Galvanic Batteries"*, 1882, p. 71. British Patent No. 3464, 21 July, 1882. (No Drawing)

Patent:

"Provisional Specifications left by John Henry Johnson at the office of the Commissioners of Patents on the 21st July 1882. A communication to him from abroad by, Henry Sutton of N° 31 Sturt Street Ballarat in the Colony of Victoria Music Seller.

John Henry Johnson of 47 Lincolns Inn Fields in the County of Middlesex Gentleman "Improvements in Secondary Batteries or in apparatus for Storing

Electricity" A communication to me from abroad by [J.] Henry Sutton, of Nº 31 Sturt Street, Ballarat, in the Colony of Victoria Music Seller.

My invention consists firstly in making porous plates cylinders or discs of Carbon having curved or angular extremities the said plates cylinders or discs being made of pulverized or ground Carbon moulded into form with a glutinous or saccharine or carbonaceous substance from which the volatile portions are driven off by heat leaving the said plates cylinders or discs capable of retaining their form.

These plates, cylinders or discs of Carbon are then electroplated with copper or other metallic electric conductor, at suitable places for the attachment of other conductors. These places of attachment are faced with lead by means of soldering to ensure preservation or partial protection from action of acids or alkalise.

These plates, cylinders or discs of porous carbon with or without heating them are then saturated with an aqueous solution of sulphuric acid whereby sulphate of lead in a finely divided state is formed or precipitated in the pores and cavities of the said plates, cylinders or discs of porous carbon.

The process of Saturation with Solution of acetate of lead and precipitation with solution of Sulphuric acid in some cases is repeated several times in order to form or precipitate a considerable quantity of sulphate of lead in the porous carbons.

Two or more of these porous plates cylinders or discs of carbon treated as above described are now taken and arranged as the positive and negative electrodes in a vessel containing dilute sulphuric acid---On a current f electricity being sent through this system the sulphate of lead in pores or cavities of positive becomes oxidized and the sulphate of lead in pores or cavities of negative electrode is reduced to spongy or finely divided lead---When this process is completed the plates, cylinders or discs are in a condition fit for the retention or holding or occlusion of gasses and are ready for use in secondary batteries.

Another part of my Invention consists in saturating the above mentioned porous plates, cylinders or discs of carbon with a solution of nitrate of lead and then heating them which leaves them in a condition to be treated as follows: Two or more of these plates, cylinders or discs of porous carbon treated with nitrate of lead are arranged as a positive and negative electrodes in a vessel containing dilute sulphuric acid a current of electricity being sent through the system. The material in pores or cavities of negative electrode is converted into spongy or finely divided lead.

The process of Saturation with nitrate of lead and then heating is performed as many times as required to produce the required quantity of material in the pores or cavities of said plates cylinders or discs of porous carbon.

Another part of my Invention relates to the form and substance of the electrodes of a Secondary Battery and consists in using carbon plates, cylinders or discs moulded or cut so as to have holes, cavities, grooves, corrugations, interstices or shelves divided carbon or alumina in fine division.

I also use electrodes of porous carbon without combination with metallic or other matter these porous carbon electrodes being made with extremely fine pores and cavities which are capable of retaining or occluding gases.

By constructing secondary batteries in the manner hereinafter described great electromotive force is obtained.

Take a number of hollow cylindrical vessels of lead or other conductor of electricity but preferably lead the hollow cylinders having either curved or angular form of aperture and of such a size as to fit one within the other leaving a space between each. These vessels are prevented from touching each other at the bottom by means of a layer of any insulating or semi-insulating substances such as vulcanite wood paraffine, india rubber, gutta percha, paraffined wood felt, asbestos paraffined felt woollen cotton or cloth.

The outer and inner vessel must be higher than the intermediate ones and form the terminals of the battery the spaces being filled with solution to within which a short or long distance from upper extremities. In the case of vessels having an elliptical form of aperture the inner one contains a plate of same metal or substance as that surrounding vessels the inner plate and outer containing vessel being the respective terminals of the battery must stand higher than the intermediate ones."

CHAPTER SEVEN:

No. 1: Photographic Patents:

Patent List Record: Complete list of all patent numbers and dates, British National Archives, Kew, "*Company Documents, Sutton's Process Syndicate Limited*", BT 31/502/35217,

Australian Patents:

Victoria: No. 5389, 20 October, 1887.

New South Wales: No. 284, 22 October, 1887.

South Australia: No. 384, 25 October, 1887.

Tasmania: No. 504-9, 25 October, 1887.

Queensland: No. 343, 6 February, 1888.

Overseas Patents:

New Zealand Patent: No. 2615, 1 November, 1887.

British Patent: No. 15722, 16 November, 1887.

French Patent: No. 187140, 22 November, 1887.
Belgian Patent: No. 79,636, 22 November, 1887.
Canadian Patent: No. 28139, 7 December, 1887.
Indian Patent or Certificate: No. 221, 28 December, 1887.
Cape Of Good Hope Patent: No. 430, 14 January, 1888.
Italian Patent: No. 489 v xliiv, 21 January, 1888.
United States Patent: No. 381091, 10 April, 1888.
Spanish Patent: No. 7616 - 12408, 27 April, 1888.

Patent Title and Record: *Specifications for registration of patent by Henry Sutton titled: Process of converting a photographic image on a gelatine surface into a relief or intaglio printing surface. United States Patent and Trademark Office: Specification forming part of Letters Patent No. 381,091, dated April 10, 1888.*

"Be it known that I, HENRY SUTTON, a subject of the Queen of Great Britain, residing at Ballarat, in the British Colony of Victoria, music-seller, have invented an Improved Process of Converting a Photograph Image on a Gelatine Surface into a Relief or Intaglio Printing Surface; and I do hereby declare that the following is a full, clear, and exact description of the invention, which will enable others skilled in the art to which it appertains to make and use the same.

It is well known that relief and intaglio printing-surfaces can be produced from photographs; but in practice these have been principally confined to the production of linear work on account of the difficulty of faithfully representing the delicate shades of what are known as the "half-tones." When such half-tones have been represented it has been due to the exercise of very great manual skill, and when produced it requires very great care and ability on the part of the printer to produce a satisfactory representation on paper.

It is well known that the half-tones of photographs may be "broken up" by dot, stipple, line hatch, grain, or other pattern.

I know of no method by which a photographic image on a gelatine surface can be automatically converted into a relief or intaglio printing surface.

Now, this invention is designed for the especial purpose of automatically converting a photographic image on a gelatine surface into a relief or intaglio printing surface which will faithfully represent the delicate lights and shades or half-tones, said printing-surface being of such kind that an ordinary printer with ordinary appliances can without difficulty utilise it for printing purposes.

In carrying out this process, I take an ordinary sensitized gelatine plate bearing a photograph broken up by any pattern or grain, as before described, and I immerse it in water, the temperature of which must be below that at which the particular gelatine used will melt or dissolve. Here I allow it to remain until all the gelatine swells by absorption of water; but the extent to which this is allowed must be determined by the character of the picture and judgment of the operator. After removing the plate from this bath I remove the free water from its surface, preferably by a slight rinse in alcohol or spirits of wine. I then subject this plate to a gradually-increasing temperature until it reaches about that at which the particular gelatine used begins to melt, and I continue this heat until all or nearly all the water or moisture is driven off.

The result of this is that those portions of the gelatine which are unaffected by the action of light and which are freely absorbent of water melt and are drawn by capillary attraction under the unmelted portions, forcing the latter up in relief, the dots, stipples, lines, grains, hatch, or other pattern by which the photographic image is broken up being graduated in size and form in precise proportion to the action of light, and of light and development, on such photographic image. This completes my process.

If the photographic image on the plate heated be a "negative" the printing surface produced will be in intaglio; if it be a "positive" the printing-surface will be in relief.

If the gelatine be on a sufficiently thick sheet of glass, or on wood or metal or other hard material, it may be used direct in a typographic press, or if on metal or other suitable surface it may be used direct in the copper-plate press; but if it be on the ordinary thickness of glass, electrotypes or stereotypes must be used for the copper-plate press. I find that electro-types and stereotypes may be made from my printing-surfaces in the same way, in the same way, and much more easily and quickly than from ordinary engravings, inasmuch as I can electrotype direct on my printing-surfaces, thus saving the usual " wax squeeze," and I can cast stereo-type metal direct on my printing-surfaces without the intervention of the paper or plaster moulds used in stereotypy.

In actual practice, I have taken an ordinary dry-plate (Fry's transparency make) and have impressed an image thereon by the action of light such image having its surface diversified or broken up by the ordinary wire gauze pattern, and have then developed and fixed the image in the usual way. I have then immersed this for five minutes in water of the temperature of 80° Fahrenheit, and have then removed the free water by dipping the plate in a bath of spirits of wine for 10 seconds, and have then removed the plate and fanned the spirits of wine dry from its surface. I have then placed it on a copper plate supported on feet or legs, and heated it by means of a Bunsen gas-burner until it became about the temperature of 212° Fahrenheit. This heat I have continued for two minutes, when I have found my object accomplished.

I do not confine myself to any particular uses to which my printing-surfaces may be applied, and I wish it to be distinctly understood that I do not claim the production of photographic pictures on gelatine plates broken up or diversified by dot, stipple, grain, or pattern of any kind, nor to the immersion of such plates in water for the purpose of causing the unaffected gelatine to swell and rise into relief, as I am aware that all this has been done before; but.

What I do claim as of my invention is;

The herein-described method of converting a photographic image on a gelatine surface into a relief or intaglio printing-surface, which consists in immersing the plate bearing the photographic image in water of a temperature below that of the melting-point of the unexposed parts of the gelatine, in order to swell the same by absorption, then taking out the plate and removing the free water from the surface thereof, then exposing the plate to a degree of heat at which the unexposed parts of the gelatine will begin to melt until the water-absorbed gelatine has been dissolved, substantially as described."

CHAPTER ELEVEN:

No. 1: Combustion Engine Patents:

Patent Title: "Improvements in and Relating to Internal Combustion Engines."

Australian Patents:

Victorian Patent: No. 15777, 7 December, 1898.

New South Wales Patent: No. 9327, 1899.

Overseas Patents:

Great Britain: No. GB189900634 (A), 25 March, 1899.

France: Patent No. 285240.

United States Patents: Divided into Two Patents:

Patent Titles: *"Explosive Engine." & "Speed Regulator for Explosive Engines."*

United States: No. 650,736, 29 May, 1900.

United States: No. 664,689, 25 December, 1900.

Patent Title and Record: National Archives of Australia: Record Title: *"Application for registration of patent by Henry Sutton titled - Improvements in and Relating to Internal Combustion Engines."* Series Number: A13149, Control Symbol: 15777, Item Barcode: 7583786. Date Range: 1898-1898.

The Oil Engine

"It is well known that there are many difficulties to be overcome in the production of a successful and efficient internal combustion or explosive engine to be operated by the explosion of a mixture generated from oil or spirit. One of the

main difficulties is that oils have a specific gravity of from .72 and upwards require extraneous heat for their vaporisation, and engines using the same could not be started whilst cold.

One of the chief objects of this invention is to provide an internal combustion engine in which this extraneous heat will not in all cases be required, the oil or spirit at its normal temperature being converted into an explosive mixture with air and the heat arising from the combustion of such mixture being subsequently used to further assist in the rapid vaporisation of the subsequent charges. For this purpose I place my mixer as near to, and when possible in contact with, the explosion chamber in order that the heat abstracted therefrom may compensate the refrigerating effect of the vaporisation.

In my improved internal combustion engine I mechanically atomise or pulverise the oil or spirit when cold and provide a cool exhaust with the loss of a minimum amount of heat energy.

The exhaust or expansion chamber of my engine is so constructed and arranged that it serves partly as a silencer and by causing the exhaust to expand to nearly normal pressure within this chamber I effect a still further cooling of the engine and at the same time make it self-contained, that is it combines a piston chamber, an explosion chamber, an exhaust expansion chamber and a vapour chamber, the vaporiser and mixer being arranged in a convenient position.

My invention also comprises an improved construction of vaporiser together with an improved method and apparatus for supplying oil or spirit thereto.

I have also devised means for regulating the power of the engine in a simple and efficient manner by varying the quantity of the charge without varying its quality. My invention also comprises improved electric firing apparatus constructed in such a way as to prevent leakage of the compressed or exploded mixture whilst the ignition can be timed as required to suit the speed of the piston.

The silencer is, according to my invention, constructed in such a way as to utilise the principle known in acoustics as the interference of sound waves.

In order that the various details of my invention may be easily understood I will describe it by reference to the accompanying drawings in which, -

Figure 1 is a plan of an oil engine constructed in accordance with my invention, whilst Figure 2 is a longitudinal central section on line A A Figure 1.

Figure 2 represents an oil or spirit supply pipe leading from a reservoir 14 attached either permanently or by a coupling to a hollow metal casing or cylinder 3 fitted with a nipple 4 and formed with air inlet openings 5 or holes around its lower end about on a level with the upper end of the nipple. This nipple 4 is formed with a small hole, the aperture or inlet to which it is regulated by a screw-threaded needle valve 6 passing through a stuffing box 7 in the upper end of the chamber 3, and projecting up into a protecting cap or cover 8, which is screwed on to the gland of the stuffing box, and serves to prevent any injudicious alteration of the needle valve after it has been set.

If preferred this needle valve might be dispensed with, a hole of predetermined size being bored through the dependent nipple of only just sufficient size to allow the requisite amount of oil or spirit to be fed into the engine. The use of this needle valve is however preferable because it provides a convenient means for regulating the charge, and having being adjusted, can remain fixed for the particular kind of oil, spirit or other liquid fuel that can be used.

The Atomiser.

The dependent nipple 4 projects into a cylindrical or other shaped tube 9 communicating through a passage or vapour chamber 10 with the, inlet valves 11. The lower end of the compartment 9 is provided with a small hole, 12 communicating with a waste pipe 13 leading direct into the atmosphere or into any suitable vessel provided to receive any surplus oil or spirit which might not be vaporised within the compartment 9. Across this latter are fixed the mixing contrivances 141. These may be constructed either of sheets or wire gauze of one uniform mesh or of sheets of gradually finer material towards the lower layer…

If preferred the area of the air passage around the nipple 4 may be restricted, whilst the body or cylinder of the vaporiser is enlarged so that the single sheet of wire gauze will not restrict the area of the air passage. If preferred the gauze may be dispensed with, as it is not essential, although useful, for assisting to start the engine.

In order that the oil or spirit may be delivered evenly and regularly by the nipple 4. it is necessary that the supply shall be maintained at a fixed uniform pressure. I accomplish this result by means of a reservoir formed of a tank 14 of any shape, but entirely enclosed in a perfectly air-tight manner, except for a small air-supply pipe 15 extending down nearly to the bottom of the reservoir and having its upper end projecting outside it. The lower end of this pipe is entirely open, but its upper end communicates with the atmosphere only through a hole 16. The inlet 17, whereby the reservoir 14 can be charged, is closed by an air-tight cap 18, the result being that the only means whereby air can enter the reservoir is through the minute opening 16 and down the pipe 15.

19, represents an ordinary needle valve, which is fitted in the upper end of the oil supply pipe 1 and is regulated by a lever 20 and rod 21 or other convenient means.

When this needle valve 19 is opened a limited supply of oil or spirit passes from the reservoir 14 down the tube 1 into the chamber 3, and thence down the nipple 4, on the lower end of which it forms a small bead. The effect of the suction stroke of either of the pistons 22 is to produce a partial vacuum in and around the nipple. A supply of air is then drawn in through the openings 5, and encircling the drop of oil or spirit on the lower end of the nipple carries it through the mixer 9 into the vapour chamber 10, and thence into one of the explosion chambers 23 through its inlet valve 11, any excess oil or spirit passing away down the waste pipe 13. The passage or chamber 10 serves to prevent any oil or spirit passing up into the explosion chamber, except such as has been properly mixed with air…

As the oil or spirit is withdrawn from the reservoir 14 a partial vacuum is formed in the upper part thereof above the surface of such oil, the effect being that the atmospheric pressure will act upon the oil or spirit in the tube 15, and will cause it to fall until the atmospheric pressure outside the reservoir 14 balances the height of oil or spirit therein plus the pressure of air on its surface. When this balance has been established the height of the oil or spirit in the pipe 15 is stationary except for the limited gravity supply to the valve 6, and this supply will be maintained at a constant pressure, and will be entirely independent of the height or quantity of oil contained in the reservoir.

The pressure of oil therefore at the dependent nipple 4 will equal the weight of the oil between its level in the pipe 15 and the nipple plus the difference of pressure due to the partial vacuum in and around the nipple and the full atmospheric pressure in the pipe 15. The gravity supply having been limited by the needle valve 6, the regulation of the ultimate delivery from the nipple will be effected automatically by the rapid or slow production of a partial vacuum in and around the nipple caused by the increased or decreased piston speed of the engine.

This condition of regulation is attained after a few strokes of the engine, and the oil or spirit continues to fall in the pipe 15 until it is below the bottom end thereof. Immediately this happens the reservoir is placed in communication with the outside atmosphere, and a bubble of air will then pass from the lower end of the tube, which is preferably bevelled off or inclined somewhat. This air bubble rises up through the oil or spirit in the reservoir 14, and lowers the partial vacuum in the upper part of the reservoir, thereby allowing a proportion of oil or spirit to pass into the pipe 15 until the balance is again effected between the oil or spirit in the pipe plus the atmospheric pressure and the partial vacuum in the reservoir plus the height of oil or spirit therein.

This intermittent supply of air to the reservoir 14 is proportionate to, and is regulated by, the piston speed of the engine, and therefore the charge of oil or spirit which is delivered to the vaporiser is automatically and accurately adjusted for every stroke of the engine in a simple manner and without the aid of pumps, floats, or other mechanical contrivances. It will be obvious that with this method of supply the only variation in the pressure of the oil or spirit supplied to the dependent nipple will be merely that caused by the slight rise and fall in the oil or spirit in the tube 15.

The Ignition.

The contrivances which I use for firing the explosive charge electrically consist of two contact pieces 24 and 25, the former being mounted rigidly upon the piston 22, as shown in fig. 1, and the latter projecting through a metal sleeve 26 having a flange 27 at about its centre. The cylinder cover is bored out as illustrated in 28 to receive a packing 29 of asbestos or other fire-proof or heat-resisting material, which is also a non-conductor of electricity. A screw plug 30 is fitted into the recess 28, in which the packing is placed so as to jam same tightly around the metal sleeve 26 having a flange 27 in order to hold the same rigidly in position and at the same time insulate it from the metal cylinder. The contact piece 25 projects into a metal cap 31, which is screwed on to the upper end of the sleeve 26, and contains a spiral spring 32 adapted to bear upon a shoulder 33 on said contact piece, so as to normally hold it in its projected position. This metal cap 31 also serves to prevent the escape of explosive mixture around the contact piece.

With this construction of igniter the piston 22 is electrically connected through the cylinder or otherwise with one pole of a battery, whilst the metal sleeve 26 is connected with the other pole, a coil of wire surrounding an iron core being inserted in the circuit to produce a current of self-induction on the circuit being broken between the contact pieces 24 and 25. It will be evident that this igniter will explode the charge only after the dead point has been passed, and consequently is only for use in slow-running engines. It is, however, simple in its construction, and by using the self-induction coil above described instead of a Ruhmkorff or secondary induction coil the spark can be produced by a current of low voltage, and the insulation can be accomplished by asbestos instead of requiring porcelain or other non-conducting medium to be employed. Moreover there is not so much danger of the insulation being injuriously affected by dust and dirt. For quick-running engines or where it is necessary to vary the time of the explosion, I prefer to arrange the igniter above described in the manner illustrated in figures 6, 7, 9 and 10.

Instead of arranging contact pieces 24 and 25 in the end of the cylinder I mount them in the explosion chamber proper as illustrated in Figures 6 and 9.

The movable contact piece 25 is in this case carried on the end of an arm 34 projecting from a spindle 35, which is passed through a metal sleeve 26, and is formed with a conical or valve-shaped collar 36 to form a tight joint against the end of the sleeve. This latter is held in position by a screwed plug 30 and packing 29 of asbestos or other heat-resisting electric non-conductor. A spring 32 serves to keep said conical collar against its seat. The outer end of the spindle 35 is fitted with a projecting arm 37 the end of which is connected to a vertically sliding rod or spindle 38 by a non-conducting block 37ª as illustrated in Figures 9 and 10. This block 37ª serves to electrically insulate the arm from the rod or spindle 38, that is it insulates the firing from the timing mechanism. The lower end of the rod or spindle 38 is arranged to bear against a curved lever having an inclined surface 40, against which a cam 41 secured upon a suitable countershaft 42 driven by a two to one gearing from the main shaft is arranged to act at the desired period of the stroke of the engine. In order that this period of ignition may be varied as required I provide means for adjusting the position of the inclined surface 40 around the shaft, so that the cam 41 acts upon it sooner or later as the case may be. For this purpose I mount the lever 39 upon a carrier 43 which is pivotally supported upon the shaft 42, and is capable of being turned around upon the same by means of a rod 44 attached to a quadrant lever or by any other convenient means.

A modification of this electric firing and timing mechanism is shown in figure 11, which represents the upper end of a vertical engine in section, together with the electric connections, etc., for timing and operating same. In this case three contact pieces 24, 24ª, and 25 are employed. The first is fixed upon the piston, and the second is carried by an arm on the spindle 35 but, unlike the contact piece 25, shown in fig. 6, it is bent upwards, and contacts with a projecting arm 25ª on the contact piece 25. This latter, projects upwardly through a metal sleeve 26 fitted with a cover 31 containing a spring 32. The metal sleeve 26 is secured in position in the same way as described by reference to the firing device illustrated in Figure 1.

The spindle 35 carrying the contact piece 24ª constructed as above described by reference to figures 7 and 9 and it is fitted on its outer end with an arm 37, which is insulated electrically from the spindle, but otherwise is rigidly secured.

It will be obvious that by coupling up one pole of an electric circuit to the metal sleeve surrounding the spindle 35 and the other pole to the sleeve 26 that the circuit will be completed through contact between the two arms 24ª and 25ª. This circuit will be broken when the pin or projection 24 on the piston comes in contact with the lower end of the spindle 25, so as to raise same, the effect of breaking such circuit being to cause a spark which therefore fires the explosive charge, which has previously been compressed within the explosion chamber in the ordinary manner.

By turning the spindle 35 the arm 24[a] can be raised, thereby raising 25, and causing the projection 24 to engage with said spindle at a later part of its movement, thus breaking the electric circuit between 24[a] and 25 at a later period.

By this arrangement the timing of the explosion of the charge can be varied before the dead point. Instead of taking the circuit through the spindle 35 and arm 25[a] the circuit might be completed through the cylinder, piston, and contact piece 24, in which case the circuit would be broken after the dead point. The exact moment at which this takes place can be regulated as before by turning the spindle 35, and thus raising the contact piece 25.

An Alternate Method of Ignition

The two methods of coupling up the electric firing appliances above described will be the readily understood by reference to fig. 11, in which 45 represents a battery or electric generator, and 46 a self-induction coil arranged to produce a current of self-induction when the circuit is broken. 47 represents an insulated switch or cut-out actuated by, but insulated from, a hand lever 48 fitted with a spring catch 49 engaging with the teeth of a quadrant rack 50. One pole of the battery is coupled to the switch 47, and the other to the metal cover 31 or to the sleeve 26.

51 and 52 are contact pieces, one of which is coupled up to the spindle 35, whilst the other is coupled to the cylinder or other convenient part of the engine in electrical communication with the piston, and therefore with the contact piece 24. The switch 47 is constructed so that it can be placed in contact with either 51 or 52 by means of the lever 48.

This lever 48 has an arm 53, which is connected by a rod 54 with the arm 37 on the spindle 35. These various contrivances are so constructed and arranged that if the lever 48 is in its lowest position the switch 47 will engage with the contact piece 51, and the circuit will therefore be through the spindle 35, arms 24[a] and 25[a], and vertically sliding spindle 25. This latter will also be in its lowest position, consequently the piston will break the circuit between 24[a] and 25[a] at a comparatively early part of its stroke; consequently, the explosion will take place considerably before the arrival of the piston at the dead point.

If the lever 48 is moved forward in the direction of the arrow, the spindle 25 will be correspondingly raised; consequently, the explosion will take place nearer but still before the dead point. These are the conditions which would be suitable for an engine running at a high velocity.

Upon the lever 48 being moved still further forward the switch 47 will be withdrawn from the contact piece 51 and will be placed in contact with 52. The circuit will then be through the piston and the contact pieces 24 and 25.

The effect will now be that as the piston rises it closes the circuit until its return, when at a certain point contact between 24 and 25 will be again broken, and an electric sparking will take place between 24 and 25, which will explode the charge. In this way a much weaker explosion is produced suitable for a slow revolution of the engine when, for instance, the car or the power transmission gear is not in use. Upon continuing the movement of the lever 48 the arm 24^a *begins to return to its lower position, because the rod 54 is connected with the arm 53 through a connection, which slides along the slot 55, because said arm 53 is at such an angle as to effect this result. This movement lowers the contact piece 25, and correspondingly lengthens the time which elapses between the dead centre being passed and the explosion taking place.*

By means of the firing mechanism above described and the appliances used for timing the firing I am enabled by the movement in one direction of a single lever to ensure the firing taking place at any desired moment either before, at, or after the dead point.

Fig. 12 shows a vertical transverse section and Figure 13^a *is a vertical longitudinal section illustrating the adaptation of the vaporiser and mixer above described for the supply of a pair of cylinders in such a way as that the one automatic supply suffices for both cylinders.*

4 represents the dependent nipple to which oil is supplied through the tube 1 leading from the reservoir 14, as shown in fig. 2

Vapour chambers or passages 10 lead from the mixer 9 to the inlet valves 11 of each cylinder. 11^a 11^a *represent the exhaust valves.*

56 represents a bell-crank lever, which is used for the purpose of adjusting the pressure on the springs 57 acting upon the suction valves 11, so as to manually vary the amount of charge drawn into the cylinders. A bell-crank lever 58 or other suitable mechanism is provided for operating the exhaust valves 11^a*, said lever being primarily operated by a cam mounted upon a spindle driven by two to one gearing from the main shaft or spindle of the engine. When the inlet and exhaust valves are arranged as illustrated in figs. 1 and 2, 3 and 4 the gear shown in figures 1 and 2 may be used for regulating them in the manner just above described, the spindles of the exhaust valves being acted upon directly by cams 59 on a spindle 60 driven by a two to one gear from the main shaft.*

60^a *represents a commutator which is inserted in the circuits of the electric firing devices for cutting out the current during the charging stroke and for switching it on to the firing appliances during the firing stroke of each cylinder.*

The Variable Inlet.

In order to adjust the quantity of the charge fed into the cylinder without affecting its quality I use the means illustrated in figs. 1, 2 and 2[a] comprising a bracket or support 61 between the two cylinders, and arrange a rod 62 so that it can slide longitudinally through it. A T-piece 63 is secured upon the end of this sliding rod 62, as illustrated in Figure 1. The ends of this T-piece are fitted with tubes 64, which slide freely upon the spindles of the inlet or suction valves 11, 11. On moving the rod 62 outwardly by means of a bell-crank lever 65 the springs of the inlet valves will be compressed, and therefore their tension increased. The T-piece 63 on the end of the rod 62 is so arranged that when said rod is pulled back to near its furthest extremity the T-piece contacts with the lower end of a lever 66 pivotally supported upon the bracket 61, and acting upon a collar 67 on each of the exhaust spindles, thereby holding said valves open and preventing any charge from being drawn into the cylinder.

By means of this mechanism it will be obvious that a movement in one direction of a single lever or its equivalent increases the tension on the springs of the inlet valves, thereby gradually diminishing the suction, whilst the continued movement will entirely stop the suction by opening the exhaust valves...

A modification of the above-described arrangement is shown in Figure 13 in which 65 represents a comparatively long lever mounted upon any convenient support, as illustrated at 68 and connected by means of a link 69. with a shorter lever 70, fulcrummed, as indicated at 71. The lower end of the lever 65 is connected with and operates a sliding block 72 acting upon the spring of the inlet valve. The lever 70 is arranged to act upon a stop piece 73 on the rod or spindle of the exhaust valve.

By moving the lever 65 in the direction of the arrow the spring of the inlet valve will be compressed, and therefore the charge drawn into the cylinder will be diminished in quantity without being effected in quality. The continued movement of the lever 65 will move the lever 70 into contact with the stop piece 73 on the spindle of the exhaust valve, which will therefore be opened, the effect, of course, being to stop the working of the engine.

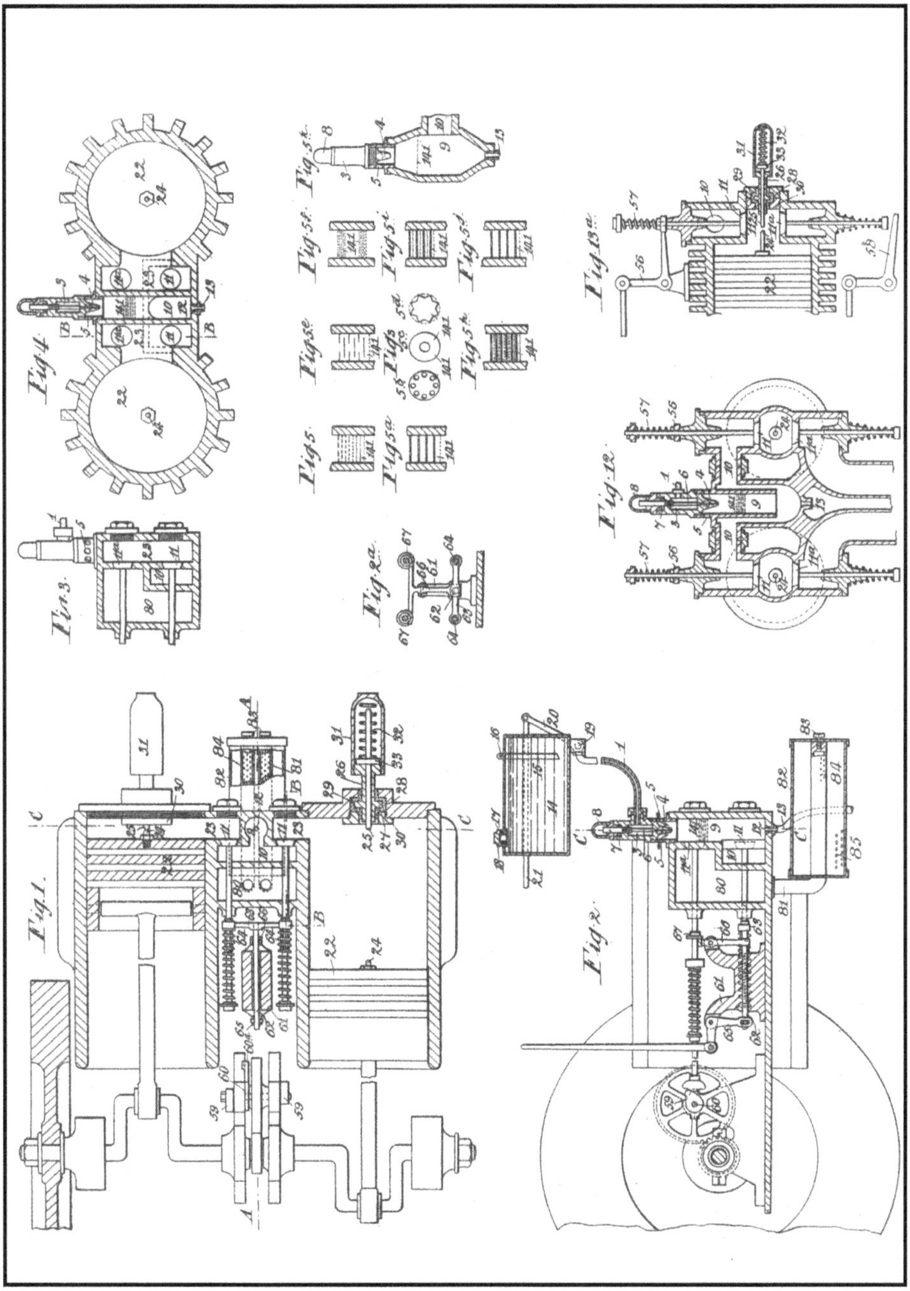

Engine Patent Drawing 1

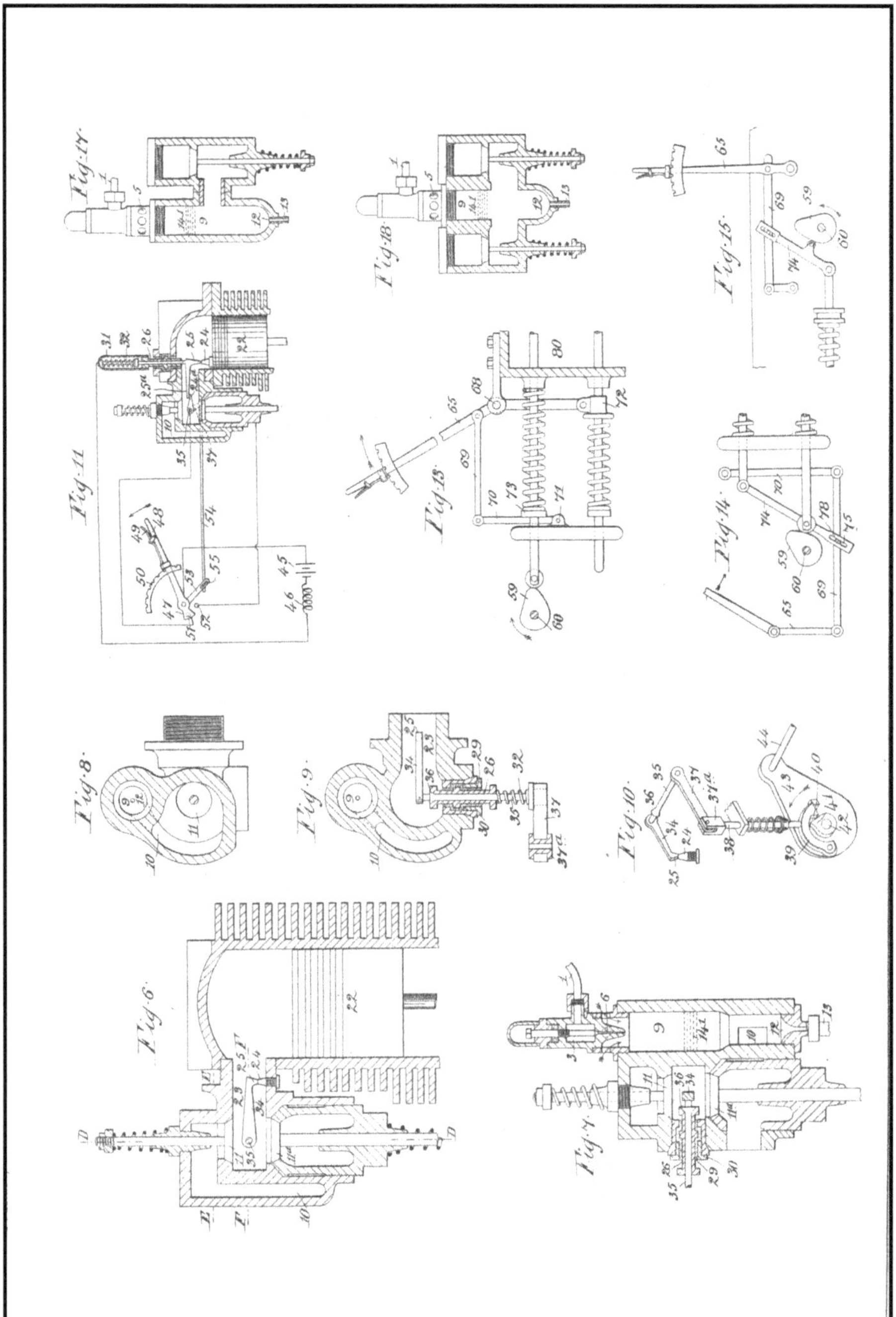

Engine Patent Drawing 2

CHAPTER FOURTEEN:

No. 1: Wireless Patent:

Provisional Patent Title and Record: Improvements relating to the production and transmission of hertzian waves. "*The Australian Official Journal of Patents, Trademarks, and Designs*", 31 December, 1909, p. 1083. Australian Patent: No. 16,633, 20 December, 1909. (No Drawing)

No. 2: Wireless Patent:

Provisional Patent Title and Record: Improved means for preventing 'arcing' of the gap in the production of high potential electrical oscillations. "T*he Australian Official Journal of Patents, Trademarks and Designs*", 22 February, 1910, p. 133. Australian Patent: No. 17,127, 15 February, 1910. (No Drawing)

No. 3: Wireless Patent:

Provisional Patent Title and Record: Improved means for detecting acoustic electric, or like waves. "*The Australian Official Journal of Patents, Trademarks and Designs*", 5 April, 1919, p. 257. Australian Patent: No. 17,433, 19 March 1910. (No Drawing)

CHAPTER FIFTEEN:

No. 1: Wireless Patent:

Patent Title and Record: An improved detector of electric oscillations for wireless telegraphy and like purposes. "*National Archives of Australia*" Series Number: A13882, Control Symbol: 002621/11, Item Barcode: 200705049, Date Range: 1911-1911. Australian Patent: No. 2621/11, 29 September, 1911. (No Drawings).

Patent:

"I, Henry Sutton, Gentleman, of 290 Bourke Street, Melbourne, Victoria, Commonwealth of Australia, hereby declare this invention and the manner in which it is performed, to be fully described and ascertained, to be fully described and ascertained in and by the following statement:-

The object of the present invention is to provide an extremely sensitive detector of electric oscillations, which will enable a current to be used direct in a telephone receiver or the like. This is attained by rectifying the alternations which property (rectification) is known to exist naturally in several crystals.

This invention utilises crystallised disulphide of iron which does not naturally rectify but by producing a conchoidal fracture in it I have found that the property of rectification is attained in a highly exalted degree.

The invention therefore consists in the use of a conchoidal fractured surface or edge of crystallised disulphide of iron (FeS_2) and an electric conductive point or surface; for the reception of wireless telegraph messages.

The crystallized disulphide of iron is of cubical conformation and the electric conductive surface in contact with its conchoidal fracture may be of gold or gold alloyed with silver or copper or silver and copper in any proportion between .001 and 24 carat.

Or the contact may be between the surface or edge of a conchoidal fracture of cubical crystallised disulphide of iron and a point or surface of silver, platinum, copper, iron, iridium, nickel or other electrically conductive metal or alloy or substance.

The contact is loose but firm and may be adjustable as to pressure in any suitable way.

The essence of the invention lies in the use of a contact consisting as aforementioned of a conchoidal fractured surface or edge of crystallised disulphide of iron (FeS_2) and an electric conductive point or surface.

Having now fully described and ascertained my invention and the manner in which it is to be performed, I declare that what I claim is:-

***1st** In an improved detector of electric oscillations for wireless telegraphy and like purposes, a contact characterized by crystallised disulphide of iron having a conchoidal fractured surface.*

***2nd** In an improved detector of electric oscillations for wireless telegraphy and like purposes, a contact characterized by cubical crystallised disulphide of iron having a conchoidal fractured surface.*

***3rd** In an improved detector of electric oscillations for wireless telegraphy and like purposes, a conchoidal fractured surface or edge of crystallized disulphide of iron, and an electric conductive point or surface in contact with said fractured surface.*

***4th** In an improved detector of electric oscillations for wireless telegraphy and like purposes, a conchoidal fractured surface or edge of cubical crystallised disulphide of iron and an electric conductive point or surface in contact with said fractured surface.*

***5th** In an improved detector of electric oscillations for wireless telegraphy and like purposes, a contact between the surface or edge of a conchoidal fracture of cubical crystallised disulphide of iron and a point or surface of gold or gold alloyed with silver or silver or copper or silver and copper.*

***6th** In an improved detector of electric oscillations for wireless telegraphy and like purposes, a contact between the surface or edge of a conchoidal fracture of cubical crystallised disulphide of iron and a point or surface of silver, platinum,* *copper, iron, iridium, nickel, or other electrically conductive metal or alloy or substance.*

No. 2: Wireless Patent:

Patent Title and Record: Improved means of producing electric oscillations for wireless telegraphy and other purposes. "*National Archives of Australia*" Series Number: A13882, Control Symbol: 003354/11, Item Barcode: 200705425, Date Range: 1911-1911. Australian Patent: No. 2621/11, 29 September, 1911. (Drawings). Australian Patent: No. 3354/11, 14 December, 1911.

Patent:

"I, Henry Sutton, Gentleman, of 290 Bourke Street, Melbourne, in the State of Victoria, Commonwealth of Australia, hereby declare this invention and the manner in which it is to be performed, to be fully described and ascertained in and by the following statement:-

The common method of producing electric oscillations is by arranging in series, a capacity or condenser, an inductance and a spark gap. The capacity is charged from electrical sources of high potential until the potential becomes great enough to cause a spark to jump across the gap, and the capacity becomes discharged, and owing to the inductance and gap this discharge takes the form of electric oscillations, when the resistance of the gap is sufficiently low. The gap is a great source of loss of energy, and in addition, it prevents the continuance of the oscillations, by virtue of losing its conductivity after the heat of the discharge is dissipated, that is, the oscillations occur only while the air is heated at the gap. The loss or stoppage of oscillations is called damping, that is, the oscillations are damped out.

A circuit of this description is usually called the closed circuit in wireless telegraphy, because it is closed, except at the gap. The presence of the gap, however, means that it is only closed while the discharge occurs, rendering the air conductive by heat. In wireless telegraphy, it is usual to associate with the gap circuit an open circuit, called the "aerial circuit." These may be coupled together by means of what is termed a transformer, in which the inductance of the gap circuit is the primary, and a coil between the air and earth wire is the secondary.

Another form of coupling is that in which the air and earth wires are directly attached to the inductance. This has been termed the autotransformer coupling.

In both these systems there only two circuits---the gap circuit in which circuit the oscillations are generated, and the open circuit in which they are radiated.

On reference to the drawings it will be seen that ***A****, is an air wire;* ***E****, an earth wire;* ***I****, an inductance;* ***C2****, a shunt capacity in parallel with* ***I, g*** *a spark gap;* ***T****, a transformer.*

My invention includes a third or intermediary circuit ***C2****, a shunt capacity in parallel with* ***I; g*** *a spark gap;* ***T****, a transformer.*

My invention includes a third or intermediary circuit **C2. I.** *placed between the gap circuit* **g. CI. I.** *and the open circuit* ***A. I. E.*** *This resonant circuit* **C2. I.** *is completely closed and has no gap.*

This intermediary circuit being of such a nature that oscillations set up in it do not rapidly die out by virtue of damping; the circuit being completely closed, any oscillations set up therein only die out gradually, and transfer to an open radiating circuit a much longer wave train or number of oscillations, than can be produced in any other way.

The third or intermediary circuit I so construct as to be in resonance with the gap circuit and also with the open radiating circuit. By this means my intermediary circuit becomes exalted in its action on the open or radiating circuit. My intermediary circuit is of such a nature that it also acts as a variable coupling between the gap circuit and the open radiating circuit.

My improvement consists in placing a capacity in parallel with the inductance of a common circuit, that is. By doing this, I constitute the inductance, and this capacity a complete closed circuit. The inductance being also in series with the gap and a reservoir capacity.

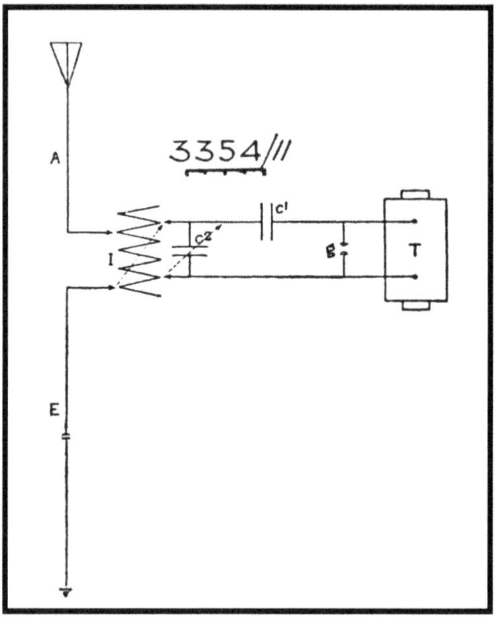

Circuit diagram from patent

The one variable or fixed inductance therefore is situated in the gap circuit, and in the open radiating circuit, and in my closed resonant intermediary circuit, the open radiating circuit being constituted by an air wire, the inductance, and an earth wire.

Having now fully described and ascertained my said invention and the manner in which it is to be performed, I declare that what I claim is:-

***1st.** A means of producing electric oscillations of a sustained character by connecting a capacity in parallel with a variable or fixed inductance, the said variable fixed inductance being also in series with a spark gap, and another capacity, the charging conductors being attached on each side of the gap.*

2nd *The combination of a closed oscillating circuit comprising a variable or fixed inductance and capacity, with an open radiating circuit comprising an air wire attached to said variable or fixed inductance, and a reservoir capacity and gap in series with said variable or fixed inductance, the charging conductors being attached on each side of the gap.*

***3rd.** The combination of a closed oscillating circuit comprising a variable or fixed inductance and capacity with an open radiating circuit comprising an air wire and an earth wire attached to said variable or fixed inductance, and a reservoir capacity and a fixed gap or rotating gap or quenched gap, in series with said variable or fixed inductance, the charging conductors being attached on each side of the gap."*

END NOTES

CHAPTER ONE:

Note 1: Site of Tent on Bakery Hill:
The Sutton tent was pitched on the site of St Paul's Church at 3 Humffray Street Bakery Hill, Ballarat East.

Note 2: James Johnson:
Little is known about James Johnson or what happened to him after the battle at the Eureka Stockade, it is only known that years later he was to survive his sister Mary as she left him money in her will.

Note 3: Henry's Birth Date:
Henry's date of birth has been popularly but wrongly believed to be 3 September, 1856, the error of the wrong date being attributed to William Bramwell Withers book, "The History of Ballarat" which was published in 1887.

Note 4: Scott's Parade Cottage:
The old miner's cottage in Scott's Parade is still there, now known as 121A, Scott Parade Ballarat.

CHAPTER TWO:

Note. 1: Mr. Taylor Australian Airships: "The Argus", Melbourne, 15 September 1909, p. 8
Sir, Without in any way detracting from the experimental work of Mr Lawrence Hargrave, of Sydney, I would point out that history does not bear out the statement in "The Argus" of today by Mr George Taylor that "the original idea of the aeroplane was invented by Mr Hargrave, of Sydney 15 years ago."

The facts are:- The original monoplane was invented by Hensen in 1843; this comprised a single plane, driven at a small angle by two propellers. In 1867 Wenham advocated the use of superimposed planes. In 1868 Stringfellow constructed a three superimposed planes' model on Wenham's plan. In Australia in 1876 I made numerous experiments in aeronautics, and embodied these in two papers which are published in the London Aeronautical Society's "Journal" for 1878. In these papers I mention Wenham and Stringfellow's work on aeroplanes. In my paper I show drawings of what I think is the first Australian-designed aeroplane, and in referring to my superimposed planes I say:- "A kite made on this principle will be found considerably more buoyant than if a single plane." This is the first suggestion for using more than one plane in a kite, and is the forerunner by many years to the box kite.

My experiments were published in 1878 (31 years ago), and I refer mainly to bird flight. Stringfellow constructed his model aeroplane 40 years ago. Wenham suggested the superimposed (box form) planes and invented the word "aeroplane" 41 years ago. Hensen invented his monoplane aeroplane 65 years ago, so Mr Taylor is in error when claiming that "the idea of the aeroplane was invented in Sydney 15 years ago."

The true conquerors of the air are Wilbur and Orville Wright. They have discovered principles that all others missed.

Yours, &c., HENRY SUTTON. Sept 14.

Note. 2: Henry Sutton, Australia and the Aeroplane: "The Argus", Melbourne, 17 September, 1909, p. 5

Sir:- Mr Henry Sutton, in "The Argus" of to-day, takes me to task for crediting Hargrave, and incidentally Australia, with giving the world the successful aeroplane, but a brief glance at the facts will substantiate my claim. It must be pointed out that I based my remarks upon practical work done, not upon theory only, for the theory of the aeroplane goes back over 4,000 years, and it's origin must be sought for in the kite.

Without in any way depreciating the pioneering work of the experimenters mentioned by your correspondent-namely, Hensen, Stringfellow, and Wenham history does not credit them with much practical success.

As far back as 1809, Sir George Cayley first pointed out that two planes at a dihedral angle formed a basis of stability, and he made a machine with a surface of 300 square feet, which collapsed before there was an opportunity of testing its effect. Thirty years after, Hensen patented an idea of a flying machine, and in his specification described the surface of the planes as measuring 4,500 square feet, the tail being 1,500 square feet extra, and a total weight of 3,000lb.! But he never completed the machine, for reasons that are apparent.

With Stringfellow in 1845, he made a model of 70 square feet, but the machine was never able to make any flight.

Wenham was the first to point out that the supporting surface could be best utilised by disposing it in such a manner that it offered a broad spread with little fore and aft depth, this being a great step forward, but Hargrave did a good deal more than theorise. In a paper read before the Royal Society of New South Wales on August 6, 1884, on his discoveries regarding the Trochoidal plane, he pointed out the track to successful mechanical flight, and his cellular kite, universally known as the Hargrave box-kite, he demonstrated its remarkable lifting powers, particularly in 1895, when he himself was lifted in the air by its aid, thus actually demonstrating the possibility of the heavier than air machine that the world was not slow to adopt.

Professor Chanute, the world's greatest authority on aerodynamics, whose monumental work, "Artificial Flight," is the standard textbook on aerial navigation, says therein:- " If there be one man more than another who deserves to fly, that man is Lawrence Hargrave, of Sydney, Australia. He began with small, cheap models, and gradually enlarged their size, and obtained flights longer than any heretofore accomplished. Hargrave's cellular kites constitute quite a new departure, and practically consist of superposed aeroplanes connected together in pairs. He may not be the first man to ride at will, but he deserves to succeed." Professor Threlfall, now of Birmingham University, thus crowns Hargrave and Australia:- "Sydney will be one day noted, not for its famous harbour, but as being the home of Hargrave, the man who invented the flying-machine."

Yours, &c., Sept 14. GEORGE A TAYLOR

Note. 3: Australia and the Aeroplane: "The Argus", Melbourne, 21 September, 1909, p. 7.

Sir:- I trust you will afford me a little space in reply to Mr. Taylor's letter of to-day. In your issue of the 15th inst, I questioned Mr. Taylor's statement that the "idea of the aeroplane was invented by Mr. Hargrave, of Sydney, 15 years ago." I am glad to see in your issue of to-day Mr. Taylor now puts the idea back to over 4,000 years. That, at any rate, shifts it from Sydney, as the kite is supposed to have originated in China.

But an aeroplane is not a kite, neither is a kite an aeroplane. They each obey perfectly defined laws that have little in common. The word aeroplane is used to define a plane, urged mechanically against the air, and this is the basis of aeroplane fight, but it is not the basis of kite flight.

Mr. Taylor says he based his remarks on "practical work done." and cites Mr.

Hargrave's paper to the Royal Society of New South Wales in 1884. This 25 years ago. Mr. Taylor's previous statement was "the idea of the aeroplane was invented by Mr. Hargrave, of Sydney, 15 years ago." Now why should the idea be invented 10 years after the "practical work done"? My first letter contended the original idea was more than 15 years old. I will now show that the practical work was done by others more than 25 years ago.

Penaud, in Paris, in the early seventies, devised, manufactured, and sold models that made successful flights.

At the annual meeting of the Aeronautical Society, held at the Society of Arts June 13, 1878, Mr J. Glashier, FRS. in the chair, flying models were exhibited. The "Journal" says:- The model of the albatross was particularly successful. This model was 12½ feet across."

Referring to other models it says:- "They transported themselves with varied velocity the length of the large hall, and were then only stopped in their further flight by the walls." This was the result of practical work done prior to 1878.

Penaud is credited with inventing the rubber propulsion in his practical models. This rubber propulsion was subsequently used by Hargrave in his first experimental models. Hargrave is credited with very many clever models propelled by compressed air, &c,. but all his work appears to be subsequent to the dates I have given for Penaud, Breary, myself and others. Neither, Professors Chanute or Threlfall's opinions as quoted by Mr. Taylor are evidence of "un fait accompli."

Yours, &c., Sept 17. HENRY SUTTON.

Note. 4: Sutton Dynamo/Electric Motor: "English Mechanic and World of Science" 15 August, 1879, p. 571.

A New Electric Motor:

"It is generally acknowledged that the primary difficulties in the way of the successful utilisation of the immense power of electro-magnets are:- 1st. Owing to the magnetic force obeying the law of inverse squares, and rendering even the most minute decrease of distance between magnetic surfaces almost in contact a terrible means of waste of power. 2nd. The creation of induced currents (to the detriment of the original current) by the approach and recession of parts (i.e. electro-magnets) and make and break of current, the main difficulties being the law of inverse squares and induction from the approach and recession of movable from fixed electro-magnets. These difficulties have been held insurmountable; and yet, with your permission, I will explain an electro motor devised by myself, in which there is no approach or recession (in the strict sense) of magnets, also no

make and break, but a continuous current circulating in coils of electro-magnets. Also the attractive and attracted, repellent and repelled surfaces never vary their distance (yet move), but may be placed so close as to exert the greatest possible force without absolutely touching. The following drawing of side elevation will explain the principle by which these results may be secured."

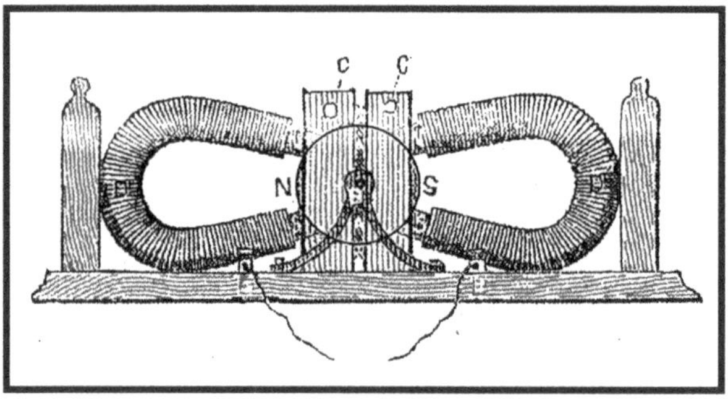

"AA.- A flat disc of soft iron, having an axis running through its centre.

B.- Bearing and support for disc AA, the disc revolving in bearing B.

CC and CC'.- Coils of insulated wire, so arranged that the disc AA is free to revolve in the coils, the disc and coils forming an electro-magnet having a core free to revolve. On a current being sent through the coils CC and CC' the disc AA becomes magnetic, having two polar surfaces N S and two neutral surfaces XX. It will now be seen that by placing the pole of any other magnet so as to act between a polar surface and one of the neutral surfaces rotation will take place, without having to break the current traversing the coils CC and CC'.

DD'.- Two electro or permanent magnets, having their poles arranged as in drawing, the dotted portions being brought within the coils CC and CC'. The action is as follows: the S pole of disc electro-magnet is attracted by the N pole of electro or permanent magnet D', and repelled by its S pole. The N pole of disc electro-magnet is attracted by the S pole and repelled by the N pole of electro or permanent magnet D. By reversing the current in coils CC and CC', and thus changing the poles of disc, the rotation is reversed. By breaking the circuit of disc-magnet the magnets D and D' act on soft iron disc AA as a brake, and stop the rotation.

Owing to the mode by which the magnet force is utilised, the motor occupies little space laterally; therefore a number of discs can be mounted on a common axis and their combined power obtained.

The motor has these advantages:-

1. The current circulating around the electro-magnets never being broken, there is no induced extra current.

2. There being no make and break of current, there are no contact-points to become destroyed or corroded.

3. The electro-magnets always having the same relative position, there is no induced current, owing to approach and recession of movable from fixed electro-magnets.

4. The relative positions being fixed, advantage may be taken of the law of inverse squares, and the great power due to diminished distances in magnetic force utilised.

The whole may be worked with simple commutator or switch for reversing or stopping. The principal point of novelty is that the motor over-comes all the difficulties hitherto held insurmountable in the 'application of electro-magnetic force'. It does not overcome that unfortunate difficulty, and perhaps greatest stumbling-block, the relative cost of zinc and coal. However we live in an age of wonders, and there is no telling where the electricity of the future may come from. Henry Sutton, 31 Sturt-street, Ballarat, Victoria, Australia.

CHAPTER FOUR:

Note. 1: Description of Vacuum Apparatus with Diagram: Henry Sutton, "Transactions of the Royal Society of Victoria", Volume XVIII, 1882, 13 December, 1881, p. 122.

"This apparatus is the outcome of several forms of vacuum apparatus I have devised for the rapid production of high vacuum suitable for electric lighting by the incandescent system, or investigations such as Mr Crookes, has recently given to science.

A mercury bottle (m) when held up, allows mercury to rise in the tube (A), filling reservoir (B), and rising in branch pipe (C) isolates lamp globe (H) from reservoir (B). The mercury passes by outlet pipe (D) filling vessel (G) till it overflows by pipe (E) into vessel (F); (G) is a glass tube of large diameter, having a glass stopper in each end, both stoppers being covered by a mercury jacket; the lower half of this tube is filled with broken pumice, soaked in sulphuric acid to absorb any moisture, the upper half being filled with pumice, gilded with gold leaf to absorb traces of mercury vapour. On lowering mercury bottle, the mercury descends in reservoir (B) and pipe (C), leaving a Torcillian vacuum in (B), the mercury in (D) being supported by atmospheric pressure. On lowering (m) sufficiently, communication is established between reservoir (B) and vessel to be exhausted (H) by means of branch pipe (C) and dissecting apparatus (G), the air in (H) then expanding over into (B); the mercury bottle (m) is then raised again, and the mercury rising in (B) drives out the expanded air by outlet pipe (D). This pipe is bent back on itself at the lower end; this is to prevent the ingress of any air that might collect on end of pipe in vessel (I), alternate raising and lowering the mercury bottle (m) producing a vacuum."

***Note. 2: Sutton's apparatus for the rapid production of high vacua: "English Mechanic and World of Science", No. 886, 17 March, 1882, p.p. 34-35**.*

"I have just received my English Mechanic for Nov. 11, 1881, in which, under the heading "Obtaining and Measuring High Vacua," there is a description of Prof. Rood's improved Sprengel pump: its limit is stated to be arrived at "when the leakage of the apparatus is equal to the amount of air that can be removed in a given time."

I send you a description and a sketch of a vacuum pump invented by myself some time ago, and which, you will see by its construction, has a higher limit still, as the amount of air that can be removed in a given time exceeds the leakage many hundred-fold; in short, the air can be withdrawn more rapidly than it can leak in. Its construction is simple; there are no stop-cocks, nor valves; it is not liable to derangement, and will do in a few minutes that which takes hours on the Sprengel principle.

It is the result of studying the defects of the Sprengel pump. It occurred to me that the mercury in a Sprengel pump, having to go through the form of numberless drops, the entire surface of the mercury is exposed to the influence of the high tension of the vacuum, and, therefore, air that would with mercury in the mass be retained cohesively, would, in the Sprengel principle, be given up in the vacuum.

Another fault I considered to be, that mercurial vapour must be formed at these high tensions, given off from the drops---a fact that can be verified by examining microscopically the junction of a Sprengel fall tube, with its connecting tube to receiver; a gentleman of high scientific eminence having told me that in conjunction with a well-known English scientific gentleman, they had observed this defect.

To overcome these objections I argued that:-

(1) The air must be removed rapidly.

(2) That the mercury must not be in the form of drops.

(3) That the mercury must not leave the pump, but must be used continuously.

I have succeeded in accomplishing this, as the following will show.

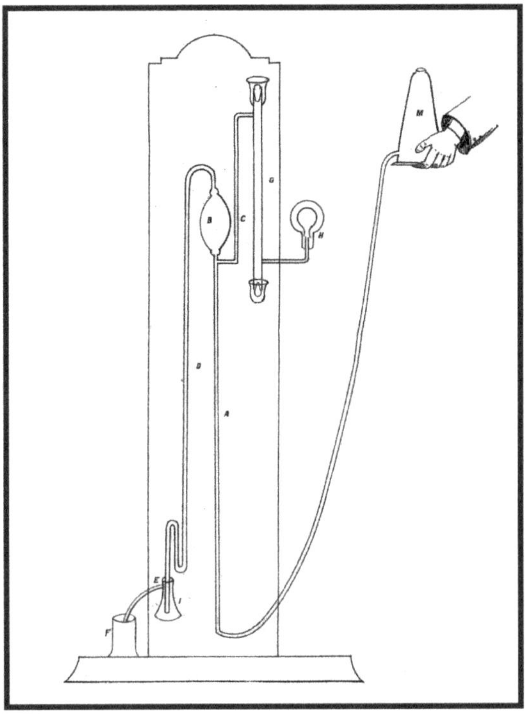

A is a glass globe; at its lower end is attached a glass tube B, 36in. long; the lower end of this tube is bent to a right angle, and attached by a short piece of rubber tube to a cylinder C; in this cylinder a piston D is worked by the lever handle E.

From the upper end of tube B, a branch tube F rises, and is bent to a right angle at the top, and joins the desiccating-tube G, which has its upper end widened into a funnel, and an India rubber stopper fitted in so as to leave the funnel surrounding it. The lower end of G, also has a rubber stopper, which is surrounded by a glass cap H.

About 2in. from lower end of G, a tube of glass I, projects, to which is attached the vessel to be exhausted, L.

From the upper portion of globe, A, a glass tube, J, of fine bore, rises, and bends down parallel to B. About half-way down tube J bends back on itself, then returns on its downward course and enters the tall conical vessel K, reaching to near the bottom.

To set the pump in action, the short rubber tube is removed from B and C, and a current of hot air passed for some time through tube B, and thence through the whole of the pump, driving out all moisture and leaving the pump warm and dry. The rubber tube is then connected and sealed between B and C, and pure mercury previously dried, placed in cylinder C, by removing the piston, which is done by taking the pin out of the junction of D with E, the piston being returned to its place.

The rubber stopper is placed tightly in lower end of desiccating tube G, and its cap H filled with mercury, making an hermetically-sealed joint. Pumice soaked in sulphuric acid, or phosphoric anhydride, is then dropped into G from its upper end till half full; the upper half then being filled with gilded pumice, the rubber stopper belonging to upper end of G is now tightly pressed in, and the funnel end filled with mercury, thus hermetically sealing this joint.

The apparatus is now ready, and is worked by pressing the lever E, and thereby forcing the mercury up into tube B, thence into globe A, the mercury also rising in branch tube F, isolates the vessel L and its connections from communication with globe A.

The mercury being still forced into A forces the air out by the fine tube J, and when the mercury begins to rise in K, the lever E is raised, and the mercury falling back from A, produces in top of A a Torricellian vacuum. The mercury in K now rises into J, until the column of mercury in J exactly balances the atmospheric pressure at K. Any air that may rise in the bottom of J cannot pass the bend, owing to the specific gravity of the air being less than the mercury.

The lever E still being raised allows the mercury to fall from A, and also from the branch-pipe F, until, having fallen below the junction of B with F, communication is restored, and the air expands over from L, and its connections, into the vacuum in A, the mercury column in J lowering at the same time.

The lever E is now pressed down again and the same action takes place as in first stroke, a mass of air equivalent to the capacity of A being removed at each stroke of the piston. The rapidity with which the vacuum is produced is proportional to the ratio of the capacity of A with the capacity of vessel to be exhausted and connections. And as the globe A can be made very large, and thereby a great bulk of air removed with each stroke, it follows that the limit of the pump is very high, much higher than it can possibly be for any form of Sprengel.

The gilded pumice is to absorb any trace of mercurial vapour, and the chemical substance in tube G to absorb moisture. In this pump there is little chance of mercurial vapour forming, as the mercury is not divided into a number of small drops and under high tension for a long period of time. It will be seen that my form is a departure from the usual mercury pumps, as the mercury is worked in the mass and without the aid of stopcocks or valves. It does its work perfectly and rapidly.

I have not yet measured the vacuum produced, as I cannot obtain the appliances here; but it is and possibly reaches the theoretical limit. In conclusion, the pump is not patented, and therefore all have the liberty of using it.

Note. 3. On a New Form of Secondary Cell for Electric Storage: Henry Sutton, "Transactions and Proceedings of the Royal Society of Victoria", 15 December, 1881, p.p. 110-114.

"The Great utility of some thoroughly practical method of conserving electric force has caused a great deal of attention to be applied to the subject. No system of electric supply can be considered perfect until some means is used to store the force generated that it may be drawn off equally and regularly and this whether the generator be on or off.

If we take as an example of electric supply the present systems of electric lighting, it is at once seen, should an accident or stoppage take place in the machinery generating the current, the whole of the apparatus, such as lamps or motive machines, would be influenced. Should there be a reservoir of electricity between the generator and the apparatus for utilising the force, this inconvenience could not occur, as the reservoir would then supply the necessary force.

All the present systems of storing electricity depend on certain chemical changes produced by electrolysis; the first effective apparatus being that discovered by M. Plante, and which consisted of two sheets of lead immersed in dilute sulphuric acid. When these plates are made the positive and negative electrodes of a few voltaic cells or other source of electricity of sufficient power, the oxygen of the decomposed solution combines with the positive electrode, and forms a coat of the insoluble peroxide of lead. On disconnecting the charging source, the plates have the power of generating a powerful current of electricity, owing to the great affinity the oxygen of the peroxide has for the hydrogen of the solution when the circuit is closed.

A recent form of this cell has been devised by M. Faure, in which the lead plates have a coat of red oxide of lead painted on, the chemical change in this case being- the red oxide on the positive electrode on charging becomes exalted to the state of peroxide, whilst the hydrogen combines with the oxygen of the negative red oxide coating and reducing metallic lead; the chemical change of discharge being precisely the same as in Plante's cell.

This red oxide cell has caused a great stir in the electrical world, owing to some exaggerated reports having gone through the press.

Although there is a great deal that is pernicious in exaggerated press reports, they have one virtue, and that is, the incentive they give to workers to improve, or if possible, discover new means of producing the same results. This was shown in the invention of the telephone; and the same will occur in connection with electrical storage.

These reports reminded me of an experiment I made some time previously on the behaviour of peroxide of lead as an element in voltaic cells. I prepared some peroxide by treating red oxide of lead with dilute hydric nitrate till the brown precipitate of peroxide fell; collected the precipitate, and made a conglomerite of it, and using it as an element in a small voltaic cell.

I have since then gone through a long series of experiments in storing electricity, and made many different forms of cell; one being a porous pot containing dilute hydric sulphate and a sheet of lead, and an outer containing vessel, having a sheet of lead immersed in a solution of acetate of lead, the plate in the porous pot being made the positive electrode. This cell had the power of storing electricity, by peroxidising the positive electrode and depositing metallic lead from the solution on the negative electrode, the acetate being then converted into a solution of acetic acid. On discharging the peroxide is reduced and the apposed element oxydised, the oxide formed combining with the acetic acid, and forming acetate of lead.

During my experiments I found that the red oxide of lead is a bad conductor of electricity; that peroxide of lead is a good conductor; and that by amalgamating lead plates with mercury a mark-ed increase was immediately manifest in polarisation effects, the plates being more uniformly and rapidly peroxidised, and local action entirely disappearing. These mercury amalgamated plates at once gave me advance over other cells.

I used them in many ways, constructing cells in which the positive electrode was amalgamated lead, the negative being coated with prepared peroxide or with the red oxide, or amalgamated, or combinations of these and solutions of various kinds. I also made cells of peroxide, and also of red oxide, formed into porous conglomerites, having a wire projecting from each end, the conglomerate being immersed in dilute hydric sulphate. I constructed cells in which the plates were parallel, and the red oxide, as also the peroxide, being filled in between the plates. In this case the red oxide is useless and peroxide efficient, owing to their relative conductive power. In all cases in which amalgamated lead was used as a positive electrode the effects were very marked.

Having thoroughly tested the amalgamated lead, and found it the most efficient positive electrode for conserving the oxygen of the decomposed solution, I investigated the behaviour of various forms of negative electrode, having always the object in view, of making the hydrogen of the decomposed solution do some work; the proper thing, of course, being to make it assist in depositing the metal from the solution on the negative electrode.

I thought by having negative electrodes, whose oxides should be soluble in the solution remaining after electrolysis, and which could be re-deposited from the solution, or by having metallic solutions from which the metal could be deposited- the resulting solution being such that on the oxidation of the deposited metal it would combine with the oxide, and again from the original solution-I should produce a perfect means of storing electric energy by conversion into chemical energy, and which would economically return the current as the chemical affinity ran down.

Having started from this foundation, the results obtained are such as to have an important practical bearing on the future of electric work.

The experiments comprised lead amalgamated with mercury as a positive electrode, with negative electrodes composed of either, zinc, iron or copper; in each case the solution between the electrodes being a salt of the metal composing the negative electrode. With zinc a solution of zinc sulphate was used; with iron, sulphate of iron was the solution; and with copper, sulphate of copper.

In all these cases the results were not only far more powerful than any form of cell I had hitherto devised, but also remarkably constant.

The cell with zinc negative electrode I discarded, owing to the necessity there would be to keep the zinc plate amalgamated to prevent local action; the iron negative cell being set

aside owing to the iron oxidising when not in use. In both the zinc and iron negative electrodes the evil effects only show when the cell is charged.

The cell having a negative electrode of copper is the one I have adopted as a thoroughly economical, lasting and practical form of storage reservoir. The chemical changes in this cell are exceedingly interesting and beautiful.

The cell is composed of a sheet of lead cleaned with dilute sulphuric acid, and amalgamated thoroughly with mercury, and a sheet of thin copper a little shorter; the two sheets are perforated with a number of holes. The copper is placed uppermost, and separated by bands of rubber, having pieces cut out every few inches, the whole being rolled into a spiral; the object of the holes in plates and pieces cut out of rubber being to allow the solution to have free access to the plates. This combination is immersed in a solution of cupric sulphate, and the amalgamated lead made the positive electrode of a suitable source of electricity. The chemical action being the oxygen of the decomposed solution combines with the lead, the hydrogen replacing the copper in the solution, and metallic copper being deposited on the negative electrode. As decomposition of the solution proceeds it becomes colourless, and more acid, until, when the whole of the copper is deposited, we have the solution transformed into dilute sulphuric acid, the amalgamated lead peroxidised, and pure copper deposited on the negative electrode.

During discharge the peroxide of the lead element is reduced, the copper element becoming oxidised; the oxide combines with the acid, and forms cupric sulphate, the solution returning to its original azure blue colour. This change in colour forms a beautiful means of knowing when the cell is fully charged. The power of this cell is very great and constant; it can be made to last for hours, the length of time being dependent on the quantity of cupric sulphate being decomposed in solution. I have, by the decomposition of one pint of solution cupric sulphate, obtained over two hours effective work as electric energy.

It will be seen from the foregoing that this method of conserving energy has a wide field before it."

NOTE: 1-- The amalgamated lead electrode will not peroxidise if free cupric sulphate crystals are present in the solution, it being essential that the solution become acid.

NOTE: 2-- The charging is accompanied with a peculiar rattling noise, and is not due to escaping gas, as the noise occurs when no gas escapes from the solution, but may be due to change of form in spiral due to deposition of copper.

Note. 4: On a New Form of Secondary Cell for Electric Storage: Henry Sutton, "Transactions and Proceedings of the Royal Society of Victoria", 15 December, 1881, p.p. 110-114. (ADDENDUM)

"The secondary, in its present form, consists of a flat copper case, having the form of the outer containing vessel of a Grove's cell. In this copper cell is a sheet of lead amalgamated with mercury, and which has a piece of flannel, or other porous material, enveloping it tightly, the object of which is to prevent the peroxide (which forms a thick layer) from falling off by the wash of the solution, or by jarring. I have removed as much as six ounces of peroxide, which has been formed in a single charge in one of my cells.

The amalgamated plate is supported in middle of copper cell by being held in a groove in a piece of parrafined wood lying at the bottom of copper cell, and also by a parrafined wood cap, which closes the top; it is charged with a solution of cupric sulphate, containing one-twelfth of hydric sulphate. The amalgamated lead is made the positive electrode, and the copper case receiving the deposited copper when charging with electricity. The electromotive force of the cell is about two volts."

CHAPTER FIVE:

Note.1: Henry Sutton (Alias Delta): Scientific Notes, Criminal Science: "The Touchstone" Ballarat, 4 January, 1884, p. 7.

"Science, like everything else, may be, and is, put to uses by no means elevating in character; the investigator, discoverer, and inventor, have each in their turn so increased its powers that at the present time the greatest mysteries of the conjurer are inferior in wonder to some of its accomplishments.

What a marvellously wonderful thing is the Photophone!, by means of which the human voice has actually been transmitted from one station to another along a beam of sunlight. Could anything be more calculated to excite our admiration than this? And yet it is done by means which, although ridiculously simple, are by no means simply ridiculous. The modus-operandi of this production of man is simply the human voice or other sound impinging on a thin glass mirror, throws the latter into vibrations, these vibrations causing a beam of light---which is being reflected by the mirror---to pulsate; these pulsations of light are received at the other station in an apparatus in which the variations of the light beam cause corresponding variations in an electric current, this varying current then causing an ordinary telephone to emit similar sounds to the original. Of course, all this has nothing to do with scientific crime, and I merely mention it to show that startling things are capable of being done.

The first process I intend explaining is that of "sweating," and I trust that no person will utilise the information given in this article to the detriment of their fellows. We all know the old process of putting a number of sovereigns in a bag and shaking them---be careful, reader, by shaking I do not imply the vulgarism, but simply the mechanical process of reducing the weight of the coin by their mutual friction---this process is old, better brains adopt the following:-

A jar is taken and filled with a solution of cyanide of potassium; in this stands a small jar of unglazed earthenware; this also contains cyanide solution. A sovereign is then suspended by a wire in one of the jars, and a copper plate in the other. The coin is connected to the positive pole of a battery, and the copper plate to the negative; on a current of electricity passing, the coin begins to dissolve, and in a short time metallic gold is deposited on the copper plate. After a number of coins have been thus treated, the crime is completed by removing the gold from the copper plate, and disposing of it. One shilling's worth of gold removed from each coin is an income of 5 per cent. on the number treated annually. Precisely the same process can be done with silver coin.

The criminal reproduction of bank notes is a simple process, in which the art of photography takes a prominent part. Bank officials consider themselves pretty safe by printing their notes in nonactinic or variable actinic colours. I must explain this technical expression to the uninitiated. The radiation we receive from the sun is characterised by the by three qualities: light, heat, and actinic or chemical power. If we decompose sunlight with a prism, separating it into various coloured rays, we find the light strongest in yellow rays, heat strongest in red rays, and chemical power strongest in violet rays, this actinic power, then, being strongest in violet, weaker in red, and of least power in yellow. From this it will be seen that if a bank note be printed in these three colours, and a photograph be taken of it, in the resulting print the violet will show as white, the red as grey, and the yellow as black. By judicious combination, a note can be so printed in colours as to render it impossible (till recently) to photograph. If we print on white paper in a blue coloured ink a photograph could not be taken; the whole would appear as a sheet of white paper, with no printed matter on it.

Notes which are not printed in colours are easily photographed and reproduced, the process being:- A note is photographed on glass, and instead of being printed on paper it is printed on a film of gelatine, sensitized with bichromate of potash; the peculiarity of gelatine when thus sensitized is, that wherever the light strikes it, it is hardened in exact proportion to the strength of the light. This property is utilised by simply placing the glass negative over the gelatine surface, and standing it in sunlight. An imprint is then produced on the surface of the gelatine, which is an exact replica of the note. This biochromatized gelatine print is then washed with water, and the soft parts swell up, the whole taking the form of an engraving. This photo-engraving, when dry, is thoroughly coated with a thin film of blacklead or other conductor of electricity.

The third part of the process includes the science and art of electrotypy. The gelatine plate is now placed in a bath of sulphate of copper (bluestone) containing a little sulphuric acid; a sheet of copper is also suspended in the bath. The negative or zinc pole of a battery is then properly connected to the gelatine plate; and the positive or copper pole to the copper plate; a current or electricity immediately passes, the copper plate dissolves, and metallic copper is reduced from the sulphate of copper and deposited on the gelatine plate in the form of a thin film, which continually increases in thickness. When a sufficient quantity of copper is deposited it is removed from the gelatine, and is ready when mounted in lead on boxwood, to print as an ordinary copperplate engraving. I stated earlier that owing to the use of nonactinic colours, bank officials stopped all this, but science has just made another advance, and now the process is known by which the so-called non-actinic colours can be photographed, showing their true shades. The new process is the invention of Messrs. Altout-Tailfer and Clayton, and is attained through treating gelatino-bromide plates with eosine. This discovery will have a very important bearing on photography in all its branches, and considering the rapid development of dry plate photography, through which any person, at a cost of a few shillings, may become an expert; a difficulty is presented to the authorities. Of course, there is a fair side to everything dark; the new process will enable persons with blue eyes or freckled skin to be taken without appearing to have white eyes or the pits of smallpox. Landscape photography also will benefit considerably, and ladies will feel no annoyance from the artist requesting them to remove blue or yellow ribbons, or any other highly-coloured ornament."

Note. 2: Henry Sutton (Alias Delta): Scientific Notes, Steam Locomotives: "The Touchstone" Ballarat, 14 December, 1883, p. 7.

"The present motto of the world seems to be, the annihilation of space and time. The matter of transit is just now occupying no inconsiderable amount of attention. The old world influence has extended out here; for not only Melbourne, but also Ballarat, is just now troubling itself about the matter. We hear so much about steam, compressed air, and electric trams that the mind is in a maze as to the best method. I purpose touching lightly and rapidly on these systems, and also saying a few words on one of the most interesting inventions of modern times.

I refer to the new fireless locomotives. The ordinary steam tram is the most powerful, but so surrounded by objections as to be practically out of the field. The main faults are, the liability to cause fire, and being a source of terror to that useful but timid animal, the horse. Compressed air is very good, but still not without danger to passengers. Fancy a reservoir of air compressed to one thousand pounds per square inch; almost as bad as having a barrel of dynamite for a travelling companion. Beaumont has done a great deal to perfect this system: the application of heat to the air outlets, in order to prevent the accumulation of snow through the refrigerative action of the expanded air, being an important improvement.

Electric trams and railways are the latest, and apparently promise great things. Electrical force, as practically understood, is in its infancy; it is, however, a very promising bantling, just powerful enough to assert itself---a force that one day will shake the world as old father steam has never done. The application of electricity to transit purposes is effected in two ways: one, understood as the electric transmission of power method; the other, the automotive, or electrical storage method. The former is adopted in electric railway systems, and the latter for ordinary street traffic. As showing the great value of electrical force as a medium for the interchange of energy, I may say that Sir W. Thomson has calculated that upwards of twenty five thousand horsepower may be electrically transmitted a distance of three hundred miles through a copper rod half an inch in diameter. The latest development in this direction is the Portrush and Bushmills Electric Railway, in Ireland, just completed, at a cost of £45,000 -- a big price for a line only six miles long.

I now come to two of the most beautiful applications of modern physics to practical usage -fireless steam locomotives, a strange expression but nevertheless true. A few words as to how it is done. Water, as we all know, boils at 212 degrees, at ordinary atmospheric pressure. The boiling-point depends on the pressure, steam rising as soon as its tension exceeds the external pressure. Now, if we increase the pressure, we can prevent the vaporisation of the water until it reaches a very high temperature. If we take water at 212 degrees, contained in a close vessel, and force steam in at any pressure above fifteen pounds per inch, the steam will be condensed and the water rise in temperature; the boiling-point rising with the pressure. If we now release the pressure by allowing some of the steam to escape or pass into the cylinder of an ordinary steam-engine, the water immediately commences to boil and give off steam; it will continue boiling and evaporating so long as the tension of the condensed steam is higher than the pressure on the water surface.

This is the principle first utilised, in 1872, by Dr Farmer, and then adopted by Francq; the actual process being: The locomotives have special portable boilers, surrounded by a non-conducting layer of felt, and containing a certain quantity of water; steam of high pressure is forced in from stationary boilers; this portion being completed, the locomotive is ready for work, steam being drawn off into a reservoir where the pressure is lowered to one hundred pounds, and then passes to the cylinder, as in the ordinary engine. So well are these boilers made that there is not more than from three to seven pounds pressure per hour lost by radiation.

Another fireless Locomotive is that invented recently by Honigmann, and though steam is the source of power it is based on quite a different principle. It is known that the absorption of steam by saline solutions heats them up to their boiling-point, which is higher than the boiling-point for water. Water boils at 212 degrees, and a solution of chloride of calcium at 355 degrees. Now, if we condense the steam which rises at 212 degrees from water with a solution of chloride of calcium, the latter would rise in temperature to 355 degrees. This higher temperature is then capable of developing a further supply of steam, which after doing work in the engine, passes into the condenser, and keeps the temperature at 355 degrees; this process going on until the chloride becomes exhausted. Honigmann uses caustic soda, it being commercially cheaper, and has a strong affinity to water; this process is far better than Francq's, as the pressure can be maintained better, and it possesses the peculiar feature that the greater the speed the greater the pressure becomes. The matter is well worthy the study of engineers and there is also a fine field open for further invention in this direction."

Note. 3: Henry Sutton (Alias Delta): Scientific Notes, Mineral Extraction: "The Touchstone" Ballarat, 21 December, 1883, p. 7.

"There is another important application of electricity force in the direction of gold saving. No really practical method has yet been devised for this purpose though a glance through the "Patent" records shows that a great deal of attention and ingenuity has been spent on the subject. All these inventors follow in one

groove, and that, the endeavour to pass a current of electricity through the water passing over the copper plates. A process of this sort, a short time back, caused a little stir in London; but to anyone conversant with electrical matters, it is self-evident that the mere passage of electricity through either the water or the plates can have little effect on the gold. The gold must be in solution in order to come within the electrolytic influence of the current. I have often wondered why inventors do not proceed on the lines originated by Davy when he decomposed the alkalies with electricity, and forced asunder compounds that had defied the powers of the chemist.

Andrew Crosse, many years ago, quietly conducted a series of experiments which, I think, contain a very valuable germ. Chemists know that both the sodium and ammonium amalgams with mercury are very active as gold-saving agents. Crosse, basing his experiments on these facts, endeavoured and succeeded in producing the ammonium amalgam by the aid of electricity, and extracted gold up to assay. This process has evidently been forgotten, as in no modern electrical work, that I have seen, have I observed any reference to it. It was only by the chance straying into my hands of a book on the Life of Crosse, written by his wife, that I stumbled across his experiments. I give the following one:- Crosse placed the powdered quartz in a clay vessel, and intimately mixed with mercury; over this he poured a solution of carbonate of ammonia, and then applied the negative pole of a voltaic battery to the mercury, and the positive pole to the solution of carbonate of ammonia. The result is, the mercury becomes very much more fluid and active, and on stirring, the finest particles of gold are held. I have myself tried Crosse's process. There will, no doubt, someday be discovered a simplified process whereby this truly great force will be utilised for the purpose of gold extraction."

Note. 4: Henry Sutton: Correspondence, The Peculiar Sunset: "The Touchstone" Ballarat, 7, December, 1883, p. 12.

"I noticed a letter in the Ballarat Courier of Wednesday, in which the writer endeavours to explain the cause of our highly-coloured sunsets. I would not call your attention to the matter, only that the knowledge of some of our most able men is incidentally denied. I refer to the writer's novel announcement to the effect that there are no skilled spectroscopists out here. Now, this is an error; for both Mr Ellery, the Government astronomer for Victoria, and Mr Russell, filling the same office in New South Wales, are highly skilled in the use of the spectroscope.

We have in our very midst a gentleman who has been under the tuition of the greatest living authorities on spectroscopy, namely, Professor Smith, of the School of Mines, who, I believe, has studied under Roscoe and Bunsen, than whom two higher exponents of both the art and science of spectroscopy it is impossible to find. It may interest the writer of the letter in "The Courier" to know that actual observation with the spectroscope led Mr Ellery to conclude the presence of watery vapour.

To the theory that we attached the remnants of Biela's comet, it is only necessary to say, the amount of matter representing the remains of Biela's comet would not refract the sun's light sufficiently. According to Proctor, millions of cubic miles of the comet's tail may be compressed into an ordinary sized hat. Starlight is not appreciably dimmed after passing through three or four millions of miles of tail.

The peculiar sunsets are seen over some sixty millions of miles of the earth's surface, so that the matter of Biela's comet, as last seen, would be too tenuous to refract the sun's light to the extent observable over this immense area.

Note. 5: Henry Sutton: Correspondence, The Strange Light: "The Touchstone" Ballarat, 7, December, 1883, p. 6.

Our evening sky for the last two or three months has been a theme for general speculation, scientific and otherwise (especially otherwise), owing to the roseate hue visible after sunset. There seems to be slight doubt, as to the true cause: Mr Ellery, favours the view that it is owing to the presence of moisture high up in the atmosphere, but considers it might be due to the earth passing through a stream of meteoric dust. In either case, the light of the sun being refracted is the cause of the colour. I think Mr Ellery, has given the true explanation in the assumption of water vapour.

I remember observing and noting down the fact on returning from a visit in the country. Immediately to my right, and in the direction of the setting sun, was a range of hills; on one of these, and situate some two or three hundred feet higher than my position, was a mining claim, from the engine stack of which steam was issuing. From my standpoint, the sun had set. My attention was, however, arrested by the beautiful manner in which the sunlight was being refracted by the steam near the orifice of the stack; the uncondensed and partially condensed steam so refracted the light as to give me the hue is now seen in the evening. At the point where the steam was completely condensed there was partial opacity, the perfect condensation preventing the thorough transmission of light; the condensed was distinctly marked off from the partially and uncondensed by the carmine or roseate hue of the latter. The presence of an abnormally excessive quantity of watery vapour in our higher atmosphere should, during its presence, have considerable meteorological influence, for warm clouds on meeting this cool layer would become condensed, and discharge their contents; the opposite effect should occur if we are passing through the dust of a shattered planet, as some conceive meteoric streams to be…

The light we receive from the sun is composed of a number of rays of light, having waves of different length. A wave of red light is one thirty-nine thousandth of an inch long, and a wave of violet light one fifty-seven thousand five hundredth of an inch: between these extremes the wave lengths of all visible colours are situate.

Now, when a ray of light strikes obliquely any transparent substance of different density from that which it is leaving, on entering this substance the ray is bent (or refracted) from its original direction; the amount of bending varies with the length of the waves of light, and as sunlight is composed of a number of rays having different wave lengths, the shorter being bent most, and longest least, the result is the sunlight is split up into a series of coloured rays (colour being dependent on length of wave). The rainbow is an instance of split up sunlight due to the refractive power of rain drops. In the case of our strange light, we see the least refracted rays, and to some extent rays of shorter wave length. The most refracted rays of shortest wavelength reach us too near sunset to be visible.

This peculiar atmospheric phenomenon seems to be very well spread, for not only is it seen over the entire continent of Australia; but I notice it is observed at the Mauritius. In India, during the latter part of August and during September the sun appeared of greenish hue; and on September the sun of greenish hue; and on September 1st, at Trinidad, the sun appeared blue. These daylight effects, it may be observed, occurred during the activity of the recent seismic disturbance at Java. Whether this disturbance and the varied atmospheric phenomena may be considered as cause and effect it is difficult to say. Some have gone so far as to consider our strange light to be due to volcanic dust ejected during the late Javan disturbance.

This I hardly conceive possible; but still the man would be bold who should deny that there is some connection between the seismic effects that at the present time are almost continuously being recorded and our atmospheric state. It is well known that outbursts of gas in coal mines are intimately associated with barometric pressure, and it is not beyond the scope of theory to imagine that volcanic outbursts may to some extent be connected with pressure; the idea that such is the case is held by some scientists. The evident cause of the blue and green coloured sun, is, that the refraction was so great as to render visible the rays of shortest wave length. During our last moon I noticed it as having a slight green tint."

Note. 6: Mr Alexander: Correspondence, The Peculiar Light: "The Touchstone" Ballarat, 28 December, 1883, p. 13.

"Mr Sutton, in his last letter, gives evidence that he has indeed an easily satisfied mind. He saw the thing "in print," and therefore he was convinced. I must have a wickedly sceptical mind, for even if I had seen Mr Ellery's reasons in print, I should still have deemed them most unsatisfactory.

Perhaps it may be news to Mr S -- at any rate from the way he writes he seems to be ignorant of the fact-- that vapour of water is always present in the atmosphere. And therefore the absorption lines due to its presence should always be found in the spectrum of the sky. No reasonable person can doubt the soundness of this position. The presence then of these lines in the spectrum is not a convincing fact (except to an easily satisfied mind), proving the composition of the strange light which has abided with us so long, and instead of that most likely is only the spectrum of the watery vapour which inevitably exists between the earth and this ethereal matter enveloping us. Mr Sutton ridicules the idea of the great star shower of 1872 being anything more than a coincidence with the position of Biela's comet. Yet Mr Proctor, in his papers on "Biela's Comet", and "A Recent Star Shower," considers the connection between star showers and comets as "perfectly demonstrated." Mr. Sutton deals rather disingenuously with a portion of my letter. He writes as if I said that there was no change in the form of the strange light.

What I did say was that there had been no material change, having in mind at the time the very material changes which hitherto, at any rate, have been continuously taking place in vapour of water suspended in the air; how one day it is totally invisible and the next heaped up and condensed into great cumulous clouds, or stretched out into long feathery streamers, or driving along as damp, wintry fogs. If this peculiar light is caused by vapour of water, why has it so strongly altered its usual behaviour.

I see Mr Norman Lockyer is not of the opinion that this strange phenomenon is caused by water. But Mr Sutton knows better than him, for he has seen it "in print," and that has convinced him. You see, then, Mr Sutton has been doing the very thing he has been charging me with, only he has been trying to "sit on" Mr Lockyer, as well as his favourite authority, Mr Proctor. I don't think either of the gentlemen have been hurt much by the process. The fact that Biela's comet does not complete its revolution till December next does not affect my mild little theory much. For Mr Proctor states that there are lagging portions of Biela's comet following in its track many millions of millions of miles behind the rest, and this earth touches that track sometime between the months of October and December each year."

Note. 7: Henry Sutton: Correspondence, The Peculiar Sunsets: "The Touchstone" Ballarat, 4 January, 1884, p. 4.

By your last issue I see Mr E. J. Alexander has satisfied himself that I have an easily satisfied mind. Now, there are certain minds which, owing to a combination of dullness and stubbornness, are never satisfied. I feel truly grateful to Mr A. for so kindly placing me out of this class. If he will just go through my letters again, he will see my reasons for being convinced. Mr A. has made a brilliant discovery, and that, my ignorance of the continual presence of aqueous vapour in the atmosphere; this he reasons out from the "Way I write." Mr A. credits me with an expression "absorption lines," and from this deduces argument, "that no reasonable person can doubt the soundness of." This is emphatic! and almost equal to his expression "I don't think I would have much difficulty in showing that Mr. Proctor contradicts himself."

But again, Mr A. must go through my letter, and he will find no reference to absorption lines; he will see simply the word "absorption." Mr A. is evidently not aware that there is such a thing as general absorption, as well as selective absorption, the latter producing the lines, the former the effect referred to in my letter, and the existence of which proved the cause of the peculiar sunsets. The word appears in a very simple passage, so there can be no excuse for Mr A. supplementing the word "lines."

Another discovery Mr A. has made, and that: that I ridicule the idea of the meteoric shower in 1872 being the disintegration of Biela's comet. If Mr A. will just run through the letter again, he will find there is no attempt at ridicule; had I used the word "absurd," then it could be construed into ridicule. Yet another discovery is made, and that, that Mr Proctor is my favourite authority; read my letters again "Mr A.," and you will find only one reference to Mr. Proctor and that on the tenuity of the comet's tail. "Mr A." evidently rivals Edison as a discoverer, having further found that I sit on both Me Proctor and Mr Lockyer. If I have done so, it is quite unintentional, but then I sin in good company, for Mr E. J. Alexander sits on everybody, including my humble self.

Why did not "Mr A" tell us that Lockyer (according to a telegram) thought the peculiar sunsets were due to volcanic dust ejected in the Straits of Sundra, but then this tail would not fit in with the comet's tail. It may be of interest to "Mr A." to know that when the peculiar sunsets were first seen here, the earth was something like seventy-one million miles off the track of Biela's comet. I must now withdraw from this discussion with the assurance to "M.A." that the spectroscope cannot lie, and that one proved fact is worth more than the universe filled with theory.

CHAPTER SIX:

Note. 1: Henry's student, David Avery:
David Avery went on to attend Melbourne University where he specialized in chemistry taking the honours examination for Bachelor of Science in 1892, and won the Kernot Research Scholarship for Chemistry.

CHAPTER EIGHT:

Note. 1: Reversal, by Henry Sutton: "The Journal of the London Camera Club" 22 October, 1891, p. 192-194.

"In opening the subject of reversal this evening, it would be well to understand that a reversal negative and a reversed negative are different things, a reversal negative being the result of what has been termed reversing action; it is produced direct from a negative, and is reversed as regards right and left. A reversed negative is reversed as regards right and left, but is produced from a positive.

We all know that over-exposure produces flatness owing to the difficulty of obtaining density in the high lights; it is easy to conceive an exposure sufficiently prolonged as to cause an entire loss of the high lights. If in such an exposure the shadows have not received sufficient light to produce a strongly developable image, the result on fixing would be an extreme case of over-exposure, as understood by a thin image; but, if during development stray light obtains access to the plate, the shadows fog, and we have reversal.

I propose making a reversal negative from an ordinary negative. When making experiments of this kind, the natural conclusion one arrives at is that dark flashes seen on photographs of lightning are the result of reversal.

In Australia we have some of the finest electric storms and there I have seen many flashes, the impression of which on the eye was of greater duration than would be due to persistence of vision. On one occasion I was fortunate in observing a brilliant thread suspended between two clouds for several seconds. I mention this to show there are discharges of exceptional duration. The light from one grain of burning magnesium is sufficient to produce reversal, as we shall see by experiment this evening. So we may consider a lightening flash having a similar actinic or photographic value should do the same. This value may be due to a discharge of exceptional duration, or to a brilliant flash caused by a maximum electromotive force with a minimum quantity of matter in the path of the flash.

One evening I was observing a display of electrical discharges in a large cloud situated on the northern horizon; these discharges were accompanied by sympathetic discharges in a cloud on the southern horizon. The northern horizon cloud was approaching, and I concluded that, should the clouds be at different altitudes, a difference of air currents might enable the one to overtake the other.

A more rapid movement of the northern cloud was soon noticed; a camera fixed up and carefully focused on the southern cloud, by aid of the local discharges. All being in readiness and the shutter of slide drawn, but cap still on the lens, I waited events. When the northern cloud reached the zenith, it began to show considerable symptoms of electric excitement at the edges. Knowing this to be my opportunity, the lens was uncapped, and in about half a minute a pair of brilliant flashes darted from the zenith to the southern cloud. I estimated the duration at over one second.

The plate was immediately developed, but not a sign of the flashes. I expected to find a pair of curly black streaks right across the plate; instead, there appeared a very fair cloud negative and the black outlines of some buildings low down in the field. These buildings enabled me to verify the position of the camera, which being attended to, proved the field must have included the flashes. I could never account for getting the clouds and not the flashes, until the matter came to mind whilst making the reversal experiments, and now think the exposure reached the transition stage described in experiment 14 of my paper in the club journal for September. I give these particulars to show every care was taken in the endeavour to photograph the pair of flashes.

There may be points which with our present limited knowledge of the dark flash, seem difficult to reconcile, still I think we are pretty safe in accepting reversal as the cause.

One of these points is stated by Professor Burton in Photography, September 17, namely, the crossing of dark and bright flashes being bright. This is probably due to an electrolytic creeping action of the deposit across the dark flash during development, and not due to the reducing action of light. If the dark flash be due to the undevelopable state produced by oxidation, then at the point of crossing the difference in condition at the junction between the reduced and oxidized flashes is probably sufficient to produce a difference of potential that may determine an electrolytic action. A plate brought to insensitive state by oxidation should form a fair couple with a plate simply reduced; it is difficult to conceive it otherwise, for where there is reduction or oxidation we invariably find electric energy, and when we find electric energy under suitable electrolytic conditions there we find reduction and oxidation."

Note. 2: Photography on Racecourses: Henry Sutton, "The Argus" Melbourne, 28 October, 1893, p. 11.

"I have read with interest Mr. Law's letter on the above in the Argus of today, and whilst agreeing with him as to the accuracy and advisability of using the camera in scientific pursuits, I do not think it will come into practice in the determination on the finish of a horse racecourse. The unreliability of light will be a main factor against it.

The accuracy of a photograph cannot be doubted, but the determination of what is accurate in a photograph is a most difficult matter.

I have seen some of Professor Boys' apparatus and results in scientific instantaneous photography, and I have seen Herr Anschultz's wonderful productions in the same direction. Among the latter the horses of some Uhlans are trotting in line. The camera was placed a little in front of the line; the effect produced is that the far horse is in front, when as a matter of fact, it is in accurate line. This is purely a matter of angle of view, but it would happen on the racecourse if the photograph was taken the fraction of a second before the finish, and the reversed effect produced if taken the fraction of a second after.

Mr Law's suggestion to have the camera overhead on an arch is the nearest way out of the difficulty, but in that case the arch would have to be over 100ft high, in order to use a long focus lens, otherwise the horse immediately under the camera would have the best of it. Still the uncertainty of light remains. No race stewards would allow anything to come into contact with the horses; therefore no automatic apparatus could be devised for discharging the shutter. To have an operator to perform the function, the question of personal equation comes in and we are no better off than at the present. There are other difficulties, the photograph may show a rider pulling his horse, and the evidence would probably be accepted, but would it be accurate evidence? The exposure may have occurred at the fraction of time during which the jockey's body was on the swing or balance. The photograph crystallizes that, and only that particular position of the body, a position held for perhaps the one thousandth of a second, and of which the jockey has no knowledge. I cannot think the camera, will ever be accepted in the determination of the complicated factors of horse racing."

Note. 3: Photography on Racecourse, Robert Law: "The Argus" Melbourne, 10 November, 1893, p. 3.

"It is with interest that Mr Sutton's letter, which appeared in The Argus of this morning, has been read by me. The difficulties raised there on the above subject are not without ground. The variability of the light is certainly a serious obstacle, were it necessary to have perfect pictures of the finish of a race no

doubt photography would be found somewhat inefficient; but is a perfect picture required? Is it not only the position of the animals which is necessary? A very poor photo, on a phenomenally dark day it would require to prevent an accurate representation of the progressive finish of a race being obtained. It is agreed that the accuracy of a photograph cannot be doubted, and also that the determination of what is accurate in "a" photograph is an extremely difficult matter sometimes, this being of course produced by the incorrect or exaggerated perspective, due to the use of an unsuitable lens or combination of lenses on the one hand, and an unfavourable angle of view at which the photo, may have been taken on the other; the angle at which the photo may have been taken being of the utmost importance, as was previously pointed out; but when a series of photos is taken, and the one position is seen merging into the other, it is a very simple matter for a correct conclusion to be arrived at, and if the letter which appeared in The Argus on Saturday be referred to it will be found that it was proposed to take a series of photos, in rapid succession, so that the position of each horse could be taken and shown at definite intervals, thus obtaining a graphic and progressive representation of the actual finish for a definite distance before the winning-post, and if need be also after the winning post had been passed.

By this method it must be obvious that the difficulty referred to by Mr Sutton as to result varying according to the time the photo, was taken, whether taken a few seconds before or after the race, can be overcome completely.

Isolate any of the photos in a series, or take any single instantaneous photo, and I admit it may be difficult, if not impossible, to determine what is exact, but take a series and see the progressive stages, and all is simple. My previous letter may have conveyed a wrong impression to some who may have read it. When I mentioned an arch it was not intended to have an erection which would in the slightest degree obscure the spectators' view of the finish, but only a structure sufficiently rigid to support the photographic apparatus; in other words a stand to replace the ordinary tripod, but to hold the camera over the course. If such an arrangement were objected to, a pole could be placed on the "off" side from the grand-stand, and the apparatus supported on an arm very similar to the method adopted for suspending electric arc-lamps.

The horses cannot be in any way interfered with, it is true, but it is not true, therefore, that no automatic arrangement could be devised for discharging the shutter. The starting of the race could be so intimately connected with a chrono-photographic arrangement as to get rid of the difficulty of in any way interfering with the horses. A person must start the horse, but in the action of starting he may also start a chrono-photographic apparatus, which will do as I have indicated. The personal element then is reduced to a minimum. I am quite prepared to state that such an arrangement can be devised. Mr Sutton writes of a photo taken of a position held from one thousandth part of a second. No photographic shutter exists which approaches anything like this speed. Photos have been taken in something approaching one-millionth part of a second, but these have been by the electric spark, which has a duration of about that time. That photography has difficulties when applied for the above purpose, as proposed, no one will deny, but they are all surmountable, and when one considers the importance of accuracy and freedom from personal element in such matters it is surely worth consideration. An unscrupulous jockey would ride fair when he knew that any unfair action was as likely as not to be detected; thus racing as a whole would be freed to a certain degree from the tricks and dodges adopted by not a few."

Note. 4: Photography on Racecourse, Henry Sutton: "The Argus" Melbourne, 24 November, 1893, p. 6.

"With reference to the above, I do not think Mr Law's letter in The Argus of today alters the matter very much from a practical point of view. It must be remembered that the horse's head decides the race. In order to obtain the relative positions on a dark day would require some background; if this background be white, the head of a white horse would not be seen; if the background be dark, the head of a dark horse would be the difficulty, hence my contention that good light is requisite in order that perfect photography be possible.

Mr Law's suggestion of chrono-photography certainly introduces fresh complications. A starter is not always successful in getting the horses away, though he would have no difficulty in starting the chrono-photographic apparatus. It would be difficult to place a series of cameras in advantageous positions, as no one could foretell the position on the course for the horses during any particular fraction of the time occupied in the race. The animals may go off in a rush, or they may go off slowly. Not so chrono-photographic cameras. They would make their exposures at intervals definitely set prior to the race. The only time during which any chrono-photographic apparatus could be used would be during the last two or three seconds of the race, and still hold the personal element would come in. Other difficulties can be pointed out, but are purely matters of opinion.

Mr Law, however, raises a question of fact when writing that no photographic shutter exists capable of making an exposure of the one thousandth of a second. Shutters do exist capable of making exposures of very much greater rapidity than the one thousandth of a second. The principle was invented by Mr B. J. Edwards, of London, who has shown me his shutter. I have heard Mr. England state that he used the same principle in 1851. The method is now adapted by Loman in

Holland, Anschutz in Germany, and Thornton and Picard in England. In 1882 I used the same principle for astronomical photography in this colony.

It is the principle of the narrow slot passing immediately in front of the sensitive plate. If we take a lens of 6in equivalent focus and aperture of 1in, have our shutter to travel immediately in front of the sensitive plate, the shutter having a slot across it 1-10th of an inch wide, the sensitive plate being 5in wide, make the shutter travel across the plate in 1-20th of a second (a comparatively slow rate) the particles of sensitive gelatino bromide of silver will then receive an exposure of exactly the one-thousandth part of a second, and the full aperture of the lens be used during the whole time of exposure. It is with shutters of this description that the most perfect instantaneous photographs are taken in the old world.

With Mr Law I agree the presence of the camera (in its present Bohemian use) will be a deterrent on the unscrupulous jockey. At the same time, as pointed out in my previous letter, there is a danger to the honest jockey. This alone, in my opinion, should condemn its acceptance."

CHAPTER TEN:

Note. 1: Don Walker:

Some readers may already know the story of Major Taylor and Don Walker as their story was featured in the book by Jim Fitzpatrick titled "*Major Taylor in Australia*" and in the Australian television series which was made in 1994 titled "*Tracks of Glory*". In the television series Don Walker was played by the Australian actor Cameron Daddo and Major Taylor, his full name was Marshall Walter Taylor was played by the American actor Phil Morris. The book and television series recounts Taylor's racing time in Australia, his friendship with Don Walker and the breaking down of the racial barriers both here in Australia and America.

CHAPTER ELEVEN:

Note. 1: Electric Motor vs. Oil Motor, Henry Sutton: "The Austral Wheel" June, 1898, p.p. 159-160.

"However desirable it may be that electricity should be the mode of energy used in automotive vehicles the time has not yet arrived for its practical application. There is only one known method of utilising electrical energy for automotive purposes and that is through the agency of the storage battery; this latter, however, has advanced but little during the past twenty years. We still adopt some method of using oxidised lead for electrodes. The crux of the whole position being

the amount of chemical, and therefore electrical energy, that maybe conserved per pound weight of lead; could we use the lightest known materials, the weight can never be reduced to less than one-fourth of what it is at present.

Outside the question of luxury, the consumer is interested mainly in the cost of the energy as developed for moving his car, and it is just here where the general reader makes his mistake. Electricity is not a source of energy, it is a mode of energy, and the cheapest way we can generate electrical energy is through the agency of heat energy converted into mechanical energy. Now, heat being the source of all energy, it is equally the source of the particular mode called mechanical energy or, to be more accurate, dynamic energy. This being so, and as it is mechanical energy (we use the phrase mechanical energy as being more expressive to general readers) we require in our automotive cars, it becomes apparent that electrical energy is superfluous, and if used must incur considerable loss when generated, stored, and applied to our mechanical purpose. That this is so, we shall endeavour to prove by facts well known to all technical engineers.

Heat is the source from which we derive all our energy. Heat has its mechanical equivalent, that is it is known exactly what amount of mechanical energy or work is available from a given quantity of heat. It is equally well known the amount of electrical energy derivable from a given amount of mechanical energy. At present the cheapest method we have of generating electrical energy from heat is by first converting the heat energy into mechanical energy, and then converting the mechanical energy into electrical energy. This is a very roundabout procedure, when it is considered that it is the mechanical energy we require, and that there can be no conversion from one form of energy to another without loss, owing to our methods of conversion being imperfect.

Electricity as a motive power, if available, would be excellent. The electric motor is a simple machine, considered as an engine. It transforms electrical into mechanical energy in a quiet, unobtrusive way, it will vary its output according to demand, and it will even act to a certain extent as a conserver of energy by developing a counter electromotive force should the motor attempt to race. It exerts its minimum and maximum powers just as required, that is, the harder the work the more it exerts itself. In short, the dynamo and electrical motors are the most perfect and economical engines for the conversion of mechanical into electric and electric into mechanical energy. But- and it is a very big but- the dynamo is not a source of energy, it is a converter of mechanical into electrical energy. There are other losses: the conversion of the original heat into mechanical energy, the conversion of electrical into chemical energy during charging of the storage battery, and the conversion of this chemical energy into electric energy during discharge of the storage battery. These losses are very serious, the practical result being that it requires about 8 cwt. of lead storage battery to propel a vehicle with two passengers 35miles [approx 56 kilometres].

Heat maybe utilised either in a steam engine or gas engine, and by gas engine must be understood vapour and oil engines. With steam the loss by conversion is great, at least 90 per cent being lost. Advanced engineers therefore look to the gas engine, where the loss is 75 per cent., and which is gradually being reduced. Even now, Diesel's results, with high compression, show a marked advance, the only difficulty, I think, being that high compression and high piston speed are difficult to maintain- I refer to Diesel's enormous compressions, which strike me as being beyond the range of commercial practice. Taking the three systems (electricity, steam and oil) in order, we find, by expending £100 in fuel to generate steam in a good steam engine, £90 worth of the original heat energy of the coal is lost, and only £10 worth available in the form of mechanical energy. We utilise this and drive a dynamo, and obtain as conversion into electrical energy, say £9. We convert this electrical energy into chemical energy in charging the so-called storage battery; a big loss at once faces us, for we can only store up £6/10/- value of our original heat energy. We now convert this chemical energy into electrical energy, and then through an electric motor into mechanical energy, and have again to meet a serious loss, this last conversion from mechanical into electrical energy only leaving us with £4 worth of our original heat energy, that is, only 4 per cent; and if we are not very careful our loss will be much greater, if we hurry the discharge the loss may be even doubled. Hence, notwithstanding its cleanliness, convenience, and adaptability, electrical energy for automotive cars is probably the worse system to adopt. Bear in mind , I speak of electrical energy applied to automotive cars, not tramways or any system wherewith the electrical energy may be transmitted from its source, through proper conductors to the electric motor. The future for electrical transmission of energy is great and full of promise. It is certain that within two decades from the present the Melbourne trams will be driven by electrically-transmitted energy.

With automotive cars the problem is quite different, and far more difficult as we are dealing a mass of matter carrying its own store of energy and which must be independent of its base of supply.

So much for the economics of electricity. There are further practical difficulties, such as the weight being a very high percentage of the total load, the weight of batteries alone being fully twice that of the passenger freight. Much as I would like to see electrical energy adopted, I am afraid a great deal has yet to be done, and the promise of the future is distinctly disappointing. Steam power as a means of mechanical energy for automotive cars has more promise; and here, I must explain that by automotive car I mean the light passenger vehicle capable of replacing the horse. I do not class traction engines, as at present understood, nor large freight vans for heavy loads, I simply mean vehicles and loads up to about 15cwt.

Steam power is subject to considerable loss, straight-away we have only £10 worth of mechanical energy for £100 worth of fuel, and this is the best we are getting out of steam after over 100 years of experience- equal to ten percent of our original heat energy.

We now consider the gas engine, and with this we couple the vapour and oil engines. We take the oil engine as being the most portable. In it we manufacture our explosive gas, as we proceed, from material carried in the car; we take £100 worth of fuel and convert it into mechanical energy inside the cylinder and immediately against the piston; we thus in a single operation convert our fuel into mechanical energy and deliver it against the piston which is practically a part of our vehicle. The result is, we obtain £25 worth of mechanical energy as a minimum, and which amount is being rapidly raised by technical engineers who have grasped the problem.

The figures do not pretend to theoretical accuracy, they merely show what the actual results are in practice.

Summing up, we also find that two pounds weight of oil [about three pints or 1.4 litres] will develop as much mechanical energy in an automotive car as will eight cwt. of storage battery.

Note. 2: Fuel for Omnibuses, Henry Sutton: "The Argus" Melbourne, 14 March, 1905, p. 8.

"Sir, In your issue of the 9th inst., referring to the above, it is stated that "the advantage in price of using cheap paraffin fuel instead of petrol far outweighed the difference in theoretical efficiency between the internal combustion engine and the steam engine."

Those in the forefront of the motor movement know the great commercial problem is fuel cost. In Australia kerosene is half the price of petrol.

From the public view, there is the disadvantage of danger from fire with petrol. Recently one of the companies in England using petrol had the misfortune to lose 12 motor omnibuses by fire, owing to vapour becoming ignited in the garage. Inflammable vapour is given off from petrol at ordinary atmospheric temperature. It is heavier than air, and is invisible. In an omnibus garage, where large quantities of petrol may be handled, it will gravitate to the lowest part, where it only waits the dropping of a match or candle to instantly catch fire, like a train of gunpowder.

In the recent submarine disaster in Ireland, where seven lives were lost, the tanks were being filled with petrol, and the heavy vapour collected below, and was fired by a spark on the electric motor.

Kerosene gives off no vapour at atmospheric temperatures, and hence none can collect. The writer of the article in "The Argus" of the 9th inst. assumes that that kerosene can be, and is, only used as a fuel to generate steam. That is not so, as kerosene can be, and is being, used in England to propel vehicles by means of internal combustion engines, with an economy far higher than any steam or petrol system. When used in an internal combustion engine, kerosene gives 20 per cent more power, and can be purchased at half the price of petrol.

The internal combustion engine, as used in motor cars, has beaten the steam engine out of the field, where speed and power combined are required. Manufacturers of steam are now turning their attention to internal combustion engines, and also turning them out in their factories. Some of the most noted steam engineers have veered round, and are designing internal combustion engined omnibuses.

The internal combustion or oil engine burn the fuel in the cylinder, and delivers the energy direct on the piston. It is both theoretically and practically cheaper and more efficient, than any steam system, wherein steam is raised in a fuel-heated boiler, whether the boiler be, fire tube, water tube, or flash system."

Yours, &c., Henry Sutton, A.I.E.E. March 10.

Note. 3: New Oil Engine: "The Argus" Melbourne, 10 July, 1902, p. 8.

"For the last seven years Mr H. Sutton, of Sutton's Music Warehouse, Bourke Street, who has produced several clever, mechanical ideas at different times, has been at work on an oil-engine, with a view to its adaption to the use of kerosene. The result of his efforts is embodied in an engine which has been in use on a motor-car for several months past, and in which Mr Sutton is considered by practical engineers to have completely solved the problem of the utilisation of kerosene as fuel for oil-engines. This, of course, when extended, will revolutionise the use of oil-engines, particularly in relation to the motor-car.

In the present type of oil-engine the vaporisation of the fuel is indispensable, but Mr Sutton has gone on the principle that far greater explosive power is contained in the oil in a finely-divided state than in a vapour, and has therefore introduced an "atomiser." which brings about conditions very much the same as in steam under expansion.

The engine, which was shown in work on Wednesday at Mr Sutton's home at Armadale, is a vertical one of very compact design, and weighing when on the car about 2½cwt. The cylinder is 4½in. diameter, with a 6in. stroke. The kerosene fuel is supplied from a tank of a simple and ingenious design, the quantity sent to the "atomiser" being automatically controlled. It is by the "atomiser" that the difficulty of using kerosene fuel has been overcome.

This device mechanically controls the "feeding" of the engine. In addition to the departure from the old method of vaporising, there is a remarkable innovation in the placing of the "atomiser" in the explosion chamber itself. The regulation of the quantity of fuel for each explosion does not affect the quality of the atomised fuel, and is controlled by the tension spring in the inlet valve. The ignition of the "atomised" material is affected by a wet battery producing a jump spark. The exhaust vapours are free from smell, and there is no smoke or heat, an important consideration in motor-car work.

Running at about 500 or 600 revolutions the engine developed about 5 b.h.p. Three quarters of a pint per h.p. per hour of the ordinary Russian Shell oil is used in the engine, and Mr Sutton states that a 20 mile journey on his motor-car can be done on 3d. worth of kerosene. The idea will be applicable to all stationary oil-engines as well as motor-cars, and can be easily adapted to existing types of engines."

Note. 4: The Sutton Heavy Oil Engine: The Automotor Journal, 4 June 1904, pages 676-679

"The interesting question of the use of kerosene for automobile engines on the "spraying" principle has been brought to the fore again recently by the work done in Australia by Mr Henry Sutton, of Melbourne, who has apparently met with a large measure of success, and has, as the result of several years' careful experiments, devised a very simple form of mechanism for adapting engines to the use of this fuel…

One of the actual kerosene attachments made for an existing petrol engine is shown in Fig. 1, and the various portions forming this apparatus are also seen separately in the same illustration. In Fig. 2 is represented a similar kerosene attachment, but in this case it is combined with an atmospheric induction valve and with an ignition plug in such a way that all these parts can be together bolted to the cylinder of an engine, and that they form a separately detachable unit in themselves; this combined mechanism has been more especially manufactured for demonstrating the Sutton principle than for actual practical use, although it is applicable in certain cases to engines, provided that the cylinders have the usual exhaust valves in them. The various parts are seen detached from the main casting in Fig. 3. In both forms of apparatus the portions that atomise the kerosene and mix it with the air are practically identical, and their arrangement in relationship to one another will perhaps be most readily understood by referring to Fig. 4, in which the pulverising apparatus is shown sectionally.

The fuel is led to the upper fitting, A, which constitutes a combined two-way cock, needle-valve control, and spray nozzle. Two separate pipes are led to it, and these are connected by unions to the nipples, A5; the one pipe is led from the main fuel tank, which contains kerosene, and the other from a small petrol tank which is only needed for starting the engine. The two way cock is formed by the conical plug, A6, which fits into the conical seating, A8, in the casting A. The plug portion of the cock has a slot, A7, cut across its face, and there are three holes leading into the conical seating, A8, in the neighbourhood of the slot, A7. These holes are in open communication with the pipe fittings, A5, and with the needle-valve, respectively, and they are so placed in relationship to the slot, A7, that the cock can be made to admit kerosene and petrol simultaneously to the needle-valve, as well as to admit either fuel alone.

Petrol is used when starting the engine because, as we have already explained, it is difficult on any but very small engines to obtain a sufficient speed by hand to pulverise the kerosene sufficiently. To start the engine, therefore, when it is cold, the two-way cock is placed in that position in which petrol only can find its way to the needle-valve. Immediately the engine has attained its normal speed the two-way cock is turned round through the various positions in which an increasing proportion of kerosene becomes mixed with the petrol until, finally, it is left in that position in which kerosene alone is being fed through it. By this arrangement the flow of fuel to the engine is uninterrupted, even momentarily, when switching over from one fuel to the other, and since the two tanks are arranged at the same level as one another there is no tendency for the different fuels to become mixed in either tank at such times.

The needle-valve, A4, is screwed down into the top of the casting, A, and its point lies inside the nozzle, A1, where it regulates the quantity of fuel that can pass to the engine. The valve is permanently adjusted to suit the kind of oil that is being used, and does not normally require any attention from the driver of the car. It is of the very greatest importance that the flow of fuel should remain constant at all times for similar working conditions of the engine, and, therefore, a specially constructed supply-tank has been devised by Mr. Sutton, by which the head of oil on the gravity feed remains the same whether the tank is full or whether it is nearly empty.

The tank is very interesting and important feature of the Sutton system and its ingenious construction will be readily understood from the diagrammatic sketch given in Fig.5. The pipe, G, at the bottom of the tank, is that through which the oil is led to the engine, and the passage of the oil to it is controlled by the needle-valve, G2, the spindle of which is carried through a stuffing box in the top of the tank. The needle-valve, G2, is subject to the control of the driver, but is not needed as a compensation for differences in the level of the oil in the tank. The required compensating action for this is obtained by fitting an air-pipe, G1, down through the top of the tank, so that it nearly reaches the bottom. Except for this pipe, G1, the tank is air-tight, and, therefore, when fuel is drawn out from it, it is necessary that the air which takes its place should enter through the pipe, G1.

The result is that no air enters immediately after the tank has been refilled until enough oil has been drawn off to form a sufficient degree of vacuum in the tank to cause the level of oil air-pipe, G1, to fall to the bottom of that pipe. When further oil is drawn from the tank, air is able to take its place by bubbling up through the fuel outside the pipe, G1. By this arrangement the actual effective level of the oil in the tank is the same at all times, the "head" in the tank being that denoted by the dimension marked, h, in our diagram.

The conical nozzle, A1, projects down into the mixing chamber, and there is a row of holes, B, arranged through side of the chamber, around the upper end of the nozzle; the whole of the air that is drawn into the engine cylinder enters through these holes. In Fig.1, the upper portion of the mixing chamber is formed by a separate casting, B1, but in Figs. 2 and 3 the whole of the mixing chamber is formed in the main casting. Beneath the nozzle, Ai, are several layers of gauze, C, placed close up against one another and lying so as to form a partition across the mixing chamber. On the underside of the gauze partition is a small metal cone, C1, which prevents any unsprayed oil from forming into drops and so dripping down unatomised below; any oil that might do so, however, would flow out through the drain hole. Just below the gauze partition, the mixing chamber is given a conical shape either by the introduction of a separate sleeve, D (as in Fig,1), or by suitably shaping the inner wall of the main casting (as in Fig. 4).

It has been found that the exact shapes of the nozzle, A1, of the cone, C1, and the conical sleeve, D, are of all importance to ensure the proper working of the apparatus, and that the exact positions of these parts in relationship to one another, and to the air hole, B, must also be most carefully arranged. Provision, too, has to be made for varying the amount of air that enters the mixing chamber, and the sliding ring, B2, is fitted for this purpose. The holes need to be almost closed when first starting the engine, apparently in order to give the entering air a sufficiently high velocity to enable it to pulverise the kerosene.

When the apparatus is in use, the oil finds its way through the small central hole, A2, in the bottom of the nozzle, A1, and would flow in a small stream on to the gauze partition, C, were it not taken up by the entering air. The gauze apparently does act as an accumulator, or equaliser, by holding in suspension the oil that flows on to it between the suction strokes of the engine, but the actual flow through the nozzle is quicker during the suction strokes than between them, because of the partial vacuum which there then is around the nozzle. The air sweeps round the nozzle at a very high speed, takes up the oil, and, dashing it through the gauze, splits it up into a finely divided spray, which is carried with it through the conical sleeve, D, and the inlet-valve into the working cylinder; the sleeve, D, which has a larger internal diameter at its upper end than at its base, is found to be essential for ensuring sufficient velocity for this purpose.

The parts which we have just described are those which are essential to the Sutton system, and it will be realised that in themselves they are remarkably simple. The difficulty, however, will be experienced of comprehending quite why it is that so extraordinarily effective an atomising effect should be obtained with them, or why---if they do give this effect---there should have been any difficulty in designing such an apparatus for this purpose before. This is just one of those points in connection with which it is difficult to give a satisfactory explanation, but it is a recognised fact that extremely small differences in the shapes and positions of any such parts as these are apt to make all the difference in the working of an atomiser, and to convert comparative success into absolute failure.

The mechanism shown in Figs. 2 and 3 has the induction valve, E, arranged in such a way that the strength of the spring tending to close it, and also the extent to which it can open, are variable. For this purpose the bell-crank lever, E2, is arranged to engage with the spring, and its other end is connected with a hand-lever for operation by the driver. The valve spring rests at its upper end against the lever, E2, and the valve-stem, E1, passes freely through a slot in the lever.

An ordinary high-tension system of ignition can be used in conjunction with engines fitted with the Sutton kerosene arrangement, and we are assured that practically no carbon is formed anywhere in the engine nor across the inside of the ignition-plug. It will be noticed in Figs. 2 and 3 that an ignition-plug, F, of the usual form, is fitted into the inspection-cover, F1, that lies immediately above the induction-valve, E. It is perhaps, not surprising that there should be an absence of carbon deposits in the cylinder, or of visible and smelling vapours in the exhaust, considering that the fuel consumption per horse-power is as low as it appears to be, for in order to be able to obtain as much power as this from the fuel, it is obvious that very complete combustion must be obtained.

The real difficulty is, of course, to obtain equally complete combustion when the engine is running light as when it is working on full load, and this is where it remains to be seen how far Mr. Sutton has practically overcome previous difficulties. He regulates the power of the engine by varying the volume of the explosive charge that is drawn into the working cylinder, and to a great extent he controls the speed by varying the time of ignition. By the former control, when the lift of the inlet-valve is reduced, the quantity of air drawn through the pulveriser is less and the suction effect on the spray nozzle is reduced."

When a person purchased one of Henry's engines or carburettors they were given an information and promotional booklet on the engine and carburettor which read;

Note 5: Henry Sutton: The Sutton Heavy Oil Engine Brochure: National Archives of Australia: "The Sutton Oil Engine" Series Number: MP84/1, Control Symbol: 1997/2/200, Item Barcode: 4795772, Date Range: Circa 1 January, 1906 – Circa 31 December, 1913.

The Sutton Heavy Oil Engine Patent:

The most important function performed in an Internal Combustion Engine is the formation and maintenance of a combustible mixture of fuel and air in such perfect proportions that at all and any time during the running of the engine, and under all circumstances of variation of power, speed, and load, the mixture shall be such as to ensure perfect combustion.

If perfect combustion be attained, then economical working and clean valves and combustion chamber are also attained. How far the modern Internal Combustion Engine has failed to procure these results is so well known as to need no recapitulation here; suffice to say that so important is the combustion in the modern engine, that it is the greatest problem in engineering invention and discovery to produce an engine that will use oils of higher specific gravity than '68 or '73 without fouling the valves, or producing bad exhaust. Once attain perfect combustion, foul valves and bad exhaust, are things of the past.

So imperfect are the modern internal combustion (oil) engines, that it is considered impossible to use oil of density '825 in an automobile, owing to the foul exhaust, the deposition of tar and unconsumed carbon on the valves, and in the combustion chamber, resulting in the engine failing to work; loss of power and other objectionable features, such as the ejection of black smoke and white vapour at the exhaust at every variation of speed of the engine, the clouds of vapour and smoke at starting; these features render the ordinary oil engine an impossibility in street traffic.

Elaborate treatises have been written by chemists and engineers showing that the ordinary lamp oils can only be utilised in oil engines by being vaporised by heat.

This entails preliminary heating of vaporisers and combustion chambers to produce the requisite vapour to be afterwards mixed with heated air in order to render it combustible. The mere fact of having to heat up the oil and air at each start has rendered the ordinary oil engine useless for automobile purposes, When heated up no liberties can be taken with hydrocarbons. If heated too much, decomposition sets in, the heavy fractions being deposited as graphite and tar on the valves and internal parts of the engine, whilst the lighter hydro-carbons inflame and assist the decomposition. If not heated quite so high the lighter hydrocarbons distil out and inflame, the heavy hydrocarbons being deposited.

The discovery by Henry Sutton, A.I.E.E., that true combustion of mixed hydrocarbons can only be produced by ignition of the mixture at a lower temperature than the distilling temperature, and in such manner that all the fractions of the hydrocarbon ignite together, and are completely oxidised before decomposition can be completed, is the key to the position and solves the great problem. By utilising this important discovery, Mr. Sutton, after many years of experiment and invention, has successfully produced an Internal Combustion Engine, in which heat is not used to vaporise the oil, and in which there is no deposition of graphite or tar, and in which the exhaust at starting and during the whole time of running at any speed and at any load is as free and clean as a spirit engine. So perfect is the combustion, and so completely are the functions of perfect combustion attained in the "Sutton" Engine, that it is now without doubt the highest development of the internal combustion or oil engine, and can be used for industrial purposes in any form in which power is required; for large powers such as railway cars the saving by using the cheap fuel is enormous as compared with spirit motors.

The "Sutton" Motor can be used on an automobile even more perfectly than the modern petrol or spirit engine. The oil, (common lamp oil '825 density) being cheap and safe and weight for weight, 25% more powerful than petrol.

The patented mechanical methods utilised in the new system are so simple and perfect in action in action as to astound engineers; no fuel feed pumps, or float feed, or valve feed mechanisms are used. No mechanical moving devices are used to regulate the fuel. Yet the great difficulty of supplying the fuel for all rates of speed of engine in mathematically perfect proportions is attained, and exactly the right quantity of fuel admitted to ensure perfect combustion irrespective of load; no air or oil governing devices being needed.

The "SUTTON" Engine, therefore, automatically maintains a perfect combustible mixture at all rates of speed and load---no valve, lever, float, rod, or anything having to be regulated. If you want the automobile to go fast, advance the ignition, the combustible mixture will instantly adapt itself to the speed. If you slow the ignition the same adaption instantly takes place and this irrespective of whether heavy oil of '825 or light oil of '680 be the fuel. These being the great characteristics of the "SUTTON" Patent Oil Engine, it follows that it must completely revolutionise the existing systems of using oils and spirits in internal combustion engines. The problem of the commercial automobile is thus solved; heavy oil may now be used in motor vehicles as easily as petrol; heavy oil is obtainable anywhere; heavy oil is not restricted by insurance or by rail companies. For launch or boat work the "SUTTON"" Heavy Oil Engine is perfect, being safe and smokeless. The "Sutton" Heavy Oil Engine is being used on Automobile, and for electric light and power work; it runs at as high a speed as a spirit engine, and starts instantly.

The SUTTON Patent Heavy Oil Engine

Will use indiscriminately--

Heavy Paraffine .. '825 Sp, G.	*Petrol .. '68 Sp, G.*
Light " .. '79	*" Gasolene '65 "*
Benzine ,, .. '73	*Benzol (Coal Tar Oil)*

Note 7: Henry's Carburettor Patents:

As to the outcome of the sale of Henry's patents and carburettor system overseas nothing is known except for a few clues that have yet been researched but in time information and further research in Canada and America may turn up some information although there are some possible clues in relation to Massey-Harris and The Ford Company and the opening of their New Jersey factory in 1905.

CHAPTER TWELVE:

Note. 1: First Minutes of the Automobile Club of Victoria, 9 December, 1903: Courtesy of the Royal Automobile Club of Victoria, Heritage Collection.

Minutes of the Meeting held at the Port Phillip Club, Hotel, Wednesday December 9th 1903.

56 Gentlemen Attending.

It was moved by Mr H. B. James, and seconded by Mr W. J. Proctor that Mr H. J. Maddox be appointed Chairman pro. tem. Carried unanimously.

Moved by Mr S. Thornley that Mr H. James act as Secretary pro.tem. Seconded by Mr. C. Rice and carried.

The business of the meeting was entered upon by Mr S Stott moving "That the time is now opportune for the formation of a club in the interests of motoring and motorists." Seconded by Mr W. J. McIntosh. Carried unanimously.

Moved by Dr Merrill, and seconded by Mr A. F. Lange that the club be called "The Automobile Club of Victoria." Carried unanimously.

Mr H. Sutton then moved that the objects of the club be-

"The promotion of a social organisation and club composed mainly of persons owning self propelled vehicles or motorcycles. To afford a means of recording the experiences of members, and others using motor cars and motorcycles. To promote investigation in their development. To co-operate in securing rational legislation; and the formation of proper rules and regulations governing the use of motorcars and motorcycles in cities, towns and country districts. To maintain the lawful rights and privileges and protect the interests of owners and users of all forms of self propelled vehicles whenever and wherever such interests, rights and privileges are menaced. To promote and encourage the improvement, construction and maintenance of roads and highways, and the development generally in the State of motoring, and to maintain a club devoted to the interests and advancement of automobilism."

Seconded by Mr T. Rand, carried unanimously.

It was moved by Mr G. Buckley "That a Provisional Committee of 7 be now appointed to draft rules and by-laws. To call a further meeting on January 15th to discuss, amend, reject or accept such rules. At the same meeting the rates of entrance and subscription fees shall be determined by those present, and other business that may be given notice of or arise thereat."

Seconded by Mr A. C. Thwaites. Carried.

The following men were duly elected members of the Provisional Committee:-

Mr H. Bagot, S. Day, E.C. Joshua, Dr Merrill, Dr Officer, T. Rand and H. Sutton. The Chairman and Hon. Sec. to act ex-officio.

After some discussion in reference to the presentation of the trade in officering the club, it was moved by Mr G. Irish and seconded by Mr Coleman that "No member of the Trade be allowed to sit on the Provisional Committee." Carried.

The business concluded by Dr Merrill moving a vote of thanks to Mr H. B. James for taking the initiative in the formation of the club.

Carried by acclimation.

Mr E. C. Joshua then moved a vote of thanks to the Chairman, and the meeting closed."

Adopted:15 January, 1904.

CHAPTER THIRTEEN:

Note. 1: Tele-photography, Henry Sutton: The Telegraphic Journal and Electrical Review, 7 November, 1890, p.p. 549-551.

"Under various names the problem of transmitting optical images by aid of the telegraph wire has at different times had attention drawn to it.

In putting forward something new in this direction, I will begin by inventing a new name, and propose calling the subject Telephany, and the electro-optic instrument the Telephane.

The art of telephony is simple compared with that of telephany. In the former we deal with a consecutive series of waves of varying rate and length, and it is the consecutive character of sound waves that lends itself so admirably to electric translation.

In telephany we are met, at the outset, with a great preliminary difficulty, having to deal with a surface or plane in which the effect appealing to the brain must be observed in all the varying character at one and the same time.

The problem stands thus: a means has to be devised whereby the varying effects on a plane surface are translated into consecutive series of currents, and by means of the consecutive series of currents to reconstruct, so to speak, a copy of the original surface; that is, we have to take an optical image, seen as a surface, translate it into a line of consecutive varying electrical currents, and by means of these produce an effect as a surface, having the characteristics of the original image.

We have here two images as surfaces having no time value, and a series of electrical currents having a time value, yet these opposing characteristics are to be presented to the brain as a momentary impression.

Before showing how this apparently impossible problem may be handled, I will explain a probable means for electrically transmitting a photograph.

If we make what photo-mechanical operators call a screen negative of a portrait, using a course screen, and from this a photo-lithographic transfer, transfer it to zinc, and transmit the result by any of the several autographic systems, we have the desired result. In fact, if we apply our knowledge of half-tone-block making to telegraphy, we are at once in possession of a means of electrically transmitting the photographic semblance of any person.

We may make a screen negative, and from that obtain a print on zinc or copper coated with sensitive albumen or bitumen, using the usual solvents, water or turpentine, as the case may be, with which to wash away the unexposed albumen or bitumen, then let the stylus of any autographic system traverse the developed image, the result at the receiving end is a facsimile portrait.

I have used the expression screen negative, as it is an understood trade name;

as a matter of fact, a screen positive would be necessary. We may go further, instead of receiving the facsimile on chemically-prepared paper, as in the Caselli autographic telegraph, we can make the receiving stylus perforate thin paper (with the electric spark) by means of a constantly-working induction coil, but only put into the receiving circuit by the transmitted current. Place this paper on a lithographic stone or zinc plate, pass a roller charged with greasy ink over it, and we have a printing surface, the portrait being transmitted and reproduced for the printer, photo-electro-mechanically.

But this is not telephany; the latter must be understood as the means of transmitting images which may be in motion, as seen in a photographic camera, but not in colours.

Having spent some years in studying the problems, I designed the following system five years ago, as my Victorian scientific friends can testify. It may be of much assistance to workers in this direction. I think it offers a fair approximation to the solution of this very difficult problem; at any rate, if in its present form it is not the actual solution, I feel sure it is in the direction indicated by my method, that the successful accomplishment of telephany will result.

Transmitter.

L (figs. 1 and 5), is a photographic objective of the rapid type, producing an intensely illuminated aerial image at A A.

D D (figs. 1,2 and 5), light metal disc revolving at a fixed rate of not less than 650 revolutions per minute under the control of La Cour's phonic wheel and fork apparatus as in the Delany multiplex system.

G (figs. 1, 3 and 5), a glass or other insulating plate, to the front surface of which is held, by binding terminals, S_1 *S, two triangular pieces of metal just separated from each other, E E.*

C (figs. 1, 3 and 5) a small piece of lamp-black, selenium or other substance, the resistance of which may be varied by heat or light. Lamp-black compressed is probably the most suitable.

The disc, D D, has a series of small holes, 1, 2, 3, 4, 5, 6, 7, 8, perforated in it, and gradually approaching its centre, as a spiral, these holes must be numerous, and yet only one at a time in the, field of the image at A A.

R L (figs, 1 and 5) is the most important part of the transmitter. This I term the regulating lens; it is a lens placed with its plane surface just to receive the aerial image from the objective L, its focal length being such as to bring all rays reaching it through the perforated disc to a point or focus at C, and therefore its function is to introduce them consecutively to the circuit comprised in S_1, E_1, *C, E, S.*

Under the influence of this regulating lens the whole image, A A, is allowed to act in consecutive manner, and therefore vary the resistance of C in accord with the lights and shades of the original. We thus solve the big problem of translating the plane image into a line of consecutively varying strength of current, and by bringing C under the influence of the whole image within one-tenth of a second, and during the same time reconstruct our image at the receiving station, persistence of vision will enable us to see the image as one impression.

Receiver.

S_1 (figs. 4 and 5), any artificial source of light, a beam of which is by means of lens, L, passed through a pair of Nicol prisms, P, A, this beam reaching lens M_1, is magnified and received by the eye-piece, M, M, and viewed by the eye, E. It is absolutely requisite this beam be received by the eye through optical means. The presence of a translucent screen at X, X, would be fatal, owing to the delicate nature of the desired effect.

D D (fig. 4), is a perforated disc, similar to and revolving synchronously with the disc in transmitter.

K K (figs. 4,5 and 6), terminals inserted in glass and having a small space between, holding a drop of bisulphide of carbon, S.

On rotating the Nicol, p, 45°, we reach the position of extinction.

The terminals, K, K, being placed in the secondary circuit of the transmitter - that is, to line - the variable electrostatic strain produced in the drop of bisulphide of carbon under the action of induced currents received from transmitter, is to produce variable rotation of the polarised beam, and, therefore, variable quantities of light reach the eye, E.

It is obvious the varying tints will be seen in similar position as in original image, owing to the synchronous movement of the discs.

The receiver is, then, based on Dr Kerr's discovery of the rotation of a plane polarised beam of light, through electric stress producing a strain in the medium.

There seems to me no question that the electric impulses will do their work within one-tenth of a second, and the point is whether the stress at O O will be sufficient to produce an observable effect, and whether this may be increased by passing the light through a bisulphide of carbon cell having a longer path of, say 1/16th inch diameter, as shown at fig. 6, instead of through the drop of bisulphide; so that, conceding the telephane is based on a rightly conceived principle, it becomes a question whether the quantitative results of the physical effects utilised in its design are sufficiently great.

I think the transmitter may be considered as near the right thing as the present state of our knowledge will admit as to reach; there is an appearance of finality about it.

With regard to the receiver it is a question of degree; the actual quantity of light required to reach the eye may be very small when received optically; in fact, so small as to have no power of illumination on a translucent screen; but a quantity of light producing no visible effect on any media, when received by means of an eye-piece shows a bright disc. Owing to being away from references, I regret my inability at the moment to give the name of the inventor of the revolving disc; with the exception of these discs the whole design is original, and devised at Ballarat, Victoria."

Note. 2: Alternate Currents of High Potential and High Frequency, Nikola Tesla: "Project Gutenberg eBook" Web Page: http:/www.gutenberg.org/files/13476/13476-h/13476-h.htm, 2014:

"This investigation, then, it goes without saying, deals with alternating currents, and, to be more precise, with alternating currents of high potential and high frequency. Just in how much a very high frequency is essential for the production of the results presented is a question which even with my present experience would embarrass me to answer.

Some of the experiments may be performed with low frequencies; but very high frequencies are desirable not only on account of the many effects secured by their use, but also as a convenient means of obtaining, in the induction apparatus employed, the high potentials, which in their turn are necessary to the demonstration of most of the experiments here complicated.

Of the various branches of electrical investigation, perhaps the most interesting and immediately the most promising is that dealing with alternating currents. The progress in this branch of applied science has been so great in recent years that it justifies the most sanguine hopes. Hardly have we become familiar with one fact, when novel experiences are met with and new avenues of research are opened.

Even at this hour possibilities not dreamed of before are, by the use of these currents, partly realised. As in nature all is ebb and tide, all is wave motion, so it seems that; in all branches of industry alternating currents—electric wave motion—will have the sway.

One reason, perhaps, why this branch of science is being so rapidly developed is to be found in the interest which is attached to its experimental study. We wind a simple ring of iron with coils; we establish the connections to the generator, and with wonder and delight we note the effects of strange forces which we bring into play, which allow us to transform, to transmit and direct energy at will. We arrange the circuits properly, and we see the mass of iron and wires behave as though it were endowed with life, spinning a heavy armature, through invisible connections, with great speed and power—with the energy possibly conveyed from a great distance.

We observe how the energy of an alternating current traversing the wire manifests itself— not so much in the wire as in the surrounding space—in the most surprising manner, taking the forms of heat, light, mechanical energy, and, most surprising of all, even chemical affinity. All these observations fascinate us, and fill us with an intense desire to know more about the nature of these phenomena.

Each day we go to our work in the hope of discovering,—in the hope that someone, no matter who, may find a solution of one of the pending great problems, - and each succeeding day we return to our task with renewed ardour; and even if we are unsuccessful, our work has not been in vain, for in these strivings, in these efforts, we have found hours of untold pleasure, and we have directed our energies to the benefit of mankind."

CHAPTER FOURTEEN:

Note. 1: How inventors are treated; Ridiculous Departmental Obstruction: The Age, 1 October, 1908, p8.

"When the Postal Commission was informed this week by Mr Hesketh, chief electrician to the post and telegraph service, that, through lack of funds, the telegraphic and telephonic systems of the States could not be put in proper order, he made it clear that his department is wallowing along in a helpless condition of muddle with what it has in hand, and is quite unable to take up fresh responsibilities. It is, therefore, rather discouraging to learn that the Postmaster-General has exerted himself to interfere in the affairs of the Defence department by claiming the right of dealing with all matters affecting electric signalling by wireless telegraphy. Such an intimation means that the defence authorities are not to be permitted to avail themselves of useful inventions on their own initiative, whilst the Postal department, even if willing, seems unable through the chaotic condition of its affairs to forward any project.

A field signalling apparatus which was invented by Mr H. Sutton for transmission of messages by wireless telegraphy had been tried by a mixed committee of military and navel men with most satisfactory results, and just at the moment when Mr Sutton was to be informed by the defence authorities whether or not it would be adopted, the Postal department asserted itself, and Mr Sutton was informed by the defence people that nothing could be done by them, as the Postmaster-General claimed jurisdiction. The same inventor had been experimenting on a large scale with a pole and apparatus at Malvern for the purpose of testing a new system of wireless telegraphy called Australian Wireless.

Providing the operator with such an installation does not set up an air disturbance of more than a certain force, and does not compete with contract systems or disturb them, he is not an offender. But if he does these things he is liable to a fine of £100, or 12 months imprisonment, on the suit of the Postmaster-General.

Of course it is impossible for an inventor to test anything by which the ethereal forces are to be made subservient unless he has a certain freedom in the use of the atmosphere. Being unwilling to run the risk of fine or imprisonment, Mr Sutton, at an early stage in his operations, made enquiries, and was told by a high officer of the Defence department to "go ahead," as it would be "all right," whilst the Postmaster-General also informed him, as he understood, that no official notice would be taken of his proceedings. Since the publication of some wonderful successes achieved by the inventor, however, such as the tapping of wireless messages which were sent from New Zealand to the American fleet in Sydney harbour, and the designing of an installation that will send wireless messages between Victoria and Tasmania at a cost much below the departmental estimate, Mr Mauger has said that he cannot now remember giving even an implied permission for the carrying on of wireless tests at Mr Sutton's "station" at Malvern, and has warned him not to break the law. The only remote justification for this seems to have been, not the interception of the American fleet messages, but the hushing up, accidentally as it happens, of some department experiments.

It appears on Sunday Mr Jenvey, of the electrical branch of the G.P.O., was having a wireless message sent from Queenscliff to his house at St. Kilda, where he had put up a pole and a receiver. The day was beautifully clear and favourable, but a most unaccountable aerial disturbance struck the communication, and knocked into what the printers call "pie." Mr Jenvey went to his dinner wondering. He only understood what had happened when, some days later, he read in "The Age" that Mr Sutton's new Australian Wireless had the capability of converting messages sent by other systems into chaos. Mr Sutton chanced to be experimenting on that Sunday, in entire ignorance of Mr Jenvey's proceedings, and created a tempest incidentally without knowing its effect until afterwards.

Yesterday the inventor invited a representative of "The Age" to an exposition of Australian Wireless as applied to military field signalling, with the view of demonstrating the bona fides of what he claims to be able to do with the apparatus, supplementary to what defence officials have testified in regard to it. He suggested that a newspaper man's unskilled appreciation of what he might do should be reinforced by the opinion of some person with at least a rudimentary knowledge of electricity, and therefore it would be well to have a telegraph operator present.

Application was made to the Postal department for the loan of a telegraph operator for the purpose. There was a prompt refusal. Mr Sutton was told that he had been illegally experimenting, that he had confused departmental messages with his Malvern apparatus, and the department would not grant him any privileges or facilities for yesterday's intended exhibition. On the other hand, he was not forbidden to demonstrate; but was specifically warned not to break the law. The result was that, Mr Sutton claims the right to experiment and demonstrate freely within the walls of his house or business premises, like any university professor, not knowing precisely the law and being intimidated by the menacing observations of the secretary of the central administration and the Postmaster-General, he decided to postpone the exhibition.

What he was going to show, and that he had invented it is undoubted, is a receiver capable not only of reproducing wireless messages audibly, but of conveying sufficient surplus power through a galvanometer to turn on an electric light, or a bell signal when a message is announced, or drive a small motor. No other such instrument, he say, has yet been produced do all these operations at once, and this fact is attested by electrical text books. From Mr. Sutton's latest experiences it is evident that he is greatly confined in the exercise of his genius as an inventor by the absurd attitude of the Postal department, which, having first "cornered" the Government rights to deal with electricity as a messenger, now apparently wants to prevent all progress. This, of course, is not merely a private, but a public wrong.

CHAPTER FIFTEEN:

Note. 1: Death of Henry's sister Elizabeth's Husband:

In 1906 the Sutton family endured another tragic loss when Henry's sister Elizabeth's husband Henry Finch Rix died suddenly on the 27th February, 1906, it was a sad blow to the Sutton and Rix families Henry Rix was a much loved member of the family and his loss was greatly felt by all.

Note. 2: Extract Dr Bell's Testimony at Royal Commission 1910:

Question: *"Do you think the Government should provide an experimental laboratory?"*

Dr Bell: *"It is very important that the Postal Department should have a laboratory to test all new ideas, and find out what is in them, and if good adopt them. An inventor may invent something which is all very well in a small exchange, but which has to be adapted to fit in with a large system; it is just one cog in a machine, and that cog has to be adapted. It is found in America that this work can be best done by having a scientific laboratory and a body of experts to test inventions. It is not only advisable to have a permanent system, but some* *means of improving that system. Government ownership places an obstacle in that direction, because it is the only buyer of new inventions. Therefore, it is more advisable under a Government system to have a testing Laboratory and staff of experts to test inventions, and give a second opinion upon them than it would be in America, where there is competition. In America if one company does not test a new invention, an opposing company may do so. Without a Government laboratory it will be difficult to get improvements except from foreign nations; you would not get many Australian improvements; that is my opinion."*

Question: *"Did you see a selector invented by Mr. Godfrey, of the Railway Department, here?*

Dr Bell: *"Yes, I was very much interested in it; it is a very ingenious instrument, and works well on a set of twelve stations. But it would require very much modification to adapt it to a telephone system. This is the case of an Australian inventor who should be encouraged in his endeavours."*

Question: *"Supposing that inventor devised an automatic system, how could it be tried?"*

Dr Bell: *"In this country the Government has a monopoly, and the invention cannot be tried unless the Government gives the inventor an opportunity. All these new ideas require a great deal of adaption before they can become commercially successful. We have a number of systems such as that of Mr Godfrey in America on the step-by-step principle. I do not know that this one has any advantage over those in the United States of America except that it is an Australian production, which should give it an advantage in your eyes. You want a proper system of encouraging investors and a department to test the value of inventions."*

Question: *"Do you consider that wireless telephony is likely to develop greatly in the future?"*

Dr Bell: *"Yes, it is in use now in America experimentally, and speech has been transmitted 12 or 14 miles. When an invention begins in that way one cannot tell how far it will go. Wireless telephony is in the same stage as wireless telegraphy was some time ago, but it has not yet entered the field of practicability. In any case, it is unlikely to enter into competition with the present system of telephones or account of the difficulty of securing privacy."*

Question: *"Did you form the opinion that Australia has not yet passed out of the infant stage?"*

Dr Bell: *"I am of opinion that the laws in Australia are not favourable to inventors. You seem to base your patent law on the English model, which is not favourable to inventors. In America no patent is granted to any other man than the first inventor, and then only after an examination has been made by the* *officials of the patent office to find out whether the supposed inventor was really the first inventor of the matter claimed. A man never knows whether he is the first to invent any instrument. He may know that the invention was original with himself, but does not know what others have done. The same thing may have been invented before he was born; therefore, the patent office makes an investigation to find that out; and only if he is the first inventor will he be given a patent. When I investigated the patent system in England some thirty years ago, I found that a man, whether the inventor or not, could obtain a patent for almost anything he chose, if he paid the fees, and after getting the patent it was not worth much in the way of protection, for the onus of proof that you were the first inventor was thrown upon you. You would have to prove to the satisfaction of the court that you were the first inventor before your patent was of much avail. No patent was of value until it had passed through a lawsuit - an expensive process. In America, the patent granted by the patent office is held by a Court of law as prima facie evidence that you are the first inventor, and the burden of proof that you are not is thrown on the other side.*

The result of the American patent law can be seen in the stimulus which it gives to inventors. I am both British and American. I was a young man when I left Great Britain, and have been in America for about forty years. I am pretty sure that the telephone invented by me would not have been developed in England on account of unfavourable surroundings. Not only do the patent laws in America afford protection to inventors but the social environment is favourable to the growth of new ideas. In Great Britain the attitude of the general public towards a new idea seems to be: "This is a new-fangled notion, let someone else try it first and we can then see how it works." In America the attitude is: "This is a new thing, and therefore probably better that the old; let us try it." That attitude is typical, and stimulates invention. It would be to the interests of Australia to stimulate invention, and protect inventors in the ownership of their creations. One of the great dangers in a country like Australia is that Government ownership is going to interfere with the improvements of inventors by stopping financial rewards that are before them. One way to improve your telephone system would be to stimulate invention; and your Telephone Department should have a Board of Experts to take up new ideas, try them, and advise the Government on them."

Question: *"Do you think it would be advisable if the Government itself were to offer such incentive that would induce young men to exercise their brains in the direction of the invention?"*

Dr Bell: *"I think such would certainly be an advantage. If you could show in your telephone system the direction in which improvement is wanted it would* *be a great advantage. For example, for thirty years people have been trying to improve the telephone instrument itself, but it has not changed. It is no use their setting inventers to work to try and do that, because it is not been done under the stimulus of thirty years' competition. We have a complicated mass of instrumentalities for connecting the telephones together; there it would pay to set inventors to work to simplify the system."*

Note. 3: Untitled Article: "The Argus" Melbourne, 31 May, 1911, p. 12.

"This city is peculiarly well situated for such a purpose, much better than either Sydney or Fremantle. Melbourne geographically has the run of the main shipping route-Fremantle to Sydney, the South African route to Sydney, the Launceston route and the whole of Bass Straits; in fact all the really big traffic routes. I am able to bear testimony to the value of the Melbourne position, having a licensed aerial, which receives Sydney clearly every night. Only last evening it received clearly in Melbourne the Levuka from the Pacific some 250 miles east of Sydney. In Melbourne is heard all that is wirelessly going on at Sydney and Adelaide, even though these two latter may not hear each other. Australia seems at last to have awakened to the value of wireless, for at almost any hour in the evening electric waves are picked up. What with the big sonorous grunt of the Marconi equipped boats and the high pitched squeal of such as use the Telefunken system, the ether is becoming like a farmyard with its diversity of squeals, crackles and grunts. The other evening a message was flashed to Sydney from a steamer stating that it was ninety miles west of Melbourne. That message probably came back to Melbourne by wire (although already caught in Melbourne by wireless). The same night a hair-raising message was flashed from Sydney to a mail steamer between Albany and Adelaide. A fine of five hundred pounds with the option of two years imprisonment and confiscation of all apparatus is the practical and efficient way Postal departments protect the secrecy of that message the world over."

Note. 4: All-Red Wireless, Henry Sutton: The Argus" Melbourne, 20 June, 1911, p. 9.

"The information this morning that the British Postmaster-General has stated to the conference that an all red wireless series of stations across the Pacific is impossible owing to the British Government being tied to the Marconi system and the over sea Dominions having adopted another system will not I think prevent the linking up of the Pacific by wireless. Wireless has nothing to do with all-red routes and the most ultra-vermilion speech cannot affect the fact that wireless refuses to be chained to any particular route, country or creed. Ordinary commercial progress will determine this matter and it is an easy prediction that

within the next five years Australia, New Zealand, Suva, Hawaii, and California will be in regular communication, the missing link at the present being Australia and New Zealand. Both Australia and New Zealand are now at work on this last link."

Note. 5: Wireless Stations, Letter to Editor by Henry Sutton, "The Argus" Melbourne, 21 September, 1911, p. 8.

"The question has several times been asked in the House of Representatives---What rights will be left to the wireless companies at present operating land stations in the Commonwealth after the Government stations are taken over? Outside territorial waters the Government has no rights between ship and ship and as there are already somewhere nearly 100 ships interstate and overseas fitted with wireless, the revenue must eventually become considerable. The owners of the systems appear to have the best of it in their arrangements with the ship owners. It is stated the operators and apparatus are hired to the ship owners, the latter bearing all the expenses and the wireless companies taking all the revenue. This is certainly a good thing. They can well afford to part with any land stations as these companies will still obtain part of the land station revenue by virtue of receiving the latter's messages. Rival companies have been known to hold each other up and it is possible for a combination to hold Government land stations up except on terms. While on this matter of wireless, I would suggest that an addition be made to the recently issued excellent new wireless regulations, and that is, it be compulsory for every land and ship station to use a "break in key." There is much loss of time, and considerable confusion, round our coast due to a number of ships not being so fitted. There are several successful keys that prevent the jamming that is going on—the Marconi key, the Balsillie key. My private station has a special automatic key of our own devising. These keys all enable the transmitting station to be stopped by the receiving station, instead of waiting for an end message. With such keys you can listen between every dot and dash of transmitter, and any other station can notify you in the middle of your message whether you are interfering."

Note. 6: Wireless Experiments, A Melbourne Inventor: "The Argus" Melbourne, 10 February, 1912, p.7.

"Probably but few people are aware that for several years Mr Henry Sutton has been carrying on important research work in wireless telegraphy at his private station in Melbourne, this station is at his residence in Erskine Street, Malvern…

… In order to give some idea of the value of Mr Sutton's latest discovery a word or two may be said first of the general principles of wireless-telegraphy. For

transmitting a message, the operator uses a transformer of very high frequency and very high electro-motive power to produce a spark or flash, which sets up very rapid oscillations or electric waves. These waves are sent up the mast and flick off so to speak at the end. They are ether waves and are transmitted in all directions from the mast growing fainter as they get farther away from it. Ether is the inelastic medium of great tenuity, which is supposed to pervade all space, including even the interstices of matter, whose atoms are, as it were bathed in. Light and heat are like electric waves transmitted by the ether. Transmission is only one part of wireless telegraphy.

After the vibrations have been sent out, a receiver or electric ear is needed to catch them, so that the message they convey may be heard at the other station. At the receiving station there will be a mast to intercept the ether waves and apparatus housed at its foot to make known to the operator there the nature of the signals made at the sending station. When the ether waves strike the second mast they reproduce electrical phenomena similar to those produced by the transmitting operator and by means of a telephone connected with a detector, the original dot and dash flashes of transmission are heard as short and long buzzes according to the Morse alphabet. Of course there are several possible variations in the method of both transmission and reception, but this brief explanation will indicate the nature of the process sufficiently for our purpose.

A number of improvements in methods of transmitting and receiving have from time to time been invented by Mr Sutton, some of which have gone into use in various navies, while others have been described in technical works. In the 1911 edition of Dr James Erskine-Murray's standard handbook of Wireless Telegraphy for example it is mentioned that a new detector constructed of two pieces of galena, was discovered independently by both Mr. Sutton and the well-known English expert Dr W. H. Eccles. After Dr Eccles had announced his discovery he learned that Mr. Sutton had previously patented the same device in Australia, having invented it a year or two before March 1910 - the date of the patent. With the disinterested courtesy of the true man of science, Dr Eccles at once gave all the credit to Mr Sutton and warmly congratulated him on his success. Mr Sutton's inventions are concerned with the production as well as with the reception of the electric waves, and in his own operating room may be seen at work an ingenious automatic system which enables the operator to listen at the same time as he sends messages. There one may also observe what appear to be effective means of protecting the detectors, this protecting apparatus also being an invention of Mr Sutton's. During the last year Mr Sutton made a discovery which seems to offer an interesting explanation of certain puzzling phenomena connected with the sensitiveness of some crystals for the purpose of electric wave detecting. This it may be mentioned goes to show that the increasing sensitiveness is due to physical rather than chemical causes and can be induced by mechanical means – a most important point from the practical and commercial no less than from the scientific point of view.

Mr Sutton has written a full account of the discovery, but is reserving it for publication in a European technical journal. Upon the discovery, however, he has based a new detector which was recently patented in the Commonwealth and is showing remarkably good results. It is a great advance on the patent of 1910 and the discovery makes its multiplication easy. From the records of its work, Mr Sutton has reason to claim that it is somewhere about three times as sensitive as the next best in use. From July to October last, tests were made with it every night at the Erskine street station by Mr A. E. Sutton, the inventor's son. The results have been tabulated, and they are eloquent of the detectors efficiency. As an interesting and suggestive sample we may take the results obtained for the half hour after midnight on the night of September 12. 1911. The following ships were then off the west and south west coasts of Australia at the distances in miles stated:- S.S. Ascaneus, 1,200; S.S. Suevic,1,400; S.S. China and S.S. Scharnhorst,1,500 and the S.S. Osterley 1,900. With the new detector, the calls of all these vessels were distinctly heard in Melbourne. Again on the previous 28th of August, at a little more than 20 minutes past midnight, the S.S. Karoola was heard 1,000 off calling the S.S. Omrah, which was about 2,000 miles from Melbourne and the Omrah was also heard answering though very faintly. The longest range call that could be definitely detected during the months of trial was one heard about 8 o'clock on the night of August 8 last. It was the Suevic calling "all stations" heard at the enormous distance (for wireless) of 2,397 miles.

The new receiver evidently does good average work at 1,000 miles, detecting without difficulty calls made by such vessels as H.M.S. Powerful from the harbour at Brisbane or Wellington. An operator at the Erskine street station may be pardoned a feeling of supremacy as he hears with the new receiver, ships vainly calling to each other. Sometimes while ships try, try, try and fail to get into touch he can note every word they send out. Two ships have been heard for a full hour endeavouring to understand each other so that a message might be sent to Sydney and thence overland to Melbourne; every word from both ships being actually received in Melbourne with the aid of the new detector. In one case two ships were between Tasmania and New Zealand striving to get one code word through. The word was repeated 15 times before the recipient managed to catch it. Yet that word was heard the 15 times in Melbourne. Wireless communication between the ships was more or less blocked by distant electrical disturbances in the atmosphere known as "static," but the acute electric ear of Mr Sutton's receiver could pick out the code word in spite of the "static." As can easily be understood, sensitiveness on the part of the detector or ear of the wireless apparatus is a main criterion of its effectiveness, and Mr Sutton seems with his new discovery to have found out how to construct an electric ear that can hear messages too faint and too distant for other receivers to note. Not only has the sensitive detector a far greater range than one less sensitive, but it records more distinctly the messages which the less sensitive can also hear. The work done in the test months was in no wise exceptional, for very similar results are to be obtained nightly with Mr Sutton's installation. Last Tuesday night for instance a visitor taking up and putting on the second telephone headpiece while Mr A. E. Sutton was listening for calls could hear at different times between 10.30 and about 11.45 such things as a conversation between H.M.S. Encounter lying at Hobart and H.M.S Cambrian which had lately left Sydney for Hong Kong; or calls from the Orsova which had just started for London from Brisbane and must have been about 800 miles off, the airway of its message to Melbourne being almost wholly overland and therefore a more difficult airway than if it had been entirely over sea. Many other calls and messages were also to be heard during the hour and a quarter including calls from the Westralia which had only just left Hobart for Sydney, from the Grantala then on her way to Sydney from Melbourne and from the Hotel Australia in Sydney. Even to one unused to listening for wireless messages the Morse dots and dashes of all these calls and talks came with wonderful clearness through the new receiver. It may be mentioned in conclusion that in addition to the delicate electric measurements necessary to the working out of the principle of the new detector, Mr Sutton made last year a number of photo-micrographs illustrating the action of electric oscillations in relation to his invention. These are to be published with the technical report of his investigations."

Note. 7: An Inventor, Mr Henry Sutton and his Invention: "The Leader" Melbourne, 14 May, 1910, p. 51.

"We publish among our illustrations this week pictures showing some of the instruments designed by Mr Henry Sutton, a most indefatigable worker in this branch of scientific research. He has kindly supplied explanatory notes which will enlighten the reader as to the nature of what is being done.

Fig. 1 represents the latest branch of Mr Sutton's experimental work. (A) is an electric arc lamp with the carbons placed between the poles of a powerful blast magnet, this magnet blows out the arc as fast as it is formed, and in conjunction with condensers and inductances produces electric oscillations.

At the back of the picture is seen a box (B), this contains a number of variable condensers. A variable inductance coil is seen on top behind the four condenser switches. With this apparatus an octave of electric notes is produced, having the tones of an organ. These are received as acoustic notes at a distance, through the intermediary of the ether. A tune may be played on the ether electrically and reproduced acoustically at a distance without the use of telegraph wires.

By placing a parabolic mirror behind the arc light, the tune may be sent along the beam of light and can be received by the ear through a special receiver. The arc tunes itself also gives out the notes and times acoustically.

(C). is a fixed inductance coil and it is the action of this and the condensers on the arc that produces the oscillations, the variable condensers and inductance in the box giving the power to vary he pitch and thus play a tune or send a different tone to each of several receivers.

(D). is a special microphone transmitter capable of taking the current of a dynamo, and therefore very distinct from the telephone of commerce. By means of this special microphone the voice is impressed on the vibrating arc light, and thus on the electric osculation, these latter set up oscillations of the same period on the ether, the ether then takes on a wave motion and in turn sets up electric oscillations in the distant receiver, which latter reproduces the voice acoustically. This apparatus combines the functions of the wireless telephone, the wireless musical tone, and the transmission of both voice and musical tones along a beam of light and is unique, there being no other apparatus with such diversified functions in the world. It was made in one of the leading physical laboratories in Europe to Mr Sutton's design. It was tested there, and found to work perfectly for wireless telegraphy and light telegraphy, both musical note and telephonic (this is the announcement to the world of this apparatus).

Fig 2. The instrument (A) on the wall is Mr Sutton's universal receiver, this also has unique qualities, and it does its work with certainty.

In the middle at bottom of board is a switch with five contacts, on placing the little lever to extreme left a wireless signal reaching the aerial instantly rings the bell and calls attention, the lever is then placed to extreme right and the sounder taps the message out loudly. It can be heard over a large room.

By placing the lever second contact from the left, instead of bell ringing, the electric lamps (below bell) instantly lights up and silently attracts attention, and remains alight till lever is put over to sounder contact, and the message is then rapped out in Morse code. This instrument never goes out of adjustment.

The flat box (B) with terminals connected to telephone is Mr Sutton's especially sensitive receiver for long distances; the internal mechanism has exquisite delicacy of adjustment.

The little box (C) with telegraph key on it is his portable testing transmitter, and though small, is a practical instrument. With it and the portable tripod receiver (E) Fig. 5, and two fishing rods for aerials, perfect wireless communication can be kept up between Port Melbourne and Williamstown.

The whole of Fig. 2 is Mr Sutton's own handiwork as well as design.

Fig. 3. This shows Mr Sutton's portable receiver in use, the little box in his left hand is a complete wireless receiving station, and without an aerial will receive at a reasonable distance the wireless messages of a war boat; attached to an aerial it receives great distances. This is the little apparatus that excited our recent Japanese visitors so that they acquired a right to use it.

More will be heard shortly of the visit of the Japanese officers to Mr Sutton, and their desire to acquire other of his inventions. Mr Sutton found them wonderfully advanced in the knowledge of wireless, not merely in practical work, but advanced in the science.

Fig. 4. This apparatus is probably the most important of his inventions, as it comprises his discovery of an entirely new principle and the invention of the apparatus to utilise this principle. This is Mr Sutton's oscillator, and it produces powerful fundamental oscillations (that is, oscillations of great power and of exact period and wave length). Tabulated tests and measurements of results have been sent to London to be examined by one of the leading physicists. Being a fundamental discovery, Mr Sutton decided it should go where it had the opportunity of being understood and appreciated. The instrument (A) is a powerful induction coil, giving a ten-inch spark. The energy of this coil passes into the square box and is converted into electric oscillations of high power and absolutely accurate wave length.

The spiral of copper rod (B) is fifty feet long, and when the plug (seen at side of spiral) is placed in top spiral, the wave emitted is exactly fifteen hundred feet long, and has seven hundred and fifty thousand vibrations a second.

When the plug is placed where shown in picture, the wave is exactly one thousand feet long (the legal wave of ships), and it has one million vibrations per second.

This is the only apparatus in existence capable of producing a true fundamental wave at high electric pressure. This is also Mr Sutton's handiwork as well as design.

Fig. 5. This is a portable military set invented by Mr Sutton. (A) is the charging coil. (B) the telegraph key. (C) is the oscillator, and it is notable that in quite a small apparatus (the box is fifteen inches high), long powerful waves are produced, having a definite wave length of 300 metres, and with nothing to go wrong, and needs no tuning.

It is absolutely certain (under any conditions of warfare), to work and send out a true wave. There are no adjustments of any kind to make, and therefore any officer who can use an ordinary telegraph key can send out perfect wireless messages.

(D)is the portable detector, by which an operator can without an aerial get near the enemy and tap the messages (an aerial gives itself away even ten miles off), but the portable detector gets within reasonable distance. (E) is the field tripod receiver.

This is the receiver Mr Sutton exhibited at work to the Military Board, 29th July, 1908, and which, by leaking into the press, caused so much departmental trouble. This tripod receiver (attached to aerial), caught all messages sent out by the American Fleet during August and September, 1908. It caught the message sent by Admiral Sperry from Cape Howe to the State Governor and the late Sir Thomas Blunt. It caught their replies sent back from the Culgoa, and it caught the Connecticut message from Albany to Melbourne, 1200 miles.

The electrical officers of the U.S.A. Fleet were highly pleased with Mr Sutton's work, and some of them have kept up a correspondence from U.S.A.

A1 this apparatus is Mr Sutton's handiwork and design.

Fig. 6. This important apparatus represents the highest degree of perfection in apparatus for research into etheric and electric waves and wireless telegraphy. (B) is termed a cymometer (wave measurer). It measures electric waves, in metres, in feet, and will measure and indicate electric oscillations, having a period of three million per second. It will measure accurately waves 100 feet long up to 3000 feet long. It is designed by the greatest expert on and exponent of electric waves, Professor Fleming, F.R.S., University College, London, the consulting engineer to the Marconi Co.

(D) In conjunction with another apparatus is a thermo galvanometer. It measures quantity of electricity to the millionth of an ampere, and the fifty thousandth of a volt. (E) measures to one thousandth of an ampere. (F) is a box of special resistance coils. (C) is a fine glass tube containing the rare atmospheric gas neon (discovered by Sir William Ramsay and Professor Dewar), and found by Fleming to be extremely sensitive to electric oscillations. It glows with a bright light when the cymometer is in resonance with any particular wave. When such is the case the length of wave, &c. can be read off the scale (B).

This beautiful apparatus was specially made for Mr. Sutton by the Marconi Co., for research work. (A) Is Professor Ruhmer's Oscillograph. When electric oscillations are sent into it, the large tube glows with a violet light, oscillating (if a 300 metre wave) one million times per second. By rapidly rotating a mirror before this light, and photographing the image in the mirror, the photograph shows the oscillations separated out, and is a permanent visible proof that the oscillations exist, for they can be counted.

This oscillograph Mr Sutton received from Berlin. Without these highly developed research instruments it is impossible for anyone to obtain a real knowledge of science that lies behind wireless telegraphy and its epoch making results."

Note. 8: Correspondence, Production of Rectifiers by Physical Means: Henry Sutton, "The Electrician", Volume 66, 16 December, 1910, p.p. 384-385.

"Sir; I thank you for giving me the opportunity of being the first to congratulate Mr Sutton on the fact that his patent for a galena detector ante-dates my Paper. I may add that it would have been a matter of regret to me if the Paper had, by prior publication, injured a meritorious inventor. Your correspondent's object was technical, mine was scientific; but while achieving a great technical success he offers an explanation of the mode of action of the galena detector. He suggests that the fall of resistance at the contact is due to the increase of mechanical pressure there through thermal expansion produced by the current. This possible effect, and other conceivable ones, I include in the "negative temperature coefficient of resistance of a contact" used in my paper, but inasmuch as I have failed to find a useful oscillation detector constituted of a contact between substances possessing positive coefficients in the bulk, I believe we always have to deal, in good detectors, with bulk negative coefficients of resistance."

When I read the paper before the Physical Society in May last I announced that if the galena detector is traversed by a suitable current while shunted with a telephone and a condenser there is heard in the telephone a singing sound whose pitch changes with change of the capacity of the condenser. I mentioned this as evidence of a falling characteristic for these (and other) detectors, analogous to the characteristic of the singing arc, and I stated that this musical sound was difficult to match on a monochord because it was marred by a background of tone which is similar to the sound heard, even when the shunt circuit is removed, in another telephone placed in the direct current circuit through the detector. It is this latter sound that heard in a telephone when a suitable direct current is passed through a circuit comprising telephone, detector and source of E.M.F. which your correspondent attributes to Trevelyan rocker action. I looked into this some time ago, and have been trying recently to obtain photographic records of these pulsating currents with a thread galvanometer. In the majority of contact detectors this phenomenon may be due to the negative bulk coefficient, but for the singing of metal contacts I have meanwhile arrived at the following explanation in preference to the "rocker" one: The heat generated by the current at the constricted pathway across the contact warms the contiguous matter, causing expansions which enlarge the area of contact. Consequently, the current density decreases, cooling follows, since there is other resistance in the circuit, and contraction re-establishes the former state of things. This cycle takes place very rapidly at a rate affected by the inductance and resistance of the telephone in the circuit. The singing phenomenon in detectors must have been noticed by everyone who has worked with the crystal detectors which came into vogue some years ago. I may mention, by the way, that a similar explanation seems to hold good for the electrolytic detector.

True Trevelyan rocker action involves complete interruption of the current, and does not occur in contacts usable as detectors. It is best exhibited at a contact of metals. For instance a contact of a pair of lead electrodes, when "rocked" by an appropriate current through it, cuts the current up into pulses. This broken current affords efficient means of generating electrical oscillations in such testing circuits as those described in my paper on Coherers (The Electrician, Vol. LXV., p. 724). One of the pieces of lead should be mounted on an iron wire, so that delicate adjustments of the contact may be made by a couple of permanent magnets."

Note. 9: Correspondence, Production of Rectifiers by Physical Means: Henry Sutton, "The Electrician", Volume 69, 19 April, 1912, p.p. 66-67.

"Sir: So far as I have been able to trace, no explanation has hitherto been found for the rectifying action that certain metallic crystals have on electric oscillations, and although the effect is largely utilised in detectors used in wireless telegraphy, the phenomenon is always treated as mysterious and of a somewhat elusive nature. Given an understanding why this valuable effect takes place, the field is opened for great improvements in wireless telegraph detectors and receivers.

Whilst investigating this peculiar property in various minerals, I discovered that the rectifying power could be mechanically induced in a mineral which in its native state had not the property. I found the rectifying property could be given in a highly exalted degree by altering the physical form of the crystalline surface.

If we take a cubic crystal of disulphide of iron, FeS_2 it will be found that in its native state the surfaces have no rectifying action on electric oscillations. I find, however, by the simple process of fracturing it across the cleavage planes in such a way as to produce a conchoidal fracture, this fracture has a number of fine ridges, which can be seen under the microscope. (I have measured these, averaging 1/2500th of an inch across.) In the vicinity of these ridges, or the edge of the conchoidal fracture, or about the basin-like form of the fracture I find rectification to a high degree takes place, whereas the adjacent native surfaces are inert and have not this property.

The discovery is of great practical importance to wireless telegraphy, as the hitherto mysterious property of crystalline rectification of oscillations is now made a matter of every day certainty and there can be mechanically produced thoroughly definite rectifying conditions, and a detector of highly exalted sensitivity.

An examination with a magnifier reveals at once the sensitive spots, and a gold point placed there immediately responds with a really wonderful sensitiveness and is yet robust. The rectifying property remains and is permanent.

Comparing the new detector with silicon, I find it fully three times sensitive and, in addition, it rectifies in such a way as to give a much greater quantitative result than silicon, and is capable of working a relay.

The photomicrograph I have made shows a mechanically produced conchoidal fracture with the ridges. The dark part marked C in the figure is the native surface, the bright part B is the special fracture. The fine surface lines (not ridges) of the native surface are also shown. No part of this native surface rectifies.

The conchoidal fracture is shown by the white shell-like appearance on the right-hand side of the photograph. The ridges are plainly visible at R, and it is in the vicinity of these ridges and on any part of the fracture that rectification to a higher degree takes place. It is not necessary that the ridges shall be always visible.

The particular fracture shown in the photograph has received from a Marconi 2 kw. ship-set at a distance of 1,200 miles without missing a dot. The whole of the native surface shown can be made equally sensitive by chipping in such a way as to produce the special fracture. No applied E.M.F. is required when using this detector.

The facts are so certain and so easily produced as to force me to the conclusion that the physical state of the surface is the real cause of rectification, and the presence of the ridges as radii to the conchoidal form is an indication as to the cause. One set of alternations disappears, is absorbed or destroyed by the surface formation, and I venture the theory that they are diffracted or bent aside sufficiently to be forced into and damped out by the resistance of the surface. It is certainly not due to any chemical state, as that is uniform both FeS_2 *on the natural surface and on the fractured surface. Simple inspection (after fracture) enables one to select the more powerful rectifying spots, such as a minute crater, shapes or near ridges, or hollows.*

For several months, daily and nightly, observations were made of incoming signals, and these give some idea of the value of the new form of detector. As an illustration of the value of this detector, the following is of interest. On September 12, 1911, between 12:5 and 12:30 a.m., the following named ships were off the west and south-west coast of West Australia; the respective distances in miles from my station at Melbourne, are given in brackets after the name:-"China"(1,500), S.S. "Ascaneus", (1,200), S.S. "Osterley", (1,930) (this latter mostly overland)

S.S. "Scharnhorst" (1,500), S.S. "Suevic", (1,400). The messages of these ships were heard at my station, which has an umbrella aerial 102 ft. high.

I have designed this aerial with a special object, and that is, that the ribs have such an inclination as to cut the wave front. Instead of the aerial being flat, the ribs have an angle of 45 deg. to the two central vertical air wires. There are 16 ribs in groups of four, the ribs being 100ft. long, and the four wires of each group arranged as a rectanglar cage with 3ft. sides. The aerial appears very efficient, both in sending and receiving. In addition to the above list, we also hear ships rounding both the north and south coasts of New Zealand, 1,200 miles away. We have heard a Marconi 2 kw. set (S.S."Marmora") at 12.30 midday off Adelaide over 400 miles. Our greatest receiving distance has been the S.S. "Suevic," 2kw set, August 9, 1911, at 2,397 miles S.W. from Melbourne.

The fact that these are obtained from a crystal which in its native state is inert, and that the result is due solely to a mechanical alteration of the surface, appears to me to be of great importance and suggests the possibility of manufacturing an rectifying surface.

The property is so pronounced that I have with a low resistance single-pivot micro-ammeter taken measurements of the rectified received current. It must be remembered it is the actual useful current that is measured and not an integrated result of both sets of alterations as would be the result by a thermo-galvanometer measurement. By measuring the rectified current after passing through the tuned receiving circuits, the useful current as used in the telephone is obtained.

The following Table gives these results:-

Experiment	Distance of ship (Miles)	Micro -amps	Remarks
1	4	120	Admiralty Ship
2	4	180	Admiralty Ship
3	200	1	Telefunken quenched spark
4	5	400	Telefunken quenched spark
5	5	120	Telefunken quenched spark
6	5	100	Telefunken quenched spark
7	4	8	White Star Ship
8	4	120	Admiralty Ship
9	4	160	Telefunken quenched spark
10	4	200	Admiralty Ship
11	5	400	Telefunken quenched spark

All the above ships were communicating with Sydney, 466 miles overland, and the varied power is remarkable. The most noteworthy point, however, is that we can now hear Nos. 1, 8 and 10 at greater distance out at sea than 4, 9 and 11.

I find that the conchoidal fractured disulphide of iron rectifies so as to suppress the alterations from the earth, and transmit the alterations from the aerial, using gold wire contact on aerial side. I find that silicon acts in the reverse way---that is, silicon suppresses the alternations from the aerial and passes on the alternations from the earth, using gold wire contact on the aerial side.

Using a gold point in each case, I find the thermo-electric current due to heating the junction is from FeS_2 *to gold---that is, reverse to the rectifying effect.*

(1)FeS_2 *takes the position of copper in the thermo-electric scale when copper is heated with iron contact. With silicon, I find the thermo-electric current due to heating the junction is from gold to silicon---that is, in the opposite direction to the rectified current. (2) Silicon takes the same position as iron in the thermo-electric scale with copper contact when heated.*

I find with a native face of disulphide of iron in contact with a gold point, that, by heating the junction, the thermo-electric current is from the native face to the gold point. This is interesting, as the native face does not rectify, yet it produces a thermo-electric current whereas the conchoidal fracture produces a thermos-electric current and rectifies in the opposite direction. This is further proof that the rectification is due to the physical form of the surface. The rectified current and the thermo-electric current oppose each other, both with silicon and gold, and conchoidal fracture of FeS_2 *and gold, but the lag between the application of heat and generation of thermo-electric current is greater in the case of* FeS_2 *and a gold contact than with silicon and a gold contact, so that in the case of* FeS_2 *the thermo-electric effect is not detrimental when receiving oscillations in practice, but may be detrimental in the case of silicon.*

I made comparative experiments on signals from the same ship---a warship situate from my station 4 miles and calling Sydney 500 miles. A selected sample of silicon was placed in the tuned receiving circuit and registered a maximum of 40 microamperes. An average sample of FeS_2 *with special fracture and gold point was then put in same circuit, and the registration was 120 microamperes that is, three times the effect. This detector has been patented in the Commonwealth under date September 29, 1911."*

Note. 10: Correspondence, Production of Rectifiers by Physical Means: Greenleaf W. Pickard, "The Electrician", Volume 69, 17 May, 1912, p. 232.

"Sir: In your issue of April 15$_{th}$ *Mr Sutton presents a very interesting letter*

describing a supposedly novel form of solid rectifier detector for wireless telegraphy. In the writer's opinion, certain of the statements in this letter require modification.

In substance, Mr Sutton considers the use of a fracture face of a mineral an important novelty, and as a result of a microscopical examination of certain fracture faces he has developed an ingenious theory of their action.

As a matter of fact, solid rectifier detectors consisting of a metal point in contact with a fracture face of a crystal have been in widespread commercial use for the past five or six years. The particular detector described by Mr. Sutton is also exactly described by the writer's U.S.A. Patent, No. 933,263 of September 7, 1909, which points out the advantages in using a fracture face of iron disulphide or pyrite in contact with a metal point.

While it is true that a fracture face of pyrite is usually a better rectifier than the natural crystal faces, it is not correct to say that these natural faces are inoperative. A clean, untarnished face of a pyrite crystal is frequently as efficient a rectifier as a fracture face.

So little is really known about solid rectifier action that it is usually difficult to criticise any theory that may be advanced. Mr Sutton, finding minute ridges on a fracture face of his pyrite crystal, considers that these ridges diffract or bend aside all of the oscillations in one direction. In view of the recent work of Nipher and others on what happens when electricity turns a very sharp corner, it must be admitted that this theory has certain plausibility.

This plausibility vanishes, however, when perfectly plane surfaces exhibit efficient rectification. The writer has ground and polished many mineral surfaces, including pyrite, to optically true planes, and has found that a contact between such a surface and a metal point forms an efficient rectifier detector.

The writer also questions the comparison made by Mr Sutton between the silicon and pyrite detectors. Pure silicon, free from iron, carbon and calcium, makes a rectifier detector of practically the same efficiency as the best samples of pyrite."

Note. 11: Production of Rectifiers by Physical Means, Henry Sutton: "The Electrician", Volume 69, 2 August, 1912, p. 708.

Sir: In your issue of May 17 Mr Greenleaf W. Pickard calls attention to my letter published in your issue of April 19. I have not had the pleasure of reading Mr Pickard's patent, No. 933,263 of September 7, 1909. My patent of September 29, 1911, is purely a date patent. Mr Pickard in his letter says: "Mr Sutton, finding minute ridges on the fracture face of his pyrite crystal, considers that these ridges diffract or bend aside all of the oscillations in one direction."

What my letter says is this:- "I find, however, by the simple process of fracturing it (the crystal) across the cleavage planes in such a way as to produce a conchoidal fracture. This fracture has a number of fine ridges."

Further on my letter says:- "It is not necessary that the ridges shall be always visible."

And further on again my letter states:- "the facts are so certain and so easily produced as to force me to the conclusion that the physical state of the surface is the real cause of rectification, and the presence of the ridges as radii to conchoidal form is an indication as to the cause."

Then my letter distinctly says:- One set of alternations disappears, is absorbed or destroyed by the surface formation, and I venture the theory that they are diffracted or bent aside sufficiently to be forced into and damped out by the resistance of the surface." In venturing this theory, I was aware that at the frequency 500,000 only a superficial penetration to the crystal could occur.

The discovery of the ridges caused me to conclude that one set of alternations was being forced between the cleavage planes a sufficient depth to be damped out by the high frequency resistance, and that this penetration was caused by the edges of the cleavage planes; naturally an optical surface produced on the crystal would not prevent the forcing in of the electrons, no matter how perfect the mechanically produced surface. The frequency, preventing a natural penetration, prevents either refraction or double refraction, and we are forced to suspect the missing electrons (in this case those from the earth) have their direction changed by diffraction and the particular alternation becomes damped out. My letter clearly stated, "It is not necessary that the ridges shall be always visible" Meaning that the diffractive effect may go on.

The theory of asymmetric conduction only tells us what we already know, that one set of alternations disappears. My theory was intended to suggest why.

As there would be no conduction (at the frequency) through the crystal, my attention was focused on the surface effects.

The connection of the conchoidal form of the fracture is that, in my experience, this fracture is only associated with a physical structure of a type that will produce ridges in FeS_2 *crystal, and not that either the conchoidal form or the ridges are the cause, both the conchoidal form and the ridges being due to the existence of cleavage planes in the crystal. If by grinding and polishing his crystal Mr Pickard exposes the cleavage planes, then he enables the diffractive effect to occur; in other words, he clears the natural surface out of the way.*

Mr Pickard, as the discoverer of a great number of sensitive crystals and combinations, no doubt is aware that any FeS_2 *crystal will not do, and that only such as have the characteristics described in my letter published in "The Electrician" of April 19 are at all effective.*

Note. 12: Scientific Notes, Telegraphing Through Crystals: "The Australasian" Melbourne, 22 June, 1912, p. 52.

"This curious behaviour of crystals has, so far, proved impossible to explain. The first suggestion towards a theory was based on thermo-electricity, but this was soon seen to be impossible. Thermo-electric currents through the crystals were measured, and found to be of the wrong order of magnitude, and, for many crystals, in the wrong direction. Neither does it appear to be likely that the rectification is due to electrolysis, as it is in the case of the Nodon valve, so largely used for rectifying alternate currents from the dynamo; for silicon, which is one of the best crystal- rectifiers, is a chemical element, and cannot be electrolysed. Some investigators have attributed it to a surface film, formed by currents, which attempt to pass one way, and destroyed by those which actually do pass in the opposite way; but the explanation does not appear altogether satisfactory, seeing that in some cases the transmission is from crystal to metal point; in others from metal point to crystal.

Some light, however, is thrown on the subject by a discovery made a few months ago by Mr Henry Sutton, of Melbourne, who found that, in one case at least, the rectifying power depended on the mechanical treatment of the crystal. A native crystal of disulphide of iron is, according to Mr Sutton's researches, devoid of rectifying power. If, however, a portion of the native surface be knocked off, and the inside material exposed, places can readily be found which, in conjunction with a gold point, give a rectifier of much greater sensitiveness than silicon and gold. The right position for the gold point is indicated by the presence, in the fracture, of ridges which are easily seen under the microscope. With such a rectifier and a telephone, Mr Sutton claims to have received signals from a ship nearly 2,400 miles (approx. 3,862 kilometres) away.

If this claim be substantiated, he has hit upon the most sensitive detector yet invented, and has incidentally discovered a fact of no small scientific interest."

CHAPTER SIXTEEN:

Note. 1: The Ballarat School of Mines Students Magazine: Obituary; Second Term, 1912, page 8.

"It is our painful duty to record the death of Henry Sutton, who was a native of Ballarat, and for some time Instructor in Electricity and Magnetism at this School. In the early eighties he was a constant attendant at all the School Society meetings, such as the Field Club and Science Society, and the Camera Club (of which he was sometime president). He was a liberal contributor to the evening entertainments connected with these societies, and at all the many enjoyable

excursions of the Camera Club he was a prominent member with his full-plate camera. A little later he gained the Photographic Annual's Gold Medal for stereoscopic photographs.

Henry Sutton had always a very active and fertile brain, and his greatest joy and almost all his leisure lay in the direction of science, and especially in its practical applications. He touched many subjects; he had many interests, and excelled in them all, whether it was drawing or photography, or pianoforte construction, or practical electricity. He read heavily and omnivorously in science, and as a very young man he used to set himself tasks to be finished in given times. He kept the flywheel of his brain spinning, and thought out his work systematically and persistently; until he had developed his subject as far as his quiet, steadily-working inventive powers would carry him. His methods of work seemed to have much in common with those of Edison.

He was always eminently practical. His ingenious mercury vacuum pump is a fine appliance. He spent a good deal of thought over artificial flight, and made some interesting experimental studies with flying birds when he was very young. He had been working at the carbonization of fibres for the glow lamp before these became in other hands accomplished facts. Some special friends in Ballarat were acquainted with his theoretical devices for seeing at a distance by electricity, and so photographing. He convinced others of the possibility of accomplishing this wonder before he went home to England, and long before it became a working success.

The storage of electricity also attracted his attention, and, after he had worked and thought and worked on it, evolved the Sutton Secondary Battery, his paper on which was read for him by the President of the Royal Society, London, at one of the meetings, and afterwards printed in the Transactions. The Sutton-type process was another of his achievements, reached after much experiment and close reasoning and highly spoken of by Mr Ellery and others, as of great service for cheap printing.

Improvements in lantern lenses took up some of his time, and when he went to England he had some unpleasant experiences when endeavouring to get his ideas worked by the mechanician. Certain improvements in connection with the bicycle brought him a steady royalty. Amongst other things on which he brought to bear his great inventive powers was the Broken Hill sulphide problem, and he had gone some way towards devising a new method on the flotation lines, but more pressing demands on his time doubtless crowded this out. But of all the many attractions that science held out for him, it was electricity and its applications that possessed him most completely and were nearest to his heart. When he took a trip home to England years ago he was well received by scientific men, and we understand that the first picture actually transmitted by wire was by the process he had invented. For many years he was a close and acute student of wireless telegraphy, devoting his time especially to improving the receiver, so as to be able to receive with greater certainty and regularity long distance messages under changing atmospheric conditions. His receiver is still attracting much attention in many quarters. Henry Sutton was granted a licence by the Commonwealth Government to erect a wireless station at his Malvern home, where he could sit at his instrument and intercept long distance messages, or make his station the centre of a violent ether storm.

As an evidence of Ballarat's appreciation of Henry Sutton's powers and further promise, it may be remembered that years ago a largely-attended public banquet was tendered to him on the eve of his departure on a trip to England. The company included the Mayor (in the chair), the members for Ballarat and district, and a long line of leading citizens.

In the passing of Henry Sutton at the comparatively early age of 58, Ballarat has lost her greatest scientific inventor and one her most distinguished sons."

Note. 2: Death Of Henry Sutton: "The Ballarat Courier" July, 1912, p.6.

"News has been received of the death of Mr Henry Sutton, at his residence, Malvern, yesterday morning. The deceased was a native of Ballarat and was 57 years of age. He was the first instructor in electricity at the School of Mines and the list of inventions he gave to the world is an extensive one. First perhaps, on the list is the mercury vacuum pump, the first pump to create a perfect vacuum. It is used extensively in London hospitals and was the original means of creating the necessary vacuum in electric glow lamps. He also discovered, coincidentally with Mr Graham Bell of Edinburgh and Mr Thomas Edison, the carbonising of manila fibre, whereby the resistance is set up in electric glow lamps that creates the electric light. He was responsible for the Sutton-type process of producing printing blocks for illustration purposes and was inventor and patentee of what is known as the most effective (longest distance) receiver for wireless telegraphy messages. At the present time wireless experts all over the world are making enquiries and investigations regarding this receiver, which has been in use by the Commonwealth Government. By the process invented by him the first actual photograph ever transmitted by telegraph was sent, with the aid of Tesla, the now famous discoverer of the Tesla wireless wave. This was done in London, with the consent and assistance of the electric telegraph authorities there. Mr Sutton was an expert draughtsman, his black and white line drawings being much admired. This talent was of great advantage to him in his scientific work. He was also an expert photographer, and when in England some years ago he received the

"Photographic Annual" gold medal award for the finest stereoscopic photography. For many years past he closely studied wireless electricity, and was granted a licence by the Commonwealth Government to erect a wireless station at his home at Malvern. He devoted his investigations chiefly to improving the receiving apparatus, so as to receive with more certainty and regularity, under any and every possible variation of atmospheric conditions, messages from long distances. It is understood his success in this direction was remarkable. He came into touch with wireless operators of the American Fleet between Pago Pago and New Zealand, at that time the longest distance over which a message had ever been received. The wireless experts of the American Fleet got at once into personal communication with Mr Sutton on their arrival at Melbourne and were much interested to find that his system of wireless did away with the necessity of the familiar high pole and outdoor apparatus; Mr Sutton being able to demonstrate to them that he could receive messages as he sat at his receiving table. He also greatly interested the American experts with a loud and sensitive telephone wireless message receiver, which they declared would be invaluable under war conditions. The British Admiralty have been also much interested in Mr Sutton's wireless experiments and many leading specialists connected with the Navy consulted with him. The Japanese authorities have also interested themselves. It is understood that Mr Sutton discovered the working of a wireless wave of remarkable power and it is believed that he entered into an understanding with persons in high authority that this was to be regarded as a secret of national value, to be used only under certain conditions. The first telephonic communication in Australia was set up by Mr Sutton. He was the eldest son of R. H. Sutton and a shareholder of Sutton's Proprietary Limited, of which he was the public officer. He leaves a widow and four sons, well provided for."

Note. 3: Obituary; The Australasian: 3 August, 1912, p. 40.

"Mr Henry Sutton a member of the well known music firm, died at his residence at Waltham, Erskine-street Malvern, on July 28. Although he was of a very retiring disposition, a wide circle of friends learnt with deep regret of his decease. Mr Sutton was all his life a keen student of electrical and magnetic science, and he obtained such proficiency that when a young man he was appointed lecturer on these subjects to the Ballarat School of Mines. One of his inventions was a detector for receiving wireless messages, and this was regarded as so sensitive and reliable that at the time of his death the Defence department were in communication with him in regard to it. A widow and four sons survive Mr Sutton."

Note. 4: Obituary; "The Mining and Engineering Review" 5 August, 1912, p. 402.

"The death of Mr Henry Sutton, which occurred in Melbourne on the 28th July, removes from our midst an enthusiastic experimenter in electrical science. Of recent years he has devoted practically all his spare time to keeping up to date on the subject of wireless telegraphy, and to do this erected and operated a very complete private wireless station at his residence near Melbourne. The value of some of his patented inventions, notably detectors, is recognised by world-wide authorities, and, as has been already mentioned in these columns, it was with one of these detectors that the Commonwealth wireless station in Melbourne was enabled to get into communication with Hobart early in the year."

Note. 5: Obituary; "The Electrician" Volume 69, 13 September, 1912, p.933.

"Henry Sutton: The death occurred at Malvern (Victoria Australia) on July 28 last of Mr Henry Sutton, electrical engineer, of Melbourne. Deceased, who was 56 years of age, took considerable interest in wireless telegraphy and for some years he was lecturer on electricity at the School of Mines Ballarat."

Note. 6: Obituary/Funeral Notice; "The Melbourne Age" 30 July, 1912, p. 7.

"The funeral took place at Brighton Cemetery yesterday of Mr Henry Sutton, a member of the firm Sutton and Co., music sellers, of Melbourne, Ballarat, Bendigo and Geelong, whose death occurred suddenly on Sunday. Mr Sutton, who was of a retiring disposition, was an exceptionally clever scientist. Years ago, when a resident of Ballarat, he had been a lecturer on electricity and magnetism at the local school of mines. He did pioneering work there in aviation also, having made some remarkably useful observations on the flight motions of birds. His inventions included a kerosene carburettor for motor engines, and a wireless telegraphy detector, but his greatest achievement was the perfection of the Sutton system of Australian wireless. Deceased, who was an associate of the Victorian Institute of Electrical Engineers was 58 years of age. He was twice married; he leaves a widow and four sons. His death will be widely regretted.

Note 6: Henry's Children:

Arthur Ernest Sutton: Arthur worked for a time at Sutton's music store in Melbourne and was an engineer who learned his engineering skills from Henry, a mechanical engineer by trade, he was also one of Australia's first motorcycle champions. Arthur married twice and had two children to his second wife, one of which became an industrial chemist. Arthur lived in Melbourne Victoria all of his life and died in 1956 aged 74.

Edward Sutton: Edward worked for a time at Sutton's music store. He caught the inventing bug but not on quite the scale as Henry and spent many hours tinkering with radios bikes and cars. He was also a talented artist. Edward married and had three daughters one of which became an electrical engineer. Edward lived in Victoria all of his life and died in 1951 aged 54.

Albert Martin Sutton: Caught the inventing bug in a big way and went onto invent a number of different things such as engines, machine guns, bombs and even a washing machine. During the Second World War, Albert was appointed to the Inventions Directorate in Melbourne and during the course of the war invented a number of bombs, guns and engines. After the war for a time Albert was Sir Lawrence Hartnett's personal engineer. Albert's washing machine went into commercial production and with a business partner AMSCO Washing Machine Company was formed. Albert married and had two children; one child died at birth. Albert's surviving son Ron had great admiration for his father's inventive talents. Later in life Ron took up the love of flying. Albert moved around a lot but eventually settled in Brisbane, Queensland and died in 1980 aged 76.

John Sutton: John worked at Sutton's Music Store in Melbourne and then moved to Adelaide and worked for a time for the P.M.G, he then worked at General Motors Holden in Adelaide as a test driver and on the production line. John married and had three children and imparted a lot of engineering knowledge and skills down to his two sons. John settled in Adelaide South Australia and died in 1973 aged 63.

Note 7: Japan Exposition: 1985:

A Model of Henry's Telephane/ Television was exhibited in the Australian Pavilion at the Japan Expo in Tsukuba in 1985.

Note 8: Canberra Street: 2004:

On 20 January, 2004, a street in the Canberra suburb of Dunlop is named in Henry Sutton's honour, the street is called Henry Sutton Circuit.

Note 9: Sutton Room: 2004:

On 26 July, 2004, a room is named after Henry Sutton at the Ballaarat Mechanics' Institute it is called The Sutton Room. In 2001, the Ballaarat Mechanics' Institute undertook a building and restoration project and in 2004, as the renovations were slowly being done, Dr Frank Hurley put forward a suggestion at a committee meeting on 28 June, 2004, that certain rooms in the Institute be named the "Humffray Room", the "Sutton Room" and the "Pound Room" in honour of the three notable men associated with the Institute. Mr John Basson Humffray was the Institutes first President, Dr James Robert Pound was a Chemistry Lecturer at the Institute from 1918-1948 and Henry Sutton who was one of the Institutes best known prodigy's. The various rooms in the Institute had not been named before now and so the suggestion was scheduled to be discussed at the next meeting. At the committees next meeting Dr Hurley's suggestion was discussed and the motion for the naming of the three rooms was put forward by Mr Clive Brooks and Dr Jill Blee, the motion was passed by the Institutes Committee Members on 26 July, 2004.

Note 10: Ballarat Heritage Weekend: 2011:

As part of the Ballarat Heritage Weekend Henry Sutton and his inventions were the main exhibit in the Mining Exchange Building. The Henry Sutton timeline which was on display at the Heritage Weekend is now mounted on the wall of one of the buildings at the Ballarat School of Mines which is now part of Federation University.

Note 11: Science and Innovation: 2013:

On Australia Day, 26 January, 2013, National Radio, and the Science Show, aired a three- part series on Science and innovation. Henry Sutton was the main subject for the first hour long series in which Professor Mark Dodgson from the University of Queensland remarked that in his opinion Henry Sutton may well have been one of the greatest inventors of all time.

Note 12: Industry Skills Building Ballarat: 2014:

The new Industry Skills Building at Federation University in Ballarat highlights Henry Sutton and his achievements as part of the buildings internal features to inspire its students and displays a timeline graphic of his life's work.

Note 13: Inaugural Henry Sutton Oration 2014:

On 14 May, 2014, The Telecommunications Society of Australia adopted Henry Sutton as their Victorian Hero and held the Inaugural Henry Sutton Oration with Robyn Williams from the ABC Science Show giving the first oration.

Note 14: Federation University Australia 2018 Alumni Awards:

On the 23 July, 2018, Federation University Posthumously honoured Mr Henry Sutton as a Distinguished Alumnus within the category of Outstanding Professional Achievement and Outstanding academic or research achievement.

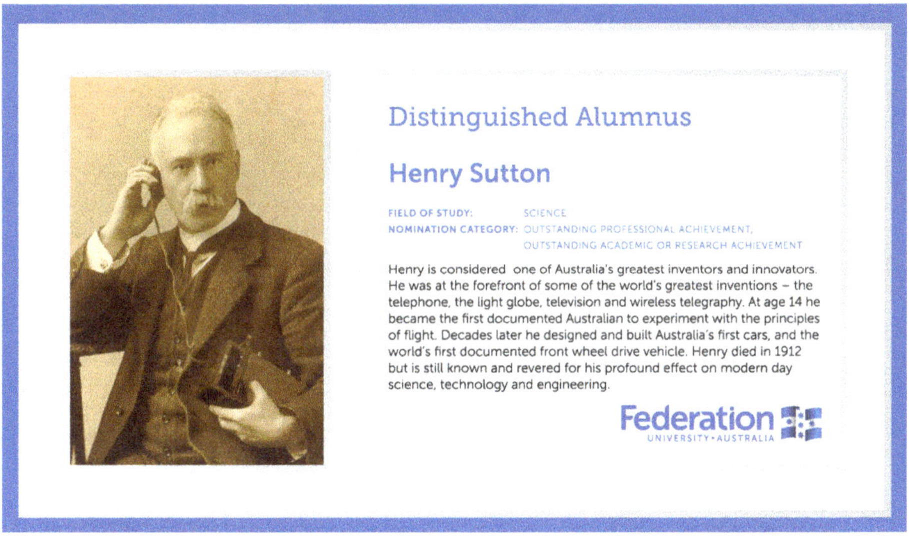

This framed picture is on permanent display at the
Caro Convention Centre, Mt Helen Campus, Federation University Australia

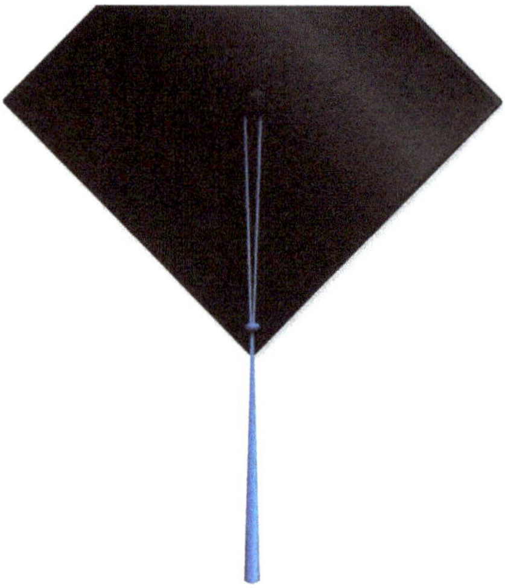

POSTHUMOUS
DISTINGUISHED ALUMNUS

AWARDED TO

Henry Sutton

ON THE TWENTY THIRD DAY OF JULY 2018

PROFESSOR HELEN BARTLETT
VICE-CHANCELLOR AND PRESIDENT

alumni | Federation
UNIVERSITY · AUSTRALIA

This Certificate was presented to the Sutton Family on 23 July 2018
by Federation University Australia.

ACKNOWLEDGEMENTS

FAMILY:

'I wish to personally thank all the Sutton and extended family members for their support and assistance in gathering family information, and a special thank you to my son Benjamin, my brother Trevor, my cousins Ron and Shirley and my aunty Betty, for their encouragement, love and support.'

PEOPLE:

Clare Gervasoni, Curator, Art and Historical Collection, Federation University Australia.
Mark Dodgson, *FASSA, FRSA.* Professor of Innovation Studies at UQ Business School and Visiting Professor at Imperial College London.
Chrystopher J. Spicer, for his treasured friendship and support.
Robyn Williams AM, ABC Science Show.
Geoff Holden, Retired Science and Technology Curator, Science Museum Victoria, for his friendship and vast knowledge.
Will McGhie, Historical and Archives Committee member, Wireless Institute of Australia, Western Australia,
Peter Wolfenden, Historical and Archives Committee Member Leader, Wireless Institute of Australia,Victoria.
Dr Andrew Davie, Technical Advisor of Mechanical Television.
Jarrod Watt, Journalist and Historian.
Professor Norman Heckenberg, University of Queensland.
Chris and Tim Clemons.
David Elms, Historian, Ballarat Fire Station.
David Burger, IEEE Historian.
The Staff at the Queensland State Library.

LIBRARIES and ARCHIVES:

Federation University Historical Collection.
The State Library of Queensland.
The State Library of Victoria.
National Library Australia.
The Australian National Archives.
The British National Archives.
The British Telecom Archives.
The Queensland Science Museum Library.
The British Science Museum Library.
The British Library.
The Bodleian Library, Oxford University, England.
St Bride Library, London.
Defence Library Service and Archives, Australian Department of Defence.
The Smithsonian Institute Library.
Nokia Bell Laboratory Archives.
The United States Naval Historical Foundation.
The Royal Aeronautical Society, National Aerospace Library.

ORGANISATIONS:

The Malvern Historical Society Inc.
The Royal Automobile Club of Victoria.
The Royal Aeronautical Society of Great Britain.
The Institute of Engineering and Technology, England.
Victorian Telecommunications Museum Incorporated.
Museum Victoria.
The Wireless Institute of Australia.
The Institute of Electrical and Electronics Engineers, Australia.
The Australian Historic Telephone Society.
The Telecommunications Society of Australia.
The Ballaarat Mechanics' Institute.
Ballarat Historical Society.
The Ballarat Fire Brigade
The Ballarat Gold Museum

OTHERS:

I would like to say a special thank you to everyone who donated towards the publication of this book and the many people and organisations who have helped with information over the years to make this book possible.

SOURCES

CHAPTER ONE:

Newspapers:

- A Valuable Suggestion: R. H. Sutton, *The Ballarat Star,* Ballarat, 13 July, 1857, p. 3.

Images:

- Photograph: Richard Henry Sutton: *Sutton Family Archives.*
- Photograph: Mary Sutton: *Sutton Family Archives.*
- Photograph: Sutton's Main Road Store: *Sutton Family Archives.*
- Image: Henry Sutton's Birth Certificate: *Victorian, Birth Deaths and Marriages Register.*
- Image: Ballaarat Mechanics Institute: *Sutton Family Archives.*

CHAPTER TWO:

Book:

- Octave Chanute, *Progress in Flying Machines*, Chapter; Wings and Parachutes Part III, January, 1892, p. 33.

Journals:

- On the Flight of Birds and Aerial Navigation and Second Paper on the Flight of Birds, Henry Sutton: *The Thirteenth Annual Report of the Aeronautical Society of Great Britain*, 1878, p.p. 30-69.
- Wing Flight- Some Researches at Cambridge: *The Automotor Journal*, 27 May, 1905, p.p. 648-651.
- Article: Henry Sutton, *The Automotor Journal*, 30 December, 1905, p. 1622.
- A New Electric Motor: Henry Sutton, *English Mechanic and World of Science*, No. 751, 15 August 1879, p. 571.

Newspapers:

- To the Editor of the Argus: Henry Sutton, *The Argus*, Melbourne, 15 September 1909, p. 8.
- Australia and the Aeroplane: George A Taylor, *The Argus*, Melbourne, 17 September 1909, p. 5.
- Australia and the Aeroplane: Henry Sutton, *The Argus,* Melbourne, 21 September 1909, p. 7.

- The Agricultural Society Prizes: Henry Sutton, *The Ballarat Courier*, 1 October 1873, p. 4.

Images:

- Drawing: Emilie Sutton, by Henry Sutton circa 1872: *Sutton Family Archives.*
- Drawing: Ornithopter, On the Flight of Birds and Aerial Navigation by Henry Sutton: *Thirteenth Annual Report of The Aeronautical Society of Great Britain*, 1878, p.35. Drawing courtesy of the *Royal Aeronautical Society of Great Britain, The National Aerospace Library.*
- Drawing: Smoked Glass Experiment, Second Paper, On the Flight of Birds: Henry Sutton: Journal, *Thirteenth Annual Report of The Aeronautical Society of Great Britain*, 1878, p.61. Drawing courtesy of the *Royal Aeronautical Society of Great Britain, The National Aerospace Library.*
- Drawing: Sutton Continuous Current Dynamo: Journal, *English Mechanic and World of Science,* 15 August 1879, p. 571.
- Images: Ballarat School of Mines: Image, Courtesy of *Federation University Historical Collection.*

CHAPTER THREE

Journal:

- Testing for Resonance: Henry Sutton, E*nglish Mechanic and World of Science*, Vol. 28, 8 November 1878, p. 222.

Newspapers:

- Untitled Article: *The Ballarat Courier*, 1 July 1875, p. 2.
- Untitled Article: *The Argus*, Melbourne, 22 February 1878, p. 5.
- Untitled Article: *The Ballarat Star*, 21 February 1878, p. 2.

Images:

- Drawing: Testing for Resonance: Henry Sutton: Journal, *English Mechanic and World of Science* Vol. 28, 8 November 1878, p. 222.
- Mr James Oddie: Photograph, Courtesy of *Federation University Historical Collection.*
- Ballarat City Fire Station: Photograph courtesy of F*ederation University Historical Collection*
- Ballarat Fire Station: Photograph Courtesy of the *Ballarat Fire Brigade Archives.*

CHAPTER FOUR

Journals:

- Description of Vacuum Apparatus with Diagram: Henry Sutton, T*ransactions and Proceedings of the Royal Society of Victoria*, Volume XVIII, 1882, 13 December 1881, p.122.

- Sutton's Apparatus for the Rapid Production of High Vacua: Henry Sutton, *English Mechanic and World of Science*, No. 886, 17 March 1882, pp. 34-35.
- On a New Form of Secondary Cell for Electric Storage: Henry Sutton, *Transactions and Proceedings of the Royal Society of Victoria*, 15 December 1881, pp. 110-114.
- A New Electric Motor: *English Mechanic and World of Science*, No. 751, 15 August, 1879, p. 571.

Newspapers:
- Scientific Notes: *The Graphic London*, 4 February 1882, p. 18.
- Mr. H. Sutton's Discovery: 14 February 1882, p. 4.
- Untitled Article: *The Ballarat Courier*, 1 March 1882, p. 2.
- Street Tramways and Electricity: Henry Sutton, *The Argus*, Melbourne, 28 August 1882, p. 4.
- The Value of Electricity to the Miner: Henry Sutton, (alias Delta), *The Touchstone*, Ballarat, 21 December 1883, p. 7.

Images:
- Drawing: Sutton Vacuum Pump: Journal: *Transactions and Proceedings of the Royal Society of Victoria*, Volume XVIII, 1882, p. 223.
- Drawing: Sutton Vacuum Pump: Journal, *English Mechanic and World of Science*, 17 March 1882, p. 34.
- Photograph: Edison-Swan United Electric Light Company, Light Globe c.1900: Photograph Courtesy *University of Queensland Physics Museum.*
- Kings of Wall Street 1882, *New York Public Library Digital Collections.* items/22059060-93fb-0130-f86a-58d385a7b928
- Copy of Henry Sutton's Admission Record of Membership to The Society of Telegraph Engineers and Electricians, London, England, 1883. Image Courtesy of, the *Institute of Engineering and Technology,* Savoy Place London.

CHAPTER FIVE:

Newspapers:
- The Microscope and Disease: Henry Sutton (alias Delta), *The Touchstone*, Ballarat, 28 December 1883, p. 8.
- Photographing the Cholera Germ: Henry Sutton, *The Argus,* Melbourne, 28 December 1885, p. 7.
- The Peculiar Sunsets: Henry Sutton, T*he Touchstone*, Ballarat, 4 January 1884, p. 4.
- The Peculiar Sunsets: Henry Sutton: *The Touchstone*, Ballarat, 7 December 1883, p. 12.

Images:

- Photograph: Emilie Sutton 1880: *Sutton Family Archives.*

CHAPTER SIX:

Book:

- Ballarat School of Mines Annual Report, 1886, p. 39.Courtesy of ***F****ederation University Historical Collection.*

Newspapers:

- One of the uses of Electricity: *The Ballarat Courier*, 22 September 1883, p. 4.
- Comet "Pons": Henry Sutton, (alias Delta), *The Touchstone*, Ballarat, 7 December 1883, p. 6.

Images:

- Photograph: Thomas Bury: Courtesy of the *Ballarat Historical Society Collection.*
- Photograph: Robert Lewis John Ellery: Courtesy of the *State Library of Victoria.*
- Photograph: Ballarat School Of Mines, Electricity Classroom: Courtesy of *Federation University Historical Collection*. Image number H24338.
- Photograph: Opening of the Baker Telescope, Ballarat Observatory, 1888, Photograph by Henry Sutton: *Government Astronomer, Robert Lewis John Ellery Collection*, Courtesy of the *Museum of Victoria.* Unique Identifier – 000050033.

CHAPTER SEVEN:

Newspapers:

- Mr. Henry Sutton's Latest Invention: *The Ballarat Courier*, 31 May 1888, p. 3.
- Ballarat, from our Correspondent: *The Argus*, Melbourne, 31 May 1888, p. 5.
- Ballarat's Farewell: Mona Marie, *Evening Post*, Ballarat, 3 February 1890, p. 4.

Images:

- Image: Mary Anderson, Electro-phototypy Advertisement. Journal, *Supplement to the Journal of the London Camera Club*, May 1891.
- Photograph: Henry Sutton's Illuminated Address, 1890: *Sutton Family Archives.*

CHAPTER EIGHT:

Journals:

- Title: Reversal, by Henry Sutton, *The Journal of the London Camera Club,* 22 October 1891, p. 192-194.
- Title: Sutton's Electro-Photypy, *The Printing World*, 1891, p. 353.
- Sutton's Process Syndicate Limited: *The Chemist and Druggist,* 3 December 1891, p.809.

Newspapers:

- Untitled Article: *The Ballarat Courier,* 10 July 1890, p. 2.

- Return of a Ballarat Inventor: *Ballarat Star*, 15 May 1893, p. 3.
- Photography on Racecourses: Henry Sutton, *The Argus*, Melbourne, 28 October 1893, p. 11.

Images:
- Image: Presenting the Address at the Library at Guildhall: *The Illustrated London News,* 18 July 1891, p.17.
- Image: Hale End, Sutton's Process Syndicate Ltd Advertisement: Photograph by Henry Sutton: The British Journal Photography Almanac,1893, p.662.
- Photograph: Henry Sutton 1891: by Barraud & Co, London: *The Austral Wheel,* May 1898, p. 130.
- Photograph: Boys At Play, black and white photograph: by Henry Sutton 1891, Photograph taken from glass plate: *Sutton Family Archives.*
- Photograph: Arthur Sutton on the Rocks at Hastings, black and white photograph: by Henry Sutton 1891, Photograph taken from glass plate: *Sutton Family Archives.*
- Photograph: Henry Sutton Colour Photograph, Study of Flowers: by Henry Sutton, photograph taken from colour glass plate: *Sutton Family Archives.*

CHAPTER NINE:

***Newspaper*:**
- What Typo Say's: The Touchstone, Ballarat, 4 January 1884, p. 15.

Images:
- Image: Otis Hydraulic Lift: *Elevator Systems of the Eiffel Tower*: By Robert M Vogel, 1889, p.13.
- Image: The Touchstone Newspaper Title: *The Touchstone,* Volume 1, Number 1, 7 December 1883, Front Page.

CHAPTER 10:

Journal:
- Article Title: Monopoly and other Topics, *The Australian Cyclist*, 7 April 1898, p.p. 46-47.

Images:
- Photograph: Sutton's Cycle Agency, Exhibition Display: *The Australian Cyclist*, 11 June 1896, p. 22.
- Photograph: Arthur Sutton on Bicycle: *The Australian Cyclist*, 17 December 1896, p.38.
- Photograph: Cyclist Snell arriving at Melbourne GPO, Melbourne: *The Australian Cyclist*, 28 January 1897, p. 31.
- Photograph: Cyclist Snell Reception at Sutton's, Bourke St, Melbourne: *The Australian Cyclist*, 28 January 1897, p. 31.

- Photograph: Alps Cycling Touring Party 1897: *The Australian Cyclist,* 29 April, 1897, p. iv.

CHAPTER ELEVEN:

Personal Papers:

- Henry Sutton's, personal papers: *Sutton Family Archives.*

Journal:

- Article Title: Electric Motor v. Oil Motor, "The Austral Wheel" June, 1898, p.p. 159-160.
- Article Title: An Australian Paraffin Motor, "The Autocar" 23 March, 1901, p. 272.
- Article Title: British Marine Oil and Spirit Engines, Edward Butler, "The Rudder" Volume 17, 1906, p.p. 325-334.
- Article Title: After a Trip Abroad, "The Scientific Australian" 20 December, 1904, p.37.
- Article Title: Monopoly and other Topics, "The Australian Cyclist" 7 April, 1898, p.p. 46-47.
- Article Title: More about the Sutton Car, "The Autocar" 31 March, 1900, p. 304.
- Article Title: The Sutton Autocar, "The Autocar" 13 January, 1900, p.p. 29-31.
- Article Title: The Sutton"Heavy Oil" Car, "The Automotor Journal" 10 January, 1903, p. 41.

Newspapers:

- Fuel for Omnibuses: Henry Sutton, *The Argus,* Melbourne, 14 March 1905, p. 8.
- New Oil Engine: *The Argus*, Melbourne, 10 July 1902, p. 8.
- Cycling: *Geelong Advertiser,* 17 September 1897, p. 3.
- The Great Inter-State Motor Contest: *Western Herald,* Bourke, New South Wales, 3 July 1901, p. 4.

Images:

- Image: Sutton Motorised Tandem Bicycle: Journal: *Tourist and Traveller*, 6 July 1899, p.19.
- Photograph: The Sutton Stationary Engine: Journal, *The Autocar"*, 23 March, 1901, p.272.
- Photographs: The Sutton Carburettor assembled and in pieces: Journal, *The Automotor Journal*, 4 June 1904, p.677.
- Photograph: Restored De Dion Beeston Motor Tricycle, photograph taken in 2012 at a museum in England: *Sutton Family Archives.*
- Drawing: Henry Sutton on Motor Tricycle: Journal, *The Australian Cyclist,* 10 June 1897, p. iii.

- Image: Sutton Brothers Cycle Agency Advertisement: Journal, *The Australian Cyclist,* 8 July 1897, p. 24.
- Photograph: Annie May Tatti on her Wedding Day, 1902, : *Sutton Family Archives.*
- Photograph: Sutton Autocar built 1899: Journal, "The Autocar", 31 March 1900, p.304.
- Drawing: Sutton Autocar built 1897: Journal, *The Cycle Age and Trade Review,* Volume XXIV, Number 14, 1 February 1900, p. 488.
- Photograph: Sutton Voiturette Pedal Car, 1900: Photograph Courtesy of *Chris and Tim Clemons.*
- Photograph: Restored Sutton Voiturette Pedal Car: Photograph Courtesy of *Chris and Tim Clemons.*
- Photograph: Sutton Tonneau built 1901: Journal, *The Automotor Journal,* Volume VIII, 10 January 1903, p. Front Cover.
- Photograph: Sutton Australian Voiturette built 1903: Newspaper, *The Melbourne Punch,* 5 February 1903, p. 22.

CHAPTER TWELVE:

Books:

- *1903 Minute Book of the Royal Automobile Club of Victoria.* Courtesy the *Royal Automobile Club of Victoria's Heritage Collection.*
- T*he Automobile Club of Victoria Rules, By-Laws and Year Book 1906-1907.* Courtesy the *Royal Automobile Club of Victoria's Heritage Collection.*

Journal:

- Article Title: After a Trip Abroad, *The Scientific Australian,* 20 December 1904, p. 37.

Images:

- Image: Automobile Club of Victoria Emblem: Image, Courtesy of the *Royal Automobile Club of Victoria's Heritage Collection.*
- Photographs: Start of the Automobile Club of Victoria's Inaugural Car Run, 20 February, 1904: Newspaper, *The Leader,* Melbourne, 27 February 1904, p.36.
- Photograph: Automobile Club of Victoria Car Race at Maribyrnong, Melbourne 1904: Journal, *The Motor Age,* Volume 5, Jan-June 1904, p. 22.
- Photograph: A. E. Sutton VMCC, 100 Mile Road Race, 1906: *Sutton Family Archives.*
- Photograph: Charles Mayman, December, 1904: Photograph Courtesy *State Library of Victoria.*

CHAPTER THIRTEEN:

Books:

- *The Romance of a Music Business*: by Alfred Sutton, 1928, pp. 6-7: *Sutton Family Archives.*
- Title: Australian Participation in International Expositions, *Documents of Australian Policy: Australian Foreign Affairs Record*, Volume 5, August 1985, p. 781.

Journals:

- Article Title: Tele-Photography, Henry Sutton, *The Telegraphic Journal and Electrical Review*, 7 November 1890, p.p. 549-551.
- Article Title: Telopy or Telephany, *The Telegraphic Journal and Electrical Review*, Volume 27, 27 November 1890, p.636.
- Article Title: Seeing by Electricity, *The Telegraphic Journal and Electrical Review*, 12 December 1890, p.782.
- Nikola Tesla: Experiments with Alternate Currents of High Potential and High Frequency, *The London Institution of Electrical Engineers,* 3 February 1892.

Newspapers:

- Forgotten Son of Ballarat Epitomised Eureka Factor: *The Age,* Melbourne, 8 August 1988, p. 15.
- Pictures by Wire: *Washington Evening Star*, 17 October 1896, p.17.
- Nikola Tesla's Great Electrical Discovery, by Henry Sutton: *The Ballarat Star*, 15 March 1892, p. 4.

Images:

- Drawing: Sutton's Telephane Design: Journal: *Scientific American Supplement*, Number 786, Volume 31, 24 January 1891, p.12559.

CHAPTER 14:

Personal Papers:

- Henry Sutton's Personal Papers: *Sutton Family Archive.*

Newspapers:

- Australian Wireless, Important New Invention: *The Argus*, Melbourne, 28 July 1908, p. 5.
- Wireless Telegraphy, An Australian Invention: *The Sydney Morning Herald,* 29 July 1908, p. 13.
- Wireless Telegraphy, No Federal Stations: *The Argus* Melbourne, 21 September 1908, p.5.
- Wireless in Australia, Mr. Hesketh's Views: *The Argus,* Melbourne, 1 October 1908, p. 8.
- How Inventors are Treated, Ridiculous Departmental Obstruction: *The Age,*

Melbourne, 1 October 1908, p. 8.

- Wireless in Australia, Mr. Hesketh's Views: *The Argus,* Melbourne, 1 October 1908, p. 8.
- Wireless Telegraphy: *The Argus,* Melbourne, 24 August 1909, p. 7.

Records:

- National Archives of Australia: Record Title, "Mr Sutton's Wireless Telegraphy Experiments" Series Number: MP84/1, Control Symbol: 1997/2/27, Item Barcode 4795750, Date Range: 1 January, 1906 - 31 December, 1913.
- National Archives of Australia: "The Sutton Oil Engine" Series Number: MP84/1, Control Symbol: 1997/2/200, Item Barcode: 4795772, Date Range: Circa 1 January, 1906 – Circa 31 December, 1913.

Images:

- Photograph: Henry Sutton with Portable Receiver 1910: *Sutton Family Archives.*
- Photograph: Sir John Quick: Photograph, Courtesy of the *National Library of Australia.*
- Photograph: Prime Minister Alfred Deakin: Photograph, Courtesy of the *National Library of Australia.*
- Photograph: Naval Captain William Rooke Creswell: Photograph, Courtesy of the *Australian War Museum.*
- Photograph: The American Great White Fleet: Photograph, Courtesy of the *United States Naval Historical Foundation.*

CHAPTER FIFTEEN:

Personal Papers:

- Henry Sutton's Personal Papers: *Sutton Family Archives.*

Journals:

- Article Title: Victorian Institute of Electrical Engineers, The Annual Dinner: *The Australian Mining and Standard, Electrical Record*, Volume XXI, Number 613, 24 August 1910, pp. 211-212.
- Article Title, Correspondence, Production of Rectifiers by Physical Means: Henry Sutton, *The Electrician*, Volume 66, 16 December 1910, pp. 384-385.
- Article Title, Correspondence, Production of Rectifiers by Physical Means: Henry Sutton, *The Electrician*, Volume 69, 19 April 1912, pp. 66-67.
- Article Title, Correspondence, Production of Rectifiers by Physical Means: Greenleaf W. Pickard,, Volume 69, *The Electrician,* 17 May 1912, p. 232.
- Article Title, Correspondence, Production of Rectifiers by Physical Means: Henry Sutton, *The Electrician*, Volume 69, 2 August 1912, p. 708.

Newspapers:

- Inventor of the Telephone, Dr. Bell in Australia: *The Age*, Melbourne, 15 June 1910, p. 6.
- Letter to Editor: George A. Taylor, *The Argus,* Melbourne, 18 November 1910, p. 7.
- Federal Affairs, Wireless Telegraphy: *Queensland Times,* Ipswich Queensland, 14 March 1911, p.2.
- Untitled Article: *The Argus,* Melbourne, 31 May 1911, p.12.
- All Red Wireless, Letter to Editor: Henry Sutton, *The Argus,* Melbourne, 20 June 1911, p. 9.
- Wireless Patents, Letter to Editor by Henry Sutton, *The Argus,* Melbourne, 14 July 1911, p. 5.
- Wireless Stations, Letter to Editor by Henry Sutton, *The Argus,* Melbourne, 21 September 1911, p. 8.
- House of Representatives, Wireless Telegraphy, *The Argus,* Melbourne, 20 December, 1911, p. 13.
- Wireless in Melbourne, Domain Station Opened: *The Argus,* Melbourne, 10 February 1912, p. 25.
- Wireless Experiments, A Melbourne Inventor: *The Argus,* Melbourne, 10 February 1912, p.7.
- An Inventor, Mr. Henry Sutton and his Invention: *The Leader*, Melbourne, 14 May 1910, p. 51.
- An Inventor, Mr. Henry Sutton and his Invention: *The Leader*, Melbourne, 14 May, 1910, p. 36.
- Scientific Notes, Telegraphing Through Crystals: *The Australasian*, Melbourne, 22 June, 1912, p. 52.

Records:

- Papers Presented to Parliament: Dr. Bell's Testimony to the Royal Commission on the Postmaster General's Department, 16 August, 1910: "The Parliament of the Commonwealth of Australia, Session 1910, Volume V".
- Records of the Proceedings and Printed Papers of the Parliament: Volume 4. 1910, p. 251.
- National Archives of Australia: Record Title, "Prime Minister's Letter Book and Correspondence", Control Symbol: Whole Series, Series A62, Volume 10, Date Range: 1910-1911, 16 October, 1911.
- National Archives of Australia: "Prime Minister's Letter Book and Correspondence", Control Symbol: Whole Series, Series A62, Volume 10, Date Range: 1910-1911, 11 October, 1911.
- National Archives of Australia: "Prime Minister's Letter Book and Correspondence", Control Symbol: Whole Series, Series Number: A62, Volume 10, Date Range: 1910-1911, 15 December 1911.

- National Archives of Australia: "The Sutton Oil Engine" Series Number: MP84/1, Control Symbol: 1997/2/200, Item Barcode: 4795772, Date Range: Circa 1 January, 1906 – Circa 31 December, 1913.
- National Archives of Australia: "Wireless- Miscellaneous Papers of 1901 and Later" Series Number: A11804, Control Symbol: 1926/427, Item Barcode: 1620418, Date Range: 1901 -1911.
- National Archives of Australia: "Wireless – Sale of Apparatus by Companies (Re. Royalties e.c.)" Control Symbol: 1912/12, Series Number: MP341/1, Item Barcode: 346012, Date Range: 1911-1912.

Images:
- Photograph: Alexander Graham Bell: Online, *Public Domain.*
- Photograph: George Augustine Taylor: Photograph, Courtesy of the *National Library of Australia.* Image record number 2088821.
- Photograph: John Graeme Balsillie: Newspaper, *The Argus,* Melbourne, 15 July 1924, p10.
- Photographs: An Inventor, Mr. Henry Sutton and his Invention: *The Leader*, Melbourne, 14 May 1910, p. 36.
- Drawing: Henry Sutton's Wireless Tower: Drawing, courtesy of *Peter Wolfenden.*
- Drawings 1 and 2: Correspondence, Production of Rectifiers by Physical Means: Henry Sutton: Journal, *The Electrician,* Volume 66, 16 December 1910, p.p. 384-385.
- Photograph: First Transistor by Bell Laboratory: Photograph, courtesy of *Nokia Corporation, Nokia Bell Laboratory Archives*, Item Number 86-301264.
- Photograph: Bardeen, Brattain and Shockley at Bell Laboratory, 1948. Photograph, courtesy of *Nokia Corporation, Nokia Bell Laboratory Archives*, Item Number 90-10567.

CHAPTER SIXTEEN:

Newspapers:
- Wireless Strength: Henry Sutton, *The Argus,* Melbourne, 13 July 1912, p. 20.
- Death of Henry Sutton: *The Ballarat Courier,* July 1912, p. 6.

Magazine:
- Henry Sutton: *The Ballarat School of Mines Students Magazine*, Second Term, 1912, page 8.

Images:
- Photograph: Henry Sutton with sons Albert and John 1912.
  Photograph Courtesy of the *Sutton Family Archives.*

INDEX

E

F

G

Q

R

T

Z

www.ingramcontent.com/pod-product-compliance
Ingram Content Group UK Ltd.
Pitfield, Milton Keynes, MK11 3LW, UK
UKHW020033121026
15625UKWH00060B/743/J

9 781925 332346